Mathematics
An Intuitive Approach for College Students

Third Edition

Leon E. Myers
South Carolina State University

Kendall Hunt
publishing company

Kendall Hunt
p u b l i s h i n g   c o m p a n y

www.kendallhunt.com
Send all inquiries to:
4050 Westmark Drive
Dubuque, IA 52004-1840

In memory of my parents, Edward and Shirley Myers, and my brothers, Charles Ervin and Robert Myers.

To my brothers, Clarence, Sonnie, Demosthenese, Bennie, my sister, Pat, and my special teachers, Mrs. Geraldyne P. Zimmerman and Mr. Frank M. Staley, Jr.

L.E.M.

Contents

Preface............ix

CHAPTER 1   INTRODUCTION TO SETS
    1.1   Sets Defined ··1
    1.2   Properties of Real Numbers (Addition and Multiplication) ······9
    1.3   Test for Divisibility ··· 17
    1.4   Prime and Composite Numbers ·························· 23
    1.5   Greatest Common Divisor or Denominator (GCD) ··········· 29
           Chapter 1 Review Exercises ··························· 35
           Chapter 1 Test ··· 37

CHAPTER 2   ESTIMATION, EXPONENTS, AND SCIENTIFIC NOTATION
    2.1   Estimating Sums, Differences, Products, and Quotients ········ 41
    2.2   Exponents ··· 49
    2.3   Symbols ··· 55
    2.4   Order of Operations ··································· 61
    2.5   Writing Numbers in Scientific Notation ·················· 65
    2.6   Converting from Scientific Notation to Ordinary Notation ····· 71
    2.7   Addition and Subtraction of Numbers in Scientific Notation ··· 75
    2.8   Multiplication and Division of Numbers in Scientific Notation 75
           Chapter 2 Review Exercises ··························· 81
           Chapter 2 Test ··· 83

CHAPTER 3   RATIONAL NUMBERS
    3.1   Meaning of a Fraction ································· 85
    3.2   Ordering Fractions ··································· 91
    3.3   Addition and Subtraction of Fractions ················· 95
    3.4   Multiplication and Division of Fractions ················101
    3.5   Complex Fractions ····································105
    3.6   Applications Involving Fractions ·······················109
    3.7   Adding and Subtracting of Decimals ·················115
    3.8   Multiplication and Division of Decimals ·················121
    3.9   Comparing and Converting Fractions to Decimals ············125
           Chapter 3 Review Exercises ·······················131
           Chapter 3 Test ··136

CHAPTER 4   RATIOS, PROPORTIONS, AND PERCENTS

4.1   Ratio and Proportion ···139
4.2   Introduction to Percents ···151
4.3   Percent Problems: Proportion Method·····························157
4.4   Percent Increase-Decrease Problems ·····························165
      Chapter 4 Review Exercises ··171
      Chapter 4 Test ···177

CHAPTER 5   FORMULAS, SEQUENCES, AND SERIES

5.1   Simple and Compound Interest·······························185
5.2   Effective Interest and Annuities ·····························195
5.3   Loan Payments, Credit Cards, and Mortgages ···········201
5.4   Arithmetic Sequences and Series·····························209
5.5   Geometric Sequences and Series ·····························216
      Chapter 5 Review Exercises ··221
      Chapter 5 Test···225
      Formula Sheet for Chapter 5 ·······································228

CHAPTER 6   INTEGERS AND PROPERTIES OF EQUATIONS AND INEQUALITIES

6.1   Uses of Integers ···229
6.2   Rules for Addition of Integers ·······························235
6.3   Subtraction of Integers ···239
6.4   Multiplication and Division of Integers ·····················243
6.5   Uses of Brackets, Parentheses, and Order of Operations ······247
6.6   Variables and Algebraic Expressions·····························253
6.7   Solving Linear Equations and Rational Equations·············261
6.8   Translating Word Problems into Algebraic Expressions ······269
6.9   Translating Sentences into Equations and Solving ···········273
6.10 Solving Word Problems ···277
6.11 Properties of Inequalities and Interval Notation ·············283
      Chapter 6 Review Exercises ··293
      Chapter 6 Test···297

CHAPTER 7   GRAPHING LINEAR EQUATIONS

7.1   Linear Equations in Two Variables   303
7.2   Graphing Linear Equations With Two Variables   309
7.3   Lines and their Slopes   315
7.4   General Form of the Equation of a Line   321
      Chapter 7 Review Exercises   327
      Chapter 7 Test   329

CHAPTER 8 FUNDAMENTALS AND APPLICATIONS OF GEOMETRY

8.1   Fundamental Concepts ·················· 333
8.2   Angle Measurements ·················· 339
8.3   Perimeter of Geometric Figures ·················· 345
8.4   Areas of Geometric Figures ·················· 349
8.5   Circles and Composite Figures ·················· 355
8.6   Solid Geometry ·················· 363
8.7   Right Triangles ·················· 369
8.8   Similar Triangles ·················· 373
        Chapter 8 Review Exercises ·················· 379
        Chapter 8 Test ·················· 381

CHAPTER 9 MEASUREMENTS

9.1   The Metric System ·················· 387
9.2   Denominate Numbers (Measurement) ·················· 393
9.3   Operations with Denominate Numbers ·················· 405
9.4   Temperature Units: Fahrenheit, Celsius, and Kelvin ·········· 411
9.5   Calibrated Scales and Scale Drawings ·················· 415
        Chapter 9 Review Exercises ·················· 423
        Chapter 9 Test ·················· 427

CHAPTER 10 STATISTICS AND PROBABILTY

10.1   Measures of Central Tendency ·················· 431
10.2   Measures of Variability ·················· 441
10.3   The Normal Distribution – Interpreting and
         Understanding the Standard Deviation ·················· 449
10.4   Probability ·················· 459
10.5   Combining Events and Odds ·················· 465
         Chapter 10 Review Exercises ·················· 473
         Chapter 10 Test ·················· 475

CHAPTER 11   STATISTICAL GRAPHS AND CHARTS

11.1   Uses of Graphs ·················· 479
11.2   Constructing Graphs ·················· 493
11.3   Scatter Plot ·················· 499
         Chapter 11 Review Exercises ·················· 507
         Chapter 11 Test ·················· 511

APPENDIXES: I.   Formulas ·················· 517
                         II.   Table of Measures ·················· 519
                         III. Table of Areas Under the Normal Curve ·················· 521

ANSWERS TO SELECTED QUESTIONS ·················· 523

INDEX ·················· 553

Preface

Objectives

The aim and purpose of this book is to provide students with the knowledge of mathematics and how to interpret graphs. I have spent much time and effort in determining how the problem-solving process should be applied for minimum success of the students.

The students are provided with every opportunity to learn methods of solving mathematical problems successfully, to develop computational skills, and to build confidence. My long experience of teaching mathematics has taught me to keep the mathematical language basic and simple.

Acknowledgements

In preparing the seventh edition of the book, I have received many feedbacks from my students.

The following people deserve special tribute: Mr. Xiaomao Liu for his detailed typing, reviewing, proofreading, and checking of all problems in the book. I would like to thank Dr. Andrew Hugine, Jr., for his past assistance.

CHAPTER 1

INTRODUCTION TO SETS

1.1 SETS DEFINED

The idea of a set is used extensively in mathematics. A set is a well-defined collection of objects. The objects in a set are called elements or members. A set can be represented by listing its elements between braces $\{\ \}$. The order in which the elements are listed is unimportant.

ILLUSTRATION 1

Represent the following sets by listing the elements.

a) The vowels $\{a, e, i, o, u\}$

b) The counting numbers less than 5. $\{1, 2, 3, 4\}$

Naming sets

Capital letters are used to name sets. To state that 3 is an element of $A = \{1, 2, 3, 4, 5, 6\}$, we write $3 \in A$. If an element is not a member of a set we use the $\notin$ symbol. For example in set A, 7 is not in set A so we write $7 \notin A$.

Two methods are commonly used to describe the contents of a set. Suppose we wish to describe the set A containing the counting numbers less than 10. Using the roster method, we list all of the elements of the set as follows: $A = \{1, 2, 3, 4, 5, 6, 7, 8, 9\}$.

Another way of representing the elements in a set is to use what is sometimes called set-builder notation. To illustrate, the set A consisting of all counting numbers less than 10 can be written in set builder notation as

$A = \{x | x$ is a counting number less than 10$\}$

which is read "A equals the set of all elements x, such that x is a counting number less than 10." In this notation x denotes a typical element of the set, the vertical bar | is read "such that" and following the bar are the conditions that x must satisfy to be a member of the set A.

ILLUSTRATION 2

Represent in set-builder notation:

a) $A = \{a, e, i, o, u\}$

The elements in the set are the vowels, therefore, $A = \{x|x$ is a vowel$\}$

b) $B = \{1, 3, 5, 7, \ldots\}$.

$B = \{x|x$ is an odd counting number$\}$

Types of Sets

Sets may be classified according to the number of their elements. The set $\{1, 2\}$ contains two elements while the number of elements in the set $\{1, 2, 3\ldots\}$ cannot be counted.

In the set $\{1, 2\}$ all of the elements of a set can be listed and we say that the set is **finite**. We typically use the roster method to specify the elements of a finite set. In the set $\{1, 2, 3, 4\ldots\}$ all of the elements cannot be listed and the set is said to be **infinite**. The elements of infinite sets are typically specified using set-builder notation.

Sets with no elements are called **empty or null sets**. The symbol ϕ or $\{\ \}$, but not $\{\phi\}$, are used to denote the empty set. An example of an empty set would be the set of all days of the week that begin with the letter "O".

Equal sets are two sets A and B that have exactly the same elements and we write $A = B$.

The set $A = \{3, 4, 5\}$ and set $B = \{4, 3, 5\}$ are equal sets since they have the same elements. Note that the order of listing the elements does not matter.

Equivalent sets are two sets A and B that have the same number of elements but not necessarily identical.

ILLUSTRATION 3

Identify each of the following sets as finite, infinite, empty, equal or equivalent

a) $A = \{1, 3, 5\}$

Finite. The set only contains 3 elements.

b) $B = \{x|x > 39\}$

Infinite. All numbers greater than 39 cannot be counted.

c) $C = \{3, 1, 7\}$

Finite. The elements in the set can be counted.

d) $A = \{x | x \text{ is a counting number between 3 and 4}\}$

Empty. There are no counting numbers between 3 and 4.

e) Sets A and C are equivalent since they don't have identical elements.

Set Relations and Set Operations

When every number of set A is also a member of another set B, we say that A is a **subset** of B and denote it $A \subset B$. For example, given the set $D = \{1, 2, 3, 4, 5, 6\}$, and $C = \{2, 3, 5\}$, since every element in set C is also in set D, we can say that $C \subset D$, read "C is a subset of D."

We use the symbol $\not\subset$ to mean that one set is not a subset of another set. For sets C and D above, $D \not\subset C$ since every member of D is not a member of C.

It is important to note that every set is a subset of itself and the empty set is a subset of every set.

Sets may also be combined by forming their union or intersection.

The **union** of two sets A and B, written $A \cup B$, is the set of all elements in A or B or in both A and B.

For example, if set $A = \{1, 2, 3\}$ and set $B = \{3, 4, 5\}$, then

$$A \cup B = \{1, 2, 3, 4, 5\}$$

The intersection of two sets A and B, denoted $A \cap B$, is the set of all elements that are in both A and B (i.e., common to both sets).

For example, if $A = \{5, 6, 7\}$, $B = \{6, 7, 8\}$, and $C = \{8, 9, 10\}$, then

$$A \cap B = \{6, 7\}, \text{ and } A \cap C = \{ \}.$$

Observe that sets A and C have no elements in common and so their intersection equals the empty set $\{ \}$, and we say that the sets are **disjoint.**

EXAMPLE 1

Determine $A \cup B$ and $A \cap B$ if $A = \{2, 5, 7\}$ and $B = \{1, 2, 3, 4, 5\}$

SOLUTION

$A \cup B = \{1, 2, 3, 4, 5, 7\}$. All the elements in both sets.

$A \cap B = \{2, 5\}$, since these are the elements common to both sets.

EXAMPLE 2

Let $A = \{x | x$ is an even number$\}$ and $B = \{x | x$ is an odd number$\}$. Find $A \cup B$ and $A \cap B$.

SOLUTION

$A \cup B = \{x | x$ is an even or odd number, i.e., 1, 2, 3, 4, 5...$\}$

$A \cap B = \{ \ \}$, there are no elements common to the sets.

THE SET OF REAL NUMBERS

Numbers are used extensively in mathematics. The language of sets may be used to describe numbers. The collection of real numbers, denoted $\boldsymbol{R}$, consists of several subsets which are described below.

Natural or counting numbers, denoted $\boldsymbol{N}$, are the numbers used for counting.

$N = \{1, 2, 3, 4, 5, ...\}$

Whole numbers, denoted $\boldsymbol{W}$, are the natural numbers and 0.

$W = \{0, 1, 2, 3, 4, ...\}$

Integers denoted $\boldsymbol{I}$, are the whole numbers and their additive inverses.

$I = \{..., -3, -2, -1, 0, 1, 2, 3, ...\}$

Rational numbers, denoted $\boldsymbol{Q}$, are numbers that can be expressed as the ratio of two integers a/b where $b \neq 0$.

Examples: $\dfrac{2}{3}, \dfrac{5}{7}, 3\left($ since 3 can be expressed as $\dfrac{3}{1}\right)$

Irrational numbers, denoted $\boldsymbol{L}$, are numbers that cannot be expressed as the ratio of two integers a/b where $b \neq 0$.

Examples: $\sqrt{3}$, $\pi = 3.14159265358...$

Decimals

Every real number can be expressed as a decimal. Rational numbers can be expressed as **terminating or repeating decimals**.

Examples of rational numbers:

6 may be expressed as 6.0

$\frac{2}{5}$ may be expressed as .4

$\frac{2}{3}$ may be expressed as .6666… or $.\overline{66}$

Note: The overbar indicates that the digit(s) continue to repeat in the same pattern.

Irrational numbers cannot be expressed as repeating or terminating decimals. As decimal representations, irrational numbers are **nonterminating and nonrepeating**.

Examples of irrational numbers

$\sqrt{2} = 1.41213562...$
$\sqrt{5} = 2.236067978...$
$\pi = 3.1415925358...$

EXAMPLE 1

List all the numbers from the set that are natural, whole, integer, rational, or irrational: $\{-9,$

$\sqrt{5}, \frac{2}{5}, 6\}$

SOLUTION

-9 Integer and rational

$\sqrt{5}$ Irrational

$\frac{2}{5}$ Rational

6   Natural, whole, integer and rational

Operations on sets, unions, intersections, and subsets, may be used to show the relationship between the subsets of real numbers. For example, if we use $\boldsymbol{R}$ to denote the set of real numbers, and $\boldsymbol{Q}$ and $\boldsymbol{L}$ to denote the rational and irrational numbers, respectively, then we can express R, the set of real numbers, as $R = Q \cup L$ in set notation.

EXAMPLE 2

Determine whether the following are true or false.

a) $W \subset N$ b) $W \subset I$ c) $Q \cap L = \phi$

d) $N \cup \{0\} = W$ e) $N \cap Q = Q$

SOLUTION

It may be helpful to list the elements of the sets and then compare the members in each set.

a) $W \subset N$

$W = \{0, 1, 2, 3, \ldots\}$
$N = \{1, 2, 3, \ldots\}$

False. All of W is not contained in N, namely 0, thus, $W \not\subset N$.

b) $I = \{\ldots, -3, -2, -1, 0, 1, 2, 3, \ldots\}$
$W = \{0, 1, 2, 3, \ldots\}$

True. All of W is in I.

c) $Q = \{$numbers that can be expressed as fractions$\}$
$L = \{$numbers that cannot be expressed as fractions$\}$

True. There are no elements in common.

d) $N = \{1, 2, 3, \ldots\} \cup \{0\}$ gives $\{0, 1, 2, 3, \ldots\}$ which is the set of whole numbers. True.

e) $N = \{1, 2, 3, \ldots\}$
$Q = \{$numbers that can be expressed as fractions$\}$

False. The elements common to both sets are the natural numbers, not the rational numbers.

EXERCISE 1.1

List the elements in each set.

1) The whole numbers between 7 and 13.

2) The days of the week that begin with the letter "T".

3) $A = \{x|x$ is a positive odd number less than 20$\}$.

4) The months of the year that begin with the letter "Z".

5) $\{x|x$ is an even natural number between 9 and 25$\}$.

Write in set builder notation.

6) $\{5, 10, 15, 20, 25, 30, ...\}$ 7) $\{0, 2, 4, 6, 8, ...\}$

8) $\{1, 3, 5, 7, ...\}$ 9) $\{1, 2, 3, 4, ..., 60\}$

If $A = \{0, 1, 2, 3, ...,10\}$, $B = \{1, 2, 3\}$, $C = \{2, 3, 4\}$, and $D = \{4, 5, 6\}$, find the following:

10) $A \cap B$ 11) $A \cup B$ 12) $B \cap \emptyset$

13) $B \cap D$ 14) $C \cap D$ 15) $A \cup \emptyset$

Which of the following are true?

16) $B \subset A$ 17) $A \subset B$ 18) $B = A$

19) $D \subset B$ 20) B is equivalent to C

If $A = \{x|x$ is greater than 4 and less than 9$\}$, and $B = \{x|x$ is greater than 1 and less than 5$\}$, and $C = \{x|x$ is between 3 and 5$\}$

21) $A \cup B$ 22) $A \cap B$

23) $A \cup (B \cap C)$ 24) $(A \cap B) \cup C$

Which of the following pairs of sets are equivalent?

25) $\{x|x$ is a natural number less than 6$\}$ and $\{m, n, o, p, q\}$

26) $\{m, a, t, h\}$ and $\{f, u, n\}$

27) $\{a, b, c, d, e, f, ...m\}$ and $\{1, 2, 3, ...13\}$

28) $\{@, \&\}$ and $\{2\}$

29) Given the set $B = \{-9, -\sqrt{3}, -\dfrac{3}{5}, 0, \sqrt{5}, \sqrt{4}, 3, 5.9, 7\}$ list all numbers from the set which are (a) natural, (b) integers, (c) irrational, (d) rational, (e) real, (f) whole.

Indicate whether each statement is true or false.

30) 0 is a natural number.

7

31) $\sqrt{3}$ is a rational number.

32) $\sqrt{16}$ is an irrational number.

33) Every integer is a rational number.

34) Every rational number is an integer.

35) If a number is a whole number, it is also a rational number.

Complete each statement with the symbol $\in$ or $\notin$.

36) 5 ___ Q 37) -6 ___ N 38) $\sqrt{7}$ ___ L

1.2 PROPERTIES OF REAL NUMBERS (ADDITION AND MULTIPLICATION)

We will now discuss several assumptions about real numbers. They will be referred to as the properties of real numbers as they apply to the operations of addition and multiplication. Each property is listed below with an explanation.

Given real numbers, a, b, and c, the following properties hold for real numbers.

> **Closure Property:** If a and b are real numbers, then $a + b$ is a real number and $a \cdot b$ is a real number.

The closure property states that the sum or product of real numbers is also a real number.

To illustrate,

a)   5 is a real number, 4 is a real number, and the sum $5 + 4 = 9$ is also a real number.

b)   3 is a real number, 2 is a real number, and $3 \cdot 2 = 6$ is also a real number.

> **Commutative Property:** For any real numbers, a and b, $a + b = b + a$ and $a \cdot b = b \cdot a$. That is, two numbers may be added or multiplied in any order.

The commutative property states that the order in which numbers are added or multiplied does not change the sum or product.

This property is shown in the following examples.

$$5 + 7 = 7 + 5 \qquad\qquad (5)\,(7) = (7)\,(5)$$
$$12 \;=\; 12 \qquad\qquad\quad 35 \;=\; 35$$
$$6 + 8 = 8 + 6 \qquad\qquad (6)\,(8) = (8)\,(6)$$
$$14 \;=\; 14 \qquad\qquad\quad 48 \;=\; 48$$

> **Associative Property:** For any real numbers, a, b, and c, $a + (b + c) = (a + b) + c$ and $a \cdot (b \cdot c) = (a \cdot b) \cdot c$, that is, when three numbers are added or multiplied, the first two may be grouped together or the last two may be grouped together, without affecting the answer.

The associative property states that the order in which real numbers are associated (grouped) does not change the sum or product.

To illustrate, note the problems below.

$$(8 + 2) + 9 = 8 + (2 + 9) \qquad\qquad (2 \cdot 3)8 = 2(3 \cdot 8)$$

$$10 + 9 = 8 + 11$$
$$19 = 19$$
$$(9 + 8) + 7 = 9 + (8 + 7)$$
$$17 + 7 = 9 + 15$$
$$24 = 24$$

$$6 \cdot 8 = 2 \cdot 24$$
$$48 = 48$$
$$9(8 \cdot 7) = (9 \cdot 8)7$$
$$9 \cdot 56 = 72 \cdot 7$$
$$504 = 504$$

Identity Property: For any real number a, $a + 0 = a$ and $a \cdot 1 = a$, that is, we can add 0 to any number or multiply any number by 1 and get the same number.

The identity property states that the sum of any number and 0 is the number and the product of any number and 1 is the same number.

The problems below illustrate the identity property.

$$5 + 0 = 0 + 5 = 5 \qquad\qquad 6 \cdot 1 = 1 \cdot 6 = 6$$

$$6 + 0 = 0 + 6 = 6 \qquad\qquad 5 \cdot 1 = 1 \cdot 5 = 5$$

Inverse Property: For any real number a, there exists a number, $-a$, such that $a + (-a) = 0$. This number is called the additive inverse. There also exists a number a, such that $a \cdot \left(\dfrac{1}{a} = 1\right)$. This number is called the reciprocal or the multiplicative inverse. In words, the inverse property states that the sum of a number and its inverse is 0 and the product of a number and its inverse is 1.

By the inverse property, the sum of a number and its additive inverse is 0 and the product of a number and its inverse is 1. The identity element for addition is 0 and the identity element for multiplication is 1. For multiplication, this number is called the **reciprocal or multiplicative inverse**.

Examples using the inverse property follow.

a)   The additive inverse of 5 is -5, since $5 + (-5) = 0$.

b)   The additive inverse of 8 is -8, since $8 + (-8) = 0$.

c)   The additive inverse of $\left(\dfrac{1}{3}\right)$ is $\left(-\dfrac{1}{3}\right)$, since $\left(\dfrac{1}{3}\right) + \left(-\dfrac{1}{3}\right) = 0$.

d)   The multiplicative inverse or reciprocal of 7 is $\dfrac{1}{7}$, since $(7) \times \dfrac{1}{7} = 1$.

e)   The multiplicative inverse or reciprocal of $\left(\dfrac{1}{2}\right)$ is 2, since $\left(\dfrac{1}{2}\right)$ x (2) = 1.

Distributive Property: For any real numbers a, b, and c, $a(b+c)=a \cdot b + a \cdot c$.   That is, we may combine the numbers within the symbols of grouping and then multiply, or we may multiply and then combine the products.

By the distributive property, we may combine numbers within the symbols of grouping first and then multiply, or we may multiply and then combine the products.

Consider the expression 3(7 + 2).

Adding within symbols or groupings first, we get, 3(9) = 27.

Now, consider again 3(7 + 2). We can find the answer as follows:

$$3(7 + 2) = 3(7) + 3(2)$$
$$= 21 + 6$$
$$= 27$$

Note that we get 27 as the answer both ways.

Cancellation Property: If $a+c=b+c$, then $a=b$.
or
$a \cdot c = b \cdot c$ and $c \neq 0$, then $a=b$.

By the cancellation property, we may simplify an expression by adding or subtracting a common term from both members of an equality or we may multiply or divide both members by any nonzero constant without affecting the quality.

Zero Factor Property $a \cdot 0 = 0$
or
If $ab = 0$, then either $a = 0$, $b = 0$
or
both a and b are zero.

By the zero factor property, if the product of two numbers is zero then either or both of the factors must be zero. For example, based on the zero factor property, the product of 8(0) = 0.

EXAMPLE 1

Identify the property which justifies each statement.

a) $5 + (-5) = 0$ b) $1 \cdot x = x \cdot 1$ c) $6 \cdot 0 = 0$

d) $x + (y + 2) = (x + y) + 2$ e) $2(5x) = (2 \cdot 5)x$

f) $8(x + 3) = 8x + 8 \cdot 3$ *or* $8x + 24$

SOLUTION

a) $5 + (-5) = 0$, additive inverse

b) $1 \cdot x = x$, identity for multiplication

c) $6 \cdot 0 = 0$, zero factor property

d) $x + (y + 2) = (x + y) + 2$, associative property for addition

e) $2(5x) = (2 \cdot 5)x$, associative property for multiplication

f) $8(x + 3) = 8x + 8 \cdot 3$ *or* $8x + 24$, distributive property

EXAMPLE 2

Identify the property which justifies each statement.

a) $3 + (4 + 7) = (3 + 4) + 7$

b) $3 + (4 + 7) = 3 + (7 + 4)$

SOLUTION

a) Associative property. Observe that the order of the numbers is the same but they are grouped differently. In the left member 4 and 7 are grouped together but on the right, 3 and 4 are grouped together.

b) Commutative. The groupings of 4 and 7 are the same in both members but the order of the 4 and 7 have changed.

EXAMPLE 3

Test each set for closure with respect to the given operations.

a) $T = (0, 1)$ for multiplication, addition, and division.

b) $O = \{x|x$ is a positive odd number$\}$ for addition and multiplication.

SOLUTION

To test a set for closure, select at least two elements from the set and perform the indicated operation. All possible combinations of the elements with respect to the operation must be formed. If we get another member of the set, then the set is closed with respect to the indicated operation.

a) $T = \{0, 1\}$

   Multiplication:

   $0 \cdot 1 = 1 \in T$ $1 \cdot 0 = 0 \in T$

   $1 \cdot 1 = 1 \in T$ $0 \cdot 0 = 0 \in T$

   The set is closed with respect to multiplication because we get another member of the set.

   Addition:

   $0 + 0 = 0 \in T$ $0 + 1 = 1 \in T$ $1 + 1 = 2 \notin T$

   Not closed since 2 is not in T.

   Division:

   $0 \div 1 = 0 \in T$ $1 \div 0$ is undefined

   Not closed since $1 \div 0$ is not in T.

b) To test closure for this set, select 3 and 5 from set O.

   Addition:

   $3 + 5 = 8 \notin O$ Not closed, 8 is not an odd number.

   Multiplication

   $3 \cdot 5 = 15 \in O$ $5 \cdot 3 = 15 \in O$

   $5 \cdot 5 = 25 \in O$ $3 \cdot 3 = 9 \in O$

   Closed. All combinations of the elements are in O.

EXERCISE 1.2

Indicate the property which makes each of the following true.

1) $6 + 15 = 15 + 6$

2) $5(15 \cdot 8) = (5 \cdot 15)8$

3) $2 + (8 + 10) = (2 + 8) + 10$

4) $5 + (-5) = 0$

5) $8 + 0 = 8$

6) $7 \times 1 = 7$

7) $4 \times \dfrac{1}{4} = 1$

8) $6(5 + 3) = 6(5) + 6(3)$

9) $5 \cdot 4$ is a whole number

10) $(2 + 5) + 7 = (5 + 2) + 7$

11) $5 \cdot 8 = 8 \cdot 5$

12) $6 + (x + 3) = (6 + x) + 3$

13) $4 + (-4) = 0$

14) $1 \cdot y = y$

Fill in the blank in each of the following based on the distributive property.

15) $7(5 + 2) = $ ___

16) ___ $= 9 \cdot 2 + 9 \cdot 8$

17) $(9 + 4) \cdot 6 = $ ___

18) ___ $(7 + 9) = 8$ ___ $+ 8$ ___

Identify the properties illustrated in each of the following:

19) $-7 + 0 = -7$

20) $(7a)b = b(7a)$

21) $8(m + 4) = 8m + 8 \cdot 4$

22) If $5x = 5y$, then $x = y$

23) If x is a real number, then $x + 4$ is a real number

24) $8 \cdot 0 = 0$

25) $(x + y) + 4 = (y + x) + 4$

1.3 TEST FOR DIVISIBILITY

We formally define the relation "divides" as follows: one number, a, divides another b, if there is another number c, such that the following illustrates this definition:

a)   5 divides 35 since there is a number 7, such that 7x5=35.

b)   10 divides 50 since there is a number 5, such that 10x5=50.

From this definition of divisibility we have two theorems which will help us to determine the divisibility of numbers.

> **Theorem 1:** If the same number divides two numbers then it divides their sum.

13 divides 26 and 13 divides 39, so, 13 divides the sum of $26 + 39$ which is 65.

4 divides 100 and 4 divides 36, therefore, 4 divides 136.

7 divides 49 and 7 divides 21, therefore, 7 divides 70.

> **Theorem 2:** If a number divides another number it divides any multiple of that number.

6 divides 12, therefore, 6 divides 24, which is 12×2; also 36, which is 12×3.

8 divides 16, therefore, 8 divides 48 (16×3); also 64, which is 16×4.

From these theorems and the definition of divisibility follow the divisibility tests for numbers.

> **Divisibility Test for 2:** A number is divisible by 2 if its units digit is 0, 2, 4, 6, or 8.

EXAMPLE 1

Determine which of the following numbers are divisible by 2:

   a) 44   b) 578   c) 49

17

SOLUTION

a) 44 is divisible by 2 because the units digit is 4: $44 = 2 \times 22$

b) 578 is divisible by 2 because the units digit is 8: $578 = 2 \times 289$.

c) 49 is not divisible by 2 because the units digit is 9, not 2, 4, 6, or 8.

> **Divisibility Test for 3:** A number is divisible by 3 if the sum of its digits is divisible by 3.

EXAMPLE 2

Determine which of the following numbers are divisible by 3:

a) 57   b) 243   c) 52

SOLUTION

a) 57 is divisible by 3 because the sum of the digits is $5 + 7 = 12$, and 12 is divisible by 3: $57 = 3 \times 19$.

b) 243 is divisible by 3 because the sum of the digits is $2 + 4 + 3 = 9$, and 9 is divisible by 3: $243 = 3 \times 81$.

c) 52 is not divisible by 3 because the sum of the digits is $5 + 2 = 7$, and 7 is not divisible by 3.

> **Divisibility Test for 5:** A number is divisible by 5 if its units digit is 0 or 5.

EXAMPLE 3

Determine which of the following numbers are divisible by 5:

a) 75   b) 9,720   c) 68

SOLUTION

a) 75 is divisible by 5 because the units digit is 5: $75 = 3 \times 25$.
b) 9,720 is divisible by 5 because the units digit is 0: $9720 = 5 \times 1944$.
c) 68 is not divisible by 5 because the units digit is 8 not 0 or 5.

> **Divisibility Test for 10:** A number is divisible by 10 if the units digit is 0.

EXAMPLE 4

Determine which of the following numbers is divisible by 10:

    a) 80   b) 7400   c) 305

SOLUTION

    a) 80 is divisible by 10 because the units digits are 0: $80 = 10 \times 8$.

    b) 7450 is divisible by 10 because the units digit is 0: $7450 = 745 \times 10$.

    c) 305 is not divisible by 10 because the units digit is 5, not 0.

> **Divisibility Test for 9:** A number is divisible by 9 if the sum of the digits is divisible by 9.

EXAMPLE 5

Determine if the following numbers is divisible by 9:

    a) 54   b) 657   c) 52

SOLUTION

    a) 54 is divisible by 9 because the sum of the digits is $5 + 4 = 9$, and 9 is divisible by 9: $54 = 9 \times 6$.

    b) 657 is divisible by 9 because the sum of the digits is $6 + 5 + 7 = 18$, and 18 is divisible by 9: $657 = 9 \times 73$.

    c) 52 is not divisible by 9 because the sum of the digits is $5 + 2 = 7$, and 7 is not divisible by 9.

> **Test for 15, 6, and 9:** We can observe that, if a number is divisible by both 3 and 5, then it is divisible by 15. If a number is divisible by both 2 and 3, then it is divisible by 6. Not all numbers that are divisible by 3 are divisible by 9; however, if a number is divisible by 9, then it is also divisible by 3.

EXAMPLE 6

Determine which of the following numbers are divisible by 15, 6, or 9:

a) 135   b) 48   c) 99

SOLUTION

a) 135 is divisible by 15 because its units digit is 5 and $1+3+5=9$ which is divisible by 3.

b) 48 is divisible by 6 because the units digit is even and $4+8=12$ which is divisible by 3. (Note that 48 is divisible by 3 but not by 9.)

c) 99 is divisible by both 3 and 9 because $9+9=18$ which is divisible by both 3 and 9.

EXERCISE 1.3

Using the techniques of this section, determine which number or numbers of the set {2, 3, 5, 9, 15} will divide exactly into each of the following numbers.

1) 98

2) 254

3) 75

4) 555

5) 471

6) 470

7) 571

8) 466

9) 897

10) 695

11) 795

12) 888

13) 45,000

14) 885

15) 4,422

16) 56

17) 4321

18) 96

19) 5,678

20) 78

PRIME AND COMPOSITE NUMBERS

When we multiply two whole numbers together, the result is called the product of the numbers. The two numbers multiplied are called factors of the product. For example, since $9 \times 4 = 36$, 9 and 4 are factors of 36. Also, since $2 \times 8 = 16$, 2 and 8 are factors of 16.

> A **prime number** is a whole number that is greater than 1 that has no factors other than itself and 1.

> A **composite number** is a whole number that can be divided by some number other than itself and 1.

> **Fundamental Theorem of Arithmetic:** Every composite number has one and only one unique prime factorization.

Illustrations of the Fundamental Theorem of Arithmetic

a)   8 is a composite number because $2 \times 4 = 8$, so that 8 has factors other than itself or 1.

b)   13 is a prime number because 1 and 13 are the only factors of 13.

c)   45 is a composite number because it has factors 3, 5, 9, and 15, other than 1 and 45.

Another important idea related to factors is that of multiples. A multiple of a given whole number is the product of that number and another whole number. Thus, 24 is a multiple of 3, since $24 = 3 \times 8$; 18 is also a multiple of 3, since $18 = 3 \times 6$. A number has many multiples. The set of multiples of 3 is

$$\{0, 3, 6, 9, 12, 15, 18, 21, 24, 27 \ldots\}$$

We will note that 0 is a multiple of every number. Any number that is a multiple of two or more numbers is called a common multiple of these numbers.

EXAMPLE 1

Determine whether 117 is prime or composite.

SOLUTION

Since $10 \times 10 = 100$ and $11 \times 11 = 121$, the square root of 117 is between 10 and 11. To determine if the number is prime, we try the prime numbers, 2, 3, 5, and 7. Applying the divisibility tests, the sum of the digits in 117, $1+1+7 = 9$, therefore, the number is divisible by 3. Dividing by 3 gives 39, therefore the number is not prime.

EXAMPLE 2

Determine whether 151 is prime or composite.

SOLUTION

12x12=144 and 13x13=169, therefore, the square root of 151 is between 12 and 13. To test for primeness, we try the prime numbers 2, 3, 5, 7, and 11. Applying the tests for divisibility, the number is not divisible by 2, 3, or 5. No tests were given for 7 or 11, so we divide the number by 7 and 11. $151 \div 11 = 13 + \text{Remainder } 9$ and $151 \div 7 = 21 + \text{Remainder } 4$. Since the number is not divisible by 2, 3, 5, or 11, the number is prime.

PRIME FACTORIZATION

A composite number may be factored into prime factors, that is, it may be factored completely by drawing a factor tree. This method is fine for small numbers, but rather tedious for large numbers.

Method 1

A factor tree is demonstrated in Figure 1.1. Notice that the branches of the tree display the prime factors of 260.

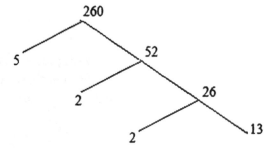

Figure 1.1. A factor tree

Method 2

There is a more systematic way of factoring a composite number completely. This method is called the consecutive primes method. We can illustrate this method by an example. Let us factor 168 completely; that is, let us find the complete factorization of 168. We begin with the least prime, 2, and decide whether or not it is a factor of 168. We see by inspection that 168 is divisible by 2, hence,

$$168 = 2 \times 84$$

Since $84 = 2 \times 42$, 2 is a factor of 84 and

24

$$168 = 2 \times 2 \times 42 .$$

Observing that 2 is also a factor of 42 since

$$42 = 2 \times 21 ,$$

We have $168 = 2 \times 2 \times 2 \times 21$.

Since 2 is not a factor of 21, we try the next prime, 3. We see that $168 = 2 \times 2 \times 2 \times 7 \times 3$.
Since all factors in this expression are primes, we have factored 168 completely. The essential results of this method can be written in this form:

$$
\begin{array}{r|r}
2 & 168 \\ \hline
2 & 84 \\ \hline
2 & 42 \\ \hline
3 & 21 \\ \hline
 & 7
\end{array}
\qquad 168 = 2 \times 2 \times 2 \times 3 \times 7
$$

EXAMPLE 1

Use the consecutive primes method to find the complete factorization of 1200.

SOLUTION

$$
\begin{array}{r|r}
2 & 1200 \\ \hline
2 & 600 \\ \hline
2 & 300 \\ \hline
3 & 150 \\ \hline
5 & 50 \\ \hline
2 & 10 \\ \hline
 & 5
\end{array}
\qquad 1200 = 2 \times 2 \times 2 \times 3 \times 5 \times 2 \times 5
$$

EXAMPLE 2

Use the consecutive primes method to find the complete factorization of 198.

SOLUTION

$$
\begin{array}{r|r}
2 & 198 \\ \hline
3 & 99 \\ \hline
3 & 33 \\ \hline
 & 11
\end{array}
\qquad 198 = 2 \times 3 \times 3 \times 11
$$

The prime factorization of 198 is $2 \times 3 \times 3 \times 11$.

EXAMPLE 3

Use the consecutive primes method to find the complete factorization of 84.

SOLUTION

$$84 = 2 \times 2 \times 7 \times 3$$

$$\begin{array}{c|c} 2 & 84 \\ \hline 3 & 42 \\ \hline 2 & 14 \\ \hline & 7 \end{array}$$

EXAMPLE 4

Use the consecutive primes method to find the complete factorization of 194.

SOLUTION

$$\begin{array}{c|c} 2 & 194 \\ \hline & 97 \end{array} \qquad 194 = 2 \times 97$$

EXERCISE 1.4

State whether each of the following numbers is prime or composite.

1) 5	2) 18	3) 101	4) 31
5) 111	6) 12	7) 49	8) 61
9) 41	10) 21	11) 13	12) 65
13) 52	14) 20	15) 65	16) 10
17) 9			

Find two sets of factors for each of the following composite numbers.

18) 38	19) 28	20) 15	21) 16
22) 50	23) 9	24) 75	25) 35
26) 68	27) 98	28) 77	29) 135
30) 96	31) 49	32) 124	

List the set of multiples for each of the following numbers.

33) 19	34) 32	35) 35	36) 14
37) 40	38) 68	39) 20	40) 43

Find the prime factorization of the following numbers.

41) 14	42) 83	43) 625	44) 500
45) 120	46) 216	47) 66	48) 75
49) 123	50) 160	51) 175	52) 343
53) 102	54) 46	55) 86	56) 51
57) 80	58) 37	59) 30	60) 61

1.5 GREATEST COMMON DIVISOR OR DENOMINATOR (GCD)

The greatest common divisor of a set of whole numbers is the greatest number that will divide into a given set of numbers.

Method 1

To find the greatest common divisor (GCD) of a set of whole numbers, list all the possible divisors of each number in the set and select the greatest number that is a divisor common to all the numbers.

EXAMPLE 1

Find the GCD for {18, 24}.

SOLUTION

List the set of divisors for each number.

Set of divisors for 18 = {1, 2, 3, 6, 9, 18}

Set of divisors for 24 = {1, 2, 3, 6, 8, 12, 24}

Observe that the largest or greatest number common to both 18 and 24 is 6. Therefore, the GCD for {18, 24} is 6.

Method 2

The method above is very time-consuming when the numbers are greater in a given set. An alternate method is to find the prime factorizations of each number in the set, and then the GCD will be the product of all the prime factors which are common to each whole number.

EXAMPLE 2

Find the GCD for {36, 18, 12}.

SOLUTION

Express each number in prime factors.

Prime factors of 36 = {2 × 2 × 3 × 3}

Prime factors of 18 = {2 × 3 × 3}

Prime factors of 12 = {2 × 2 × 3}

Notice that the product of the prime factors 2 and 3 are common to each number, therefore, the GCD = 2 × 3 = 6.

EXAMPLE 3

Find the GCD for {100, 75, 50}

SOLUTION

$$100 = 2 \times 2 \times 5 \times 5$$

$$75 = 5 \times 5 \times 3$$

$$50 = 2 \times 5 \times 5$$

$$GCD = 5 \times 5 = 25$$

We will note that if the GCD of two numbers is one (that is, there are no common prime factors), then the two numbers are said to be **relatively prime**. The numbers themselves may be prime or may be composite numbers.

EXAMPLE 4

Find the GCD of {10, 9}

SOLUTION

$$10 = 2 \times 5$$

$$9 = 3 \times 3$$

$$GCD = 1$$

10 and 9 are, therefore, **relatively prime**.

LEAST COMMON MULTIPLE (LCM)

The least common multiple (LCM) of two or more whole numbers is the smallest number that is divisible by a given set of numbers.

> **Method 1**
> To find the LCM, we list the multiples of each number and then use the smallest multiple common to each of the numbers.

EXAMPLE 1

Find the least common multiple of {4, 8, 12}.

SOLUTION

The multiples of 4 are {4, 8, 12, 16, 20, 24, ...}.

The multiples of 8 are {8, 16, 24, 32, ...}.

The multiples of 12 are {12, 24, 36, 48, ...}

We see that of the set of multiples we have, the smallest common multiple is 24. Therefore, the LCM = 24 for the numbers {4, 8, 12}.

Other basic or useful method for finding the LCM is illustrated by the following example.

EXAMPLE 2

Find the LCM of {25, 18, 20, 6}.

SOLUTION

a) Write the numbers horizontally and find a prime number that will divide into more than one number. Divide by that prime and write the quotients beneath the dividends. Re-write any numbers that were not divided.

b) Continue this process until no two numbers have a common prime divisor.

c) The LCM is the product of all the prime divisors and the last set of quotients.

$$
\begin{array}{r|cccc}
2 & 25 & 18 & 20 & 6 \\
5 & 25 & 9 & 10 & 3 \\
3 & 5 & 9 & 2 & 3 \\
\hline
 & 5 & 3 & 2 & 1
\end{array}
$$

Therefore, the LCM of {25, 18, 20, 6} = $2 \times 5 \times 3 \times 5 \times 3 \times 1 \times 2$

31

EXERCISE 1.5

Find the GCD for each of the following sets of numbers:

1) {12, 8} 2) {16, 28} 3) {85, 51} 4) {20,75}

5) {20, 30} 6) {42, 48} 7) {15, 21} 8) {27, 18}

9) {18, 24} 10) {77, 66} 11) {182, 184} 12) {110, 66}

13) {8, 16, 64} 14) {121, 44} 15) {28, 52, 56} 16) {98, 147}

17) {60, 24, 96} 18) {33, 55, 77} 19) {25, 50, 75} 20) {30, 78, 60}

21) {17, 15, 21} 22) {520, 220} 23) {14, 55} 24) {210, 231, 84}

25) {140, 245, 420}

Which of the following pairs of numbers are relatively prime?

26) {35, 24} 27) {11, 23} 28) {14, 36} 29) {72, 35}

30) {42, 77} 31) {16, 51} 32) {20, 21} 33) {8, 15}

34) {66, 22} 35) {10, 27}

Find the LCM for each of the following sets of numbers:

36) {8, 12} 37) {3, 5, 7} 38) {4, 6, 9} 39) {6, 8, 27}

40) {144, 216} 41) {40, 25} 42) {40, 75} 43) {98, 28}

44) {72, 36, 54} 45) {15, 10, 35} 46) {13, 26, 169} 47) {121, 33, 66}

48) {8, 15, 13} 49) {125, 45, 150} 50) {51, 54, 34} 51) {8, 15, 10}

52) {24, 15, 10} 53) {8, 10, 120} 54) {6, 15, 30} 55) {45, 18, 6, 27}

56) {228, 12, 95} 57) {63, 98, 45} 58) {56, 40, 196} 59) {135, 125, 225}

60) {99, 363, 143}

Find the GCD and the LCM for each of the following sets of numbers:

61) $\{3, 5, 9\}$ 62) $\{2, 5, 11\}$ 63) $\{4, 18, 14\}$ 64) $\{6, 15, 12\}$

65) $\{49, 25, 35\}$ 66) $\{45, 145, 290\}$ 67) $\{40, 48, 56, 24\}$ 68) $\{81, 54, 108\}$

69) $\{135, 75, 45\}$ 70) $\{169, 637, 845\}$

CHAPTER 1 REVIEW EXERCISES

Indicate which of the following is true/false:

1) Every set has at least two subsets.

2) The set consisting of the letters in the word "sentences" is equal to the set consisting of the letters in the word "cents."

3) Zero (0) is a natural number.

4) An infinite set is a set with a large number of elements such as the population of the United States.

5) The only difference between the set of whole numbers and the set of natural numbers is that the set of whole numbers contain negative numbers.

List the elements in the following sets.

6) Set A which contains the integers between 5 and 11.

7) Set B containing the months of the year that begin with the letter A.

Let $A = \{1, 3, 4, 5, 7\}$, $B = \{2, 3, 6, 8\}$, and $C = \{9, 10\}$.

Find:

8) $A \cup B$ 9) $B \cup C$ 10) $A \cap B$ 11) $A \cap C$

12) $B \cap \phi$

Let $R = \{$real numbers$\}$, $N = \{$natural numbers$\}$, $A = \{O\}$, $W = \{$whole numbers$\}$, $T = \{$integers$\}$, $Q = \{$rational numbers$\}$, and $L = \{$irrational numbers$\}$.

Find:

13) $Q \cup L$ 14) $N \cup A$ 15) $Q \cap L$ 16) $Q \cap I$

17) Which of the following is true?

a) $R \subset I$ b) $N \subset W$ c) $I \subset Q$

Identify the property which makes each statement true.

18) $5 + 6$ is a whole number.   19) $5 \cdot (6) = 6 \cdot (5)$ 20) $8 + (7 + 16) = (8 + 7) + 16$

21) $1 \cdot 7 = 7$ 22) $4 + (-4) = 0$ 23) $2(5 + 3) = 2 \cdot 5 + 2 \cdot 3$

Answer true or false for each statement.

24) The number 0 is rational.

25) Some whole numbers are not integers.

26) Not every rational number is positive.

27) Some real numbers are not rational.

28) Some irrational numbers are negative.

29) Every whole number is positive.

30) Every integer is positive.

31) Some integers are not real numbers.

32) Every integer is a rational number.

33) Every rational number is a real number.

Perform the indicated operations.

34) List the elements in the set, $\{ x \mid x$ is a whole number greater than 6 and less than 15$\}$.

35) Test the number 17,265 for divisibility by $\{2, 3, 5, 6, 9, 10, 15\}$.

36) Find the prime factorization of 320.

37) List the prime numbers you would try to test the number 203 for primeness.

38) If $A = \{1, 2, 3, 4, 5, 7\}$ and $B = \{2, 3, 6, 8\}$ find
   a) $A \cap B$ b) $A \cup B$.

39) Given the set $\{ \sqrt{2}, 0, 4, 5.8, \dfrac{3}{4}, \sqrt{16}, -4 \}$, list

   a) Natural Numbers   b) Whole Numbers,
   c) Integers, and   d) Irrational Numbers.

40) If $A = \{1, 2, 3, 4\}$ and $B = \{a, b, c, d\}$, then sets A and B are said to be _____ sets.

CHAPTER 1 TEST

1) Given the set $A=\{0, 1\}$ and the operation of addition. Adding 1 and 1, elements of the set, gives 2, which is not a member of the set. Therefore, the _____ property does not hold for this set.
   a) Associative   b) Identity   c) Closure   d) Addition

2) Which set of numbers is correct?
   a) $N=\{0, 1, 2\}$ b) $W=\{0, 1, 2\}$
   c) $Q=\{1.333 .., \sqrt{3}, \frac{4}{7}\}$ d) $L=\{3.434343 ..., \sqrt{5}\}$

3) If $A=\{1, 2, 3, 4, 5, 6, 7\}$ and $B=\{5, 6, 7\}$, which of the following is correct?
   a) $A \cap B=\{1, 2, 3, 4, 5, 6, 7\}$ b) $A \cup B=\{5, 6, 7\}$
   c) $A \subset B$ d) $A \cap B=\{5, 6, 7\}$

4) The number 951,810 is divisible by
   a) 2, 3   b) 5, 10, 15   c) 2, 3, 6, 9, 10, 15   d) 2, 3, 5, 6, 10, 15

5) The prime factorization of 630 is
   a) 10×63 b) $5 \times 2 \times 9$ c) $5 \times 2 \times 3 \times 3 \times 7$ d) 9×70

6) $10m + (5h + 9) =$
   a) $(10m + 5h) + 9$ b) $(10m + 5h) + (10m + 9)$
   c) $(10m + 5h) + (10m + 9)$ d) $(10 + 5) \times (m \times h) + 9$

7) $(6 \times 74) + (6 \times 13) =$
   a) $6 \times (74 + 13)$ b) $6 + (74 + 13)$
   c) $(6 \times 6) \times (74 \times 13)$ d) $(6 \times 6) + (74 \times 13)$

8) The set of all numbers between 3 and 4 is an example of
   a) a finite set   b) an infinite set
   c) an empty set   d) an equal set

9) As a number, $\sqrt{5}$ is an example of a
   a) rational number   b) integer
   c) irrational number   d) whole number

10) Which quantity is not equal to $(84 + 16)(35)$
    a) $35 \times 16 + 35 \times 84$ b) $35(84 + 16)$
    c) $(84 + 16)35$ d) $(35 \times 16) + 84$

11) Which of the following does not hold true for the commutative property?
    a) $6 - 4 = 4 - 6$ b) $(6)(4) = (4)(6)$
    c) $6 + 4 = 4 + 6$ d) all statements are true

12) If $A = \{1, 3, 4\}$ and $B = \{4, 5, 6\}$ then $A \cup B =$
a) $\{1, 3, 4\}$ b) $\{1, 4, 6\}$
c) $\{1, 3, 4, 5, 6\}$ d) $\{4, 5, 6, 7\}$

13) Which of the following is a composite number?
a) 119   b) 107
c) 97   d) none of these

14) In determining whether the number 151 is a prime number, we need only try
a) all prime numbers up to 15   b) the prime numbers 2,3,5
c) the prime number 13   d) the prime numbers 2,3,5,7, and 11

15) A collection of objects is called a
a) number   b) set
c) property   d) none of the above

16) The set of all men 20 feet tall is an example of
a) a finite set   b) an infinite set
c) an empty set   d) an equal set

17) Which property makes the statement that follows true: $25 + (75 + 35) = (25 + 75) + 35$?
a) identity   b) associative
c) inverse   d) distributive

18) If $A = \{1, 2, 4\}$ and $B = \{4, 5, 6, 7\}$, then
a) $A \cap B = \{1, 2, 4\}$ b) $A \cup B = \{4, 5, 6\}$
c) $A \cap B = \phi$ d) $A \cap B = \{4\}$

19) According to the distributive property, the missing number in the equality
$6(8 + 7) = \_(8) + 6(7)$ is
a) 6   b) 8
c) 7   d) cannot be determined

20) Given two sets A and B, if $A \subset B$ and $B \subset C$, we can say that
a) set $A \neq$ set B b) set $A =$ set B
c) $A \cap B = \phi$ d) b and c are correct

21) As a number, $\sqrt{3}$ is an example of a
a) Rational number   b) Integer
c) Irrational number   d) Whole number

22) The number -6 belongs to all of the following subsets.
a) Natural, whole, and rational   b) Whole, integer, and irrational
c) Whole, integer, and rational   d) Integer and rational

23) Of the pairs of statements and properties below, which pair is correct?
   a) $9 + 6 = 6 + 9$ Identity
   b) $6 \times 1 = 1 \times 6$ Commutative
   c) $2(6 + 3) = 2 \times 6 + 2 \times 3$ Distributive
   d) Every pair is correct.

24) Given the sets $A = \{3, 4\}$ and $B = \{5, 6\}$, the intersection of set A and set B is equivalent to the intersection of the set
   a) of real and rational numbers
   b) of irrational and rational numbers
   c) of whole and natural numbers
   d) of natural numbers and integers

25) Which quantity is not equal to $35(16 + 84)$?
   a) $35 \times 16 + 35 \times 84$
   b) $35(84 + 16)$
   c) $(84 + 16)35$
   d) $(35 \times 16) + 84$

26) If Q is the set of rational numbers and L is the set of irrational numbers, then $Q \cup L$ is equal to the set of
   a) whole numbers
   b) prime numbers
   c) real numbers
   d) natural numbers

27) Find all the factors of 28.
   a) 1, 2, 14, 28
   b) 1, 2, 4, 14, 28
   c) 1, 2, 4, 7, 14, 28
   d) 2, 4, 7, 28

28) Find the prime factorization of 120.
   a) $2 \times 2 \times 2 \times 3 \times 3 \times 5$
   b) $2 \times 2 \times 3 \times 5$
   c) $2 \times 2 \times 2 \times 2 \times 5$
   d) $2 \times 2 \times 2 \times 3 \times 5$

29) Find the LCM of 30 and 12.
   a) 12   b) 30   c) 60   d) 90   e) 180

30) Find the QCD for 80 and 32.
   a) 32   b) 4   c) 8   d) 16

31) Find the GCD for 42 and 14.
   a) 14   b) 3   c) 21   d) 42

32) Find the LCM for 20 and 12.
   a) 60   b) 240   c) 40   d) 30

33) Find the LCM of $\{15, 9, 24\}$
   a) 12×9 b) $4 \times 3 \times 9$ c) $5 \times 3 \times 3 \times 2 \times 2 \times 2$

34) The prime factorization of 424 is
   a) 4×106 b) $2 \times 2 \times 2 \times 53$ c) $2 \times 2 \times 101$ d) $2 \times 2 \times 101$

35) 3,242 is evenly divisible by
   a) 4   b) 3 and 5   c) 2 and 3   d) 3   e) 2

36) 2,445 is evenly divisible by
   a) 2   b) 2 and 3   c) 9   d) 3 and 5   (e) 10

37) Determine the LCM of {32, 24, 6}.
   a) 4608   b) 6   c) 96   d) 32
   e) None of these

38) Write 108 as a product of prime factors.
   a) 12 × 9   b) 4 × 3 × 9   c) 2 × 3 × 3 × 3   d) 2 × 3 × 3 × 3 × 3
   e) None of these

39) Which is the greatest common divisor of 12, 18, and 42?
   a) 21   b) 9   c) 7   d) 6
   e) None of these

40) The smallest number which is evenly divisible by 8, 16, and 24 is which of the following numbers?
   a) 24   b) 32   c) 48   d) 72
   e) None of these

41) Which number is a multiple of 60?
   a) 213   b) 350   c) 540   d) 666
   e) 1,060

CHAPTER 2

ESTIMATION, EXPONENTS, AND SCIENTIFIC NOTATION

You are buying some grocery items and you have only $5.00. The milk costs $2.35 and the box of cereal costs $1.65. Do you have enough money?

To answer this question we do not need to know the exact total of the items, but only an estimate. In this chapter, we will explain how estimates are found. We will also discuss techniques for multiplying when we have repeated factors (exponents) and techniques for handling extremely large and small numbers.

 ## 2.1 ESTIMATING SUMS, DIFFERENCES, PRODUCTS, AND QUOTIENTS

When an exact solution to a problem is not necessary, we can estimate the sum, difference, product, or quotient by rounding the number to a given place value position (e.g., nearest whole, tens, thousandths, etc.) and then performing the operation.

EXAMPLE 1

At a restaurant you pay the bill for yourself and two friends. The checks are $3.59, $2.78, and $4.10. Estimate the total bill to the nearest dollar.

SOLUTION

Exact cost	Round to the nearest dollar
$3.59	$4.00
$2.78	$3.00
$4.10	$4.00

The total bill was approximately $11.00

Rules for estimating

When the position place value to which the number is to be rounded is not given, then the following "rules of thumb" should be used.

EXAMPLE 2

Estimate the sum: 197 + 248

SOLUTION

No place value is given, so the numbers are rounded so that the sum is easily obtained. Estimating, 197 becomes 200, and 248 becomes 250, giving an estimated sum of 450.

	Exact	Estimate
	197	200
	248	250
Sum	445	450

> To estimate a sum or difference, round the smaller number to the first non-zero digit, then round all other numbers to that place value position and add or subtract the rounded numbers.

EXAMPLE 3

You have $12.53. You purchase items totaling $4.60. After paying for these items, estimate how much money you would have left.

SOLUTION

By the rounding rule for sums or differences, $4.60 rounds to $5.00, and $12.53 would round to the nearest dollar as well. Rounding and subtracting,

$12.53 becomes	$13.00
$ 4.60 becomes	−5.00
Amount of money left	$ 8.00

ESTIMATING A PRODUCT

> To estimate a product, round each factor to its greatest place value position and then multiply.

EXAMPLE 1

Tickets to the circus cost $9.75 each. Estimate the cost of 4 tickets.

SOLUTION

Rounding each number by the rule for estimating a product, $9.75 rounds to $10.00. Multiplying by the number of tickets, $4 \times \$10.00 = \$40.00$.

EXAMPLE 2

John's Datsun 300ZX gets 33 miles per gallon of gasoline. The tank holds 19 gallons of gasoline. Estimate the distance that John can drive on a tank of gas.

SOLUTION

Rounding each number, 33 miles becomes 30 and 19 gallons becomes 20.

To find the number of miles, we multiply the miles per gallon, 30, by the capacity of the tank, 20 gallons. The result is 600 miles.

ESTIMATING A QUOTIENT

> To estimate a quotient, round the divisor to its greatest place value position. Then, round the dividend so that the quotient is easy to estimate.

EXAMPLE 1

By using a calculator, you get the following result: $175,770 \div 378 = 465$. Check your result by estimating.

SOLUTION

a) Rounding the divisor to its greatest place value position, 378 become 400.
b) Rounding the dividend so that the quotient is easy, 175,770 rounds to 200,000.
c) Dividing, $200,000 \div 400 = 500$

The estimate makes 465 seem reasonable.

Estimates are also useful in checking the placement of decimal position in numbers.

EXAMPLE 2

Use a rough estimate to select the correct answer in the following problem:

$(20.2 - 4.0702) \div 8.961 =$

a) .018 b) 1.8 c) .18.

SOLUTION

a) Rounding each number,

20.2	becomes 20
4.0702	becomes 4
8.961	becomes 9

b) Simplify the rounded expression.

$(20 - 4) \div 9$	Subtract within parentheses
$16 \div 9$	Dividing

The estimated answer is closest to the correct answer b) 1.8.

EXERCISE 2.1

Estimate the sum or difference to the nearest ten. Then find the actual answers.

1) $346 + 784$

2) $4591 + 1685 + 5091$

3) $932 - 317$

4) $945 - 309$

5) $7075 + 2982$

6) $378 - 89$

Estimate the sum or difference to the nearest hundred. Then find the actual answers.

7) $532 + 371$

8) $3577 + 2238$

9) $1483 + 1371$

10) $9108 - 1794$

11) $19,580 - 10,946$

12) $6,007 - 184$

Estimate the sum or difference to the nearest thousand. Then find the actual answers.

13) $3244 + 2745$

14) $17,942 + 36058$

15) $4138 + 9263$

16) $19580 - 11046$

17) $358,008 - 180,210$

18) $7188 - 5068$

Estimate the product or quotient, then find the exact answer.

19) 215×78

20) 473×81

21) 6340×43

22) 47.85×32

23) 416×785

24) 7.32×328

25) 893×437

26) 5391×645

27) $563 \div 5$

28) $3,867 \div 6$

29) $95 \div 14$

30) $295 \div 27$

31) $45,702 \div 137$

32) $25,862 \div 468$

33) $357,506 \div 785$

Estimate the answers to each of the following problems.

34) Tires regularly priced at $60 are on sale at Joe's for one-fourth off. Henry's offers the same tire for $45.95 at regular price. Which place has the better buy?

35) Timothy bought a rebuilt motor for $226. It cost $112 for installation. A new motor costs $610.32. How much did he save?

36) Two quarts of oil cost $4; an oil filter costs $16; and the labor costs $11. What is the cost of putting in an oil filter and 5 quarts of oil?

Estimate each of the following sums by rounding each addend to the nearest thousand.

37)	38)	39)	40)
9137	11,548	2238	3678
2315	3,874	3925	4215
7643	14,435	+ 5217	+ 2032
+ 3092	+ 5,398		

Estimate each of the following sums by rounding each addend to the nearest hundred.

41)	1378	42)	3145	43)	379	44)	967
	519		889		1215		2365
	792		259		+ 528		544
	+ 2041		692				+ 738
			+ 2518				

45) Pat went to the local supermarket and purchased the following items: milk, $2.89; butter, $1.75; bread, $1.10; orange juice, $1.25; cereal, $3.95; and coffee, $3.80. Approximate the total cost of these items.

46) Edna went shopping for clothes. She bought a sweater for $32.95, a scarf for $9.99, boots for $68.29, a coat for $125.90, and socks for $18.15. Estimate the total amount of Edna's purchases.

47) The attendance at a football game was 2345. The cost of admission was $12 per person. Estimate the total gate receipts for the game.

48) A company can manufacture 45 sleds per day. Approximately how many can this company make in 128 days?

49) There are 52 mathematics classes with 28 students in each class. Estimate the total number of students in the mathematics classes.

50) A theater has its seats arranged so that there are 42 seats per row. The theater has 48 rows. Estimate the number of seats in the theater.

Estimate each of the following products by rounding each factor to the nearest hundred.

51)	729	52)	212	53)	179
	× 481		× 278		× 431

Estimate each of the following products by rounding each factor to the nearest ten.

54)	74	55)	93	56)	27	57)	36
	× 57		× 48		× 34		× 23

Solve the following applications.

58) Bennie has $275 to spend on shirts. If the cost of a shirt is $23, estimate the number of shirts that Bennie can buy.

59) Pat purchased a used car for $1850 by paying $275 down and the rest in equal monthly payments over a period of 18 months. Estimate the amount of her monthly payments.

60) There is $365 left in the budget to purchase pens. If each box of pens costs $18, estimate the number of boxes of pens that can be ordered.

61) Twelve people are to share equally in an estate totaling $26,875. Estimate how much money each person will receive.

62) Charles Ervin can build a house in 27 days. Estimate how many houses he can build in 265 days.

63) Kirk drove 279 miles on 18 gallons of gas. Estimate his mileage. (Hint: Find the number of miles per gallon.)

2.2 EXPONENTS

In Chapter 1 we found prime factors of numbers. For example, the prime factors of the numbers below are as follows:

Number	Prime factors
9	3×3
27	$3 \times 3 \times 3$
81	$3 \times 3 \times 3 \times 3$

We can write these prime factors using a shorthand method as shown below.

Product	Can be written	Read
3×3	3^2	3 to the second power or 3 squared
$3 \times 3 \times 3$	3^3	3 to the third power or 3 cubed
$3 \times 3 \times 3 \times 3$	3^4	3 to the fourth power

EXPONENTIAL NOTATION

The numbers above are written in exponential notation. The numbers, 2, 3, and 4 are called exponents and 3 is called the base. The exponent tells us the number of times the base is used as a factor. For example, in 4^3, 3 is the exponent and 4 is the base. Therefore, by definition, $4^3 = 4 \times 4 \times 4$ or 64.

EXAMPLE 1

Write the following numbers in exponential form.

a) $4 \times 4 \times 4 \times 4 \times 4$ b) $6 \times 6 \times 7 \times 7 \times 7$

SOLUTION

a) 4^5, since the base, 4, is used as a factor 5 times (exponent).

b) $6^2 \times 7^3$

EXAMPLE 2

Evaluate the following

a) 2^4 b) $\left(\dfrac{2}{3}\right)^2$ c) $4^3 \times (.2)^2 \times 5$ d) $\left(\dfrac{1}{2}\right)^3 + \left(\dfrac{1}{3}\right)^2$

49

SOLUTION

a) $2^4 = 2 \times 2 \times 2 \times 2 = (2 \times 2) \times (2 \times 2)$

$$= 4 \times 4$$
$$= 16$$

b) $\left(\dfrac{2}{3}\right)^2 = \left(\dfrac{2}{3}\right) \times \left(\dfrac{2}{3}\right) = \dfrac{4}{9}$

c) $4^3 \times (.2)^2 \times 5 = (4 \times 4 \times 4) \times (.2 \times .2) \times 5$

$$= 64 \quad \times \quad (.04) \quad \times 5$$
$$= 64 \quad \times \quad .20$$
$$= 12.80$$

d) First, evaluate each factor separately.

$$\left(\dfrac{1}{2}\right)^3 = \left(\dfrac{1}{2}\right) \times \left(\dfrac{1}{2}\right) \times \left(\dfrac{1}{2}\right) = \dfrac{1}{8}$$

$$\left(\dfrac{1}{3}\right)^2 = \left(\dfrac{1}{3}\right) \times \left(\dfrac{1}{3}\right) = \dfrac{1}{9}$$

Now, combine the two fractions.

$$\left(\dfrac{1}{8}\right) + \left(\dfrac{1}{9}\right) = \left(\dfrac{9}{72}\right) + \left(\dfrac{8}{72}\right) = \dfrac{17}{72}$$

From the definition of exponents we can develop rules for multiplying and dividing exponents. For example, $3^2 \times 3^3 = (3 \times 3) \times (3 \times 3 \times 3)$

$$= 3 \times 3 \times 3 \times 3 \times 3 = 3^5$$

The same result could be obtained by adding the exponents. Thus, $3^2 \times 3^3 = 3^{2+3} = 3^5$.

Product Rule For Exponents: In general, for any positive integer exponents, x and y, where the bases are the same, $a^x \times a^y = a^{x+y}$

EXAMPLE 3

Use the product rule for exponents to find each product.

a) $5^2 \times 5^6$
b) $\left(\dfrac{1}{3}\right)^2 \times \left(\dfrac{1}{3}\right)^4$
c) $(.7)^4 \times (.7)$

SOLUTION

a) By the product rule, $5^2 \times 5^6 = 5^{2+6} 5^8$

b) $\left(\dfrac{1}{3}\right)^2 \times \left(\dfrac{1}{3}\right)^4 = \left(\dfrac{1}{3}\right)^{2+4} = \left(\dfrac{1}{3}\right)^6$

c) $(.7)^4 \times (.7) = (.7)^{4+1} = .7^5$

Note: Every number is understood to be raised to the first power if no exponent is written.

From the definition of exponents, the quotient rule may be derived as well. For example,

---

Quotient Rule For Exponents: In general, for any integer exponents, x and y, where the bases are the same, $\dfrac{a^x}{a^y} = a^{x-y}$

EXAMPLE 4

Use the quotient rule for exponents to find each quotient.

a) $\dfrac{5^6}{5^2}$ b) $\dfrac{6^2 \times 4^5}{6 \times 4^2}$ c) $\dfrac{10^7 \times 10^9}{10^{10}}$

SOLUTION

a) By the quotient rule, $\dfrac{5^6}{5^2} = 5^{6-2} = 5^4$

b) $\dfrac{6^2 \times 4^5}{6 \times 4^2}$ Divide each similar factor separately by subtracting the exponents.

$\dfrac{6^2}{6} = 6^{2-1} = 6$

$\dfrac{4^5}{4^2} = 4^{5-2} = 4^3$

$6 \cdot 4^3$ Express the factors as a product.

c) $\dfrac{10^7 \times 10^9}{10^{10}}$ Multiply the factors in the numerator (add exponents)

$\dfrac{10^{16}}{10^{10}} = 10^{16-10}$ Divide by subtracting the exponents.

10^6

51

From the quotient rule, the definition of the zero exponent may be derived. To illustrate, find the quotient of $\dfrac{3^2}{3^2}$.

By definition, $\dfrac{3^2}{3^2} = \dfrac{3 \times 3}{3 \times 3} = 1$. Using the quotient rule, $\dfrac{3^2}{3^2} = 3^{2-2} = 3^0$. Since the quotient of a number and itself is one, then 3^0 must equal 1. It follows then, for any real number x, $x^0 = 1$.

EXAMPLE 5

Use a calculator to evaluate each of the following:

a) 5^3 b) 18^3 c) 5^{-4}

SOLUTION

a) To compute 5^3, using a calculator, enter 5, press either of the following exponential keys $[y^x]\,[x^y]\,[\wedge]$ depending upon your calculator, then 3, press 3, display reads 125.

b) Keystrokes: enter 18 press $\wedge$ 3 then equal, display reads 5832.

c) Enter [5], press [$\wedge$] [$-$] [4] then equal, display reads .0016.

If your calculator has a [+/$-$] key, use the following sequence:
Enter [5], press [y^x] [4] [+/$-$] then equal, display reads .0016.

EXERCISE 2.2

Write each expression in exponential form using a single exponent.

1) $5 \times 5 \times 5 \times 5 \times 5$
2) $7 \times 7 \times 7 \times 8 \times 8$
3) $5^3 \times 5^4$

4) $4^5 \times 5^5$
5) $25^7 \div 5^3$
6) 27×3^5

7) $7^{12} \div 7^3$
8) $\dfrac{1}{6 \times 6 \times 6 \times 6}$
9) $\dfrac{1}{2 \times 2 \times 2 \times 3 \times 3 \times 4 \times 4}$

Evaluate each expression.

10) $4^3 + 4^2$
11) $2^3 - 2^2$
12) $4^0 + 7^0$
13) 6^2

14) 5^3
15) $2^3 \times (\dfrac{7}{2})^2$
16) $(\dfrac{3}{2})^2$
17) $5^3 \times 10^3$

18) $\dfrac{2^8 \times 3^4}{4^3}$
19) $\dfrac{9^3 \times 12^2}{3^8}$
20) $\dfrac{6^8 \times 15^3}{6^3 \times 15^2}$
21) $\dfrac{4 \times 10^5 \times 2^3}{4 \times 10^3 \times 2}$

Use a calculator to evaluate the following.

22) 5^6
23) 38^6
24) 7^{10}
25) 9^{-3}

26) 13^{-2}
27) $3^6 + 2^5$
28) $3^4 + 5^3$
29) $2 \times 5^7 - 3 \times 3^3$

30) $\dfrac{9^8 \times 3^{10}}{2^{14}}$

2.3 SYMBOLS

Thus far, the symbols of arithmetic, such as +, −, x (or •), and ÷, have been used. Another common symbol is the one for equality, =, which shows that two numbers are equal. This symbol with a slash through it, ≠, mean "is not equal to." For example,

$$6 \neq 7$$

Indicates that 6 is not equal to 7.

If two numbers are not equal, then one of the numbers must be smaller than the other. The symbol < represents "is less than," so that "6 is less than 7" is written

$$6 < 7.$$

Also, write "7 is less than 8" as $7 < 8$.

The symbol > means "is greater than." Write "5 is greater than 2" as

$$5 > 2.$$

The statement "16 is greater than 11" becomes $16 > 11$.

Keep the symbols < and > straight by remembering that the symbol always points to the smaller number. For example, write "9 is less than 15" by pointing the symbol toward the 9.

$$9 < 15.$$

In mathematics, it is often necessary to convert word phrases to symbols. The next example shows this.

EXAMPLE 1

Write each word statement in symbols.

    a)   Twelve equal ten plus two.

    b)   Fifteen is not equal to seventeen.

    c)   Eight is less than twelve.

    d)   Nine is greater than four.

SOLUTION

    a) $12 = 10 + 2$

    b) $15 \neq 17$

    c) $8 < 12$

    d) $9 > 4$

The two symbols, ≤ and ≥, also represent the idea of inequality. The symbol ≤ means, "is less than or equal to," so that

$$7 \leq 9$$

means "7 is less than or equal to 9." This statement is true, since 7 < 9 is true. If either the < part or the = part is true, then the inequality ≤ is true.

The symbol ≥ means "is greater than or equal to." Again,

$$9 \geq 7$$

is true because 9 > 7 is true. Also, 8 ≤ 8 is true since 8 = 8 is true. But it is not true that 11 ≤ 9 because neither 11 < 9 nor 11 = 9 is true.

EXAMPLE 2

Tell whether or not each statement is true.

    a) $11 \leq 20$ b) $25 \geq 30$ c) $10 \geq 10$

SOLUTION

    a)   This statement $11 \leq 20$ is true, since 11 < 20.

    b)   Both 25 > 30 and 25 = 30 are false. Because of this, $25 \geq 30$ is false.

    c)   Since 10 = 10 this statement is true.

Any statement with < can be converted to one with >, and any statement with > can be converted to one with <. Do this by reversing the order of the numbers and the direction of the symbol. For example, the statement 6 < 10 can be written with > as 10 > 6. Similarly, the statement 4 ≤ 10 can be changed to 10 ≥ 4.

EXAMPLE 3

The following list shows the same statement written in two equally correct ways.

        a) $9 < 11$ $11 > 9$

        b) $5 > 2$ $2 < 5$

        c) $4 \leq 8$ $8 \geq 4$

        d) $12 \geq 5$ $5 \leq 12$

Here is a summary of the symbols discussed in this section.

Symbols of Equality and Inequality	
= is equal to	≠ is not equal to
< is less than	> is greater than
≤ is less than or equal to	≥ is greater than or equal to

EXERCISE 2.3

Insert < or > to make the following statements true.

1) 6 _____ 9   2) 17 _____ 9   3) 25 ___ 12   4) 8 _____ 10

5) 12 ___ 15   6) 5 _____ 3   7) 6 _____ 9   8) 41 ___ 72

9) $1\frac{5}{8}$ ___ 1   10) $\frac{2}{3}$ _____ 0   11) $\frac{3}{4}$ _____ 1   12) $3\frac{7}{9}$ ___ 2

Insert ≤ or ≥ to make the following statements true.

13) 8 _____ 28   14) 10 _____ 15   15) 35 ___ 42   16) 51 ___ 62

17) 12 ___ 17   18) 28 _____ 42   19) 16 ___ 14   20) 39 _____ 17

Which of the symbols <, >, ≤, and ≥ make the following statements true?

21) $3\frac{1}{2}$ _____ 4   22) .5 ___ .499   23) .609 ___ .61   24) $\frac{2}{3}$ _____ $\frac{5}{8}$

25) $\frac{1}{4}$ _____ $\frac{2}{5}$ 26) 5 _____ 3   27) 16 _____ 10   28) 100 ___ 1000

29) 48 _____ 0   30) 10 _____ 10   31) 5 _____ 5   32) 0 _____ 12

33) 51 _____ 50   34) 18 _____ 12   35) 6 _____ 9

Write the following word statements in symbols.

36) Six is less than or equal to six.

37) Zero is greater than or equal to zero.

38) Fifteen does not equal sixteen.

39) Twelve is not equal to five.

40) Five equals ten minus five.

41) Three is less than the quotient of fifty and five.

42) Nine is greater than the product of four and two.

43) Seven equals five plus two.

Tell whether each statement is true or false.

44) $5 \le 5$

45) $12 \ge 12$

46) $26 \ge 50$

47) $8 \le 0$

48) $1.95 \ge 1.96$

49) $2.13 < 2.13$

50) $15 < 21$

51) $6 \ne 5 + 1$

52) $15 \le 32$

53) $9 < 0$

54) $\dfrac{18}{5} < \dfrac{5}{4}$

55) $\dfrac{25}{3} \ge \dfrac{19}{2}$

56) $3\dfrac{2}{5} < 6\dfrac{1}{4}$

57) $1\dfrac{2}{3} + 2\dfrac{3}{4} = \dfrac{53}{12}$

58) $16 \ge 10$

59) $0 < 15$

60) $45 < 45$

61) $12 \ge 10$

62) $8 \ne 9 - 1$

63) $8 + 2 = 10$

2.4 ORDER OF OPERATIONS

Many problems involve more than one operation of arithmetic. For example, to simplify the expression $5 + 2 \times 3$ to a single number, should we add first, or multiply first? One way to make the order of operations clear is to use grouping symbols.

The most useful way to work problems with more than one operation is to use the following **order of operations**. This is the order used by calculators and computers.

If grouping symbols are present, simplify within them, innermost first (and above and below fraction bars separately), in the following order.

Step 1 Apply all exponents.

Step 2 Do any multiplications or divisions in the order in which they occur, working from left to right.

Step 3 Do any additions or subtractions in the order in which they occur, working from left to right.

If no grouping symbols are present, start with Step 1.

A dot has been used to show multiplication; another way to show multiplication is with parentheses. For example, 3(7), (3)7, and (3)(7) each mean 3×7 or 21. The next example shows the use of parentheses for multiplication.

EXAMPLE 1

Evaluate $6 + 3[(12 \div 4) + 5]$.

SOLUTION

$$
\begin{aligned}
6 + 3[(12 \div 4) + 5] &= 6 + 3[3 + 5] \\
&= 6 + 3(8) \\
&= 6 + 24 \\
&= 30
\end{aligned}
$$

EXAMPLE 2

Evaluate $(4 \div 2) + 4(5 - 2)^2$

SOLUTION

$$
\begin{aligned}
(4 \div 2) + 4(5 - 2)^2 &= 2 + 4(3)^2 \\
&= 2 + 4 \cdot 9 \\
&= 2 + 36 \\
&= 38
\end{aligned}
$$

The following example shows how to decide where to insert grouping symbols so that an expression equals a particular number.

EXAMPLE 3

Insert parentheses so that the following are true.

   a) $9 - 3 - 2 = 8$
   b) $9 \cdot 2 - 4 \cdot 3 = 6$

SOLUTION

   a) This statement would be true if parentheses were inserted around $3 - 2$.

$$9 - (3 - 2) = 8$$

   It is not true that $(9 - 3) - 2 = 8$, since $6 - 2 \neq 8$.

   b) Since $9 \times 2 - 4 \times 3 = 18 - 12 = 6$, no parentheses are needed here. If desired, parentheses may be placed as follows.

$$(9 \times 2) - (4 \times 3) = 6$$

EXERCISE 2.4

Find the values of the following expressions

1) $6 \cdot 5 + 3 \cdot 10$

2) $2 \cdot 20 - 8 \cdot 5$

3) $3 \cdot 8 - 4 \cdot 6$

4) $16 - 3 \cdot 5$

5) $12 - 5 \cdot 2$

6) $9 + 3 \cdot 4$

7) $4 + 6 \cdot 2$

8) $\dfrac{4^2 - 8}{15 - 3^2}$

9) $\dfrac{8^2 + 2}{5 - 2^2}$

10) $\dfrac{9(7-1) - 8 \cdot 2}{4(6-1)}$

11) $\dfrac{2(5+3) + 2 \cdot 2}{2(4-1)}$

12) $(7-1)[9 + (6-3)]$

13) $(6-3)[8 - (2+1)]$

14) $9[(14+5) - 10]$

15) $5[8 + (2+3)]$

First simplify both sides of each inequality. Then tell whether the given statement is true or false.

16) $7 \le \dfrac{3(8-3) + 2(4-1)}{9(6-2) - 11(5-2)}$

17) $3 \ge \dfrac{2(5+1) - 3(1+1)}{5(8-6) - 4 \cdot 2}$

18) $\dfrac{7(3+1) - 2}{3 + 5 \cdot 2} \le 2$

19) $\dfrac{3 + 5(4-1)}{2 \cdot 4 + 1} \ge 3$

20) $2 \cdot [7 \cdot 5 - 3(2)] \le 58$

21) $[3 \cdot 4 + 5(2)] \cdot 3 > 72$

22) $55 \ge 3[4 + 3(4+1)]$

23) $45 \ge 2[2 + 3(2+5)]$

24) $10 \le 13 \cdot 2 - 15 \cdot 1$

25) $0 \ge 12 \cdot 3 - 6 \cdot 6$

26) $9 \cdot 3 + 4 \cdot 5 \ge 48$

27) $5 \cdot 11 + 2 \cdot 3 \le 60$

28) $6 \cdot 5 - 12 \le 18$

29) $9 \cdot 3 - 11 \le 16$

Insert parentheses in each expression so that the resulting statement is true. Some problems require no parentheses.

30) $6^2 - 2 \cdot 5 = 170$

31) $8 - 2^2 \cdot 2 = 8$

32) $3^3 - 2 \cdot 4 = 100$

33) $6 + 5 \cdot 3^2 = 99$

34) $2^2 + 4 \cdot 2 = 16$

35) $4096 \div 256 \div 4 = 4$

36) $100 \div 20 \div 5 = 25$

37) $3 \cdot 5 + 2 \cdot 4 = 68$

38) $100 \div 20 \div 5 = 1$

39) $2^3 + 4 \cdot 2 = 24$

40) $3 \cdot 5 + 2 \cdot 4 = 84$

41) $3 \cdot 5 + 2 \cdot 4 = 23$

42) $3 \cdot 5 - 4 = 11$

43) $3 \cdot 5 - 4 = 3$

44) $3 \cdot 5 + 7 = 36$

45) $3 \cdot 5 + 7 = 22$

46) $16 - 4 - 3 = 15$

47) $10 - 7 - 3 = 6$

2.5 WRITING NUMBERS IN SCIENTIFIC NOTATION

Scientific notation is useful when we are working with extremely large numbers or extremely small numbers. It is quite useful when using a slide rule to perform calculations on ordinary numbers. Scientific notation is a very useful concept to know when you study the topic of logarithms in a later course in mathematics. To understand what we mean by writing a number in scientific notation, let us look at the definition.

> A number is written in **scientific notation** if the number is expressed as a number with one and only one nonzero digit to the left of the decimal point times 10 to some power.

Let us examine some numbers that are written in both ordinary notation and scientific notation to help us understand the meaning of the above definition.

$$729. = 7.29 \times 10^2 \text{ in scientific notation}$$

The number 729 in ordinary notation is written as a number with one and only one nonzero digit (the digit 7) to the left of the decimal point times 10 to the second power as is required by the definition.

$$0.0014 = 1.4 \times 10^{-3} \text{ in scientific notation}$$

RULES FOR CONVERTING FROM ORDINARY NOTATION TO SCIENTIFIC NOTATION

> To convert an ordinary number to scientific notation, move the decimal point so that one nonzero digit is to the left of the decimal point and express the result times 10 to some power. If the decimal point is moved to the left, the power of 10 is positive and has as its value the number of places the decimal point is moved. If the decimal point is moved to the right, the power of 10 is negative and has as its value the number of places the decimal point is moved.

EXAMPLE 1

Express the number 739 in ordinary notation as a number in scientific notation.

SOLUTION

$$739 = 7.39 \times 10^2$$ Move the decimal point two places to the left, and

$$= 7.39 \times 10^2$$ then multiply 7.39 by 10 to the second power.

EXAMPLE 2

Express the number 26,800 in ordinary notation as a number in scientific notation.

SOLUTION

Step 1.   2.6800.   Move the decimal point to the left so that there is one and only one nonzero digit to the left of the decimal point.

Step 2. 2.68×10^4 Multiply 2.68 by 10 to some power. To determine the power of 10, notice that we move the decimal point four places to the left, thus the power of 10 is +4.

EXAMPLE 3

Express the number 0.0016 in ordinary notation as a number in scientific notation.

SOLUTION

0.001.6   Move the decimal point 3 places to the right and
$= 1.6 \times 10^{-3}$ multiply 1.6 by 10 to the −3 power.

EXAMPLE 4

Express the number 0.000084 in ordinary notation as a number in scientific notation.

SOLUTION

Step 1.   0.00008.4   Move the decimal point to the right so that there is one and only one nonzero digit to the left of the decimal point.

Step 2. 8.4×10^{-5} Multiply 8.4 times 10 to some power. To determine the power of 10, notice that we moved the decimal point 5 places to the right. Thus the power of 10 is minus 5.

EXERCISE 2.5 A

Write the following in scientific notation.

1) 67,900 2) 400 3) 2180 4) 845,000

5) 6,280,000,000,000,000,000 6) .005 7) .0000749

8) .00002 9) .00000892 10) .0000329

Write the number in each of the following in scientific notation.

11) The sun is about 93,000,000 miles from the earth.

12) Light travels about 186,200 miles per second.

13) The wave length of a yellow light is about 0.0000228 inches.

14) One meter equals 10,000,000 microns.

15) One nanometer equals .000000001 meter.

16) Thirty-nine thousand, eight hundred twenty-four.

17) Sixty-six million.

18) Seven billion fourteen million.

19) One hundred sixty-four thousand.

20) Thirty-six thousandths.

21) Forty-seven and four tenths.

22) Eight and twenty-five thousandths.

23) 1,000,000 24) 849,000 25) 685,000
26) 12,000 27) 11,000 28) 3000

EXERCISE 2.5 B

Choose the one alternative that best completes the statement or answers the question.

Write the number in scientific notation.

1) 0.008×10^8
   a) 8×10^6 b) 8×10^5 c) 8×10^{11} d) 0.8×10^6

2) 79,000,000
   a) 7.9×10^7 b) 7.9×10^{-7} c) 7.9×10^8 d) 7.9×10^{-8}

3) 0.0009
   a) 9×10^{-4} b) 9×10^{-3} c) 9×10^{-5} d) -9×10^4

4) 0.00000651
   a) 6.51×10^{-7} b) -6.51×10^6 c) 6.51×10^{-5} d) 6.51×10^{-6}

5) 7,975,060
   a) 7.97506×10^{-6} b) 7.97506×10^6 c) 7.97506×10^1 d) 7.97506×10^7

6) 121.9
   a) 1.219×10^3 b) 1.219×10^2 c) 1.219×10^{-3} d) 1.219×10^{-2}

7) 74,000
   a) 7.4×10^{-4} b) 7.4×10^{-5} c) 7.4×10^5 d) 7.4×10^4

8) 7744
   a) 7.744×10^1 b) 7.744×10^{-3} c) 7.744×10^4 d) 7.744×10^3

9) 0.09×10^6
   a) 9×10^4 b) 0.9×10^5 c) 9×10^{-5} d) 9×10^8

10) 160.4
   a) 1.604×10^3 b) 1.604×10^{-2} c) 1.604×10^{-3} d) 1.604×10^2

11) 17,000
   a) 1.7×10^3 b) 1.7×10^{-4} c) 1.7×10^{-3} d) 1.7×10^4

12) 86,000,000
   a) 8.6×10^6 b) 8.6×10^{-6} c) 8.6×10^7 d) 8.6×10^8

13) 0.0000000379
   a) -3.79×10^8 b) 3.79×10^{-7} c) 3.79×10^{-9} d) 3.79×10^{-8}

14) 0.004
a) -4×10^8 b) 4×10^4 c) 4×10^3 d) 4×10^{-2}

15) 136,000
a) 3.6×10^4 b) 3.6×10^{-4} c) 3.6×10^3 d) 3.6×10^{-3}

16) 0.002×10^{-7}
a) 2×10^{10} b) 2×10^4 c) 0.2×10^5 d) 2×10^{-10}

17) 315.7
a) 3.157×10^2 b) 3.157×10^{-1} c) 3.157×10^{-2} d) 3.157×10^1

18) 517
a) 5.17×10^1 b) 5.17×10^2 c) 5.17×10^3 d) 5.17×10^{-2}

19) 0.04
a) 4×10^{-3} b) 4×10^{-2} c) -4×10^2 d) 4×10^{-1}

20) 4,700,000
a) 4.7×10^6 b) 4.7×10^{-6} c) 4.7×10^{-5} d) 4.7×10^5

21) 0.00000978
a) 9.78×10^{-5} b) 9.78×10^{-7} c) 9.78×10^{-6} d) -9.78×10^6

2.6 CONVERTING FROM SCIENTIFIC NOTATION TO ORDINARY NOTATION

We now understand how to convert a number from ordinary notation to scientific notation. Let us now learn how to convert a number written in scientific notation to ordinary notation.

> To convert from scientific notation to ordinary notation, the power of 10 tells how many places to move the decimal point and the sign of the exponent tells which direction to move it. If the power of 10 is positive, move the decimal point to the right the same number of places as the value of the exponent. If the power of 10 is negative, move the decimal point to the left the same number of places as the value of the exponent.

EXAMPLE 1

Convert the number 7.6×10^{-4} in scientific notation to a number in ordinary notation.

SOLUTION

7.6×10^{-4} The power of 10 is -4.

$= 0.\underline{0007}.6$ Thus, move the decimal point to the left.

$= 0.00076$

EXAMPLE 2

Convert the number 2.8×10^5 in scientific notation to a number in ordinary notation.

SOLUTION

2.8×10^5 The power of 10 is plus 5.

$= 2.\underline{80000}.$ Thus, move the decimal point 5 places to the right.

$= 280,000$

EXAMPLE 3

Convert the number 2.67×10^3 in scientific notation to a number in ordinary notation.

SOLUTION

$2.\underline{670}. = 2670$ The power of 10 is plus 3, which tells us to move the decimal point three places to the right. We must place the digit 0 to the right of the digit 7 to have three places.

EXAMPLE 4

Convert the number 1.5×10^{-1} in scientific notation to a number in ordinary notation.

SOLUTION

1.5×10^{-1} The power of 10 is minus 1.

$= 0.1.5$

$= 0.15$ Thus, move the decimal point 1 place to the left.

EXAMPLE 5

Convert the number 36.8×10^2 to a number in ordinary notation.

SOLUTION

36.8×10^2 Notice that 36.8×10^2 is not in scientific notation because there are

$= 36.80.$ two digits to the left of the decimal point. But the power of 10 tells us

$= 3680$ to move the decimal point 2 places to the right, and the desired result is 3680.

EXERCISE 2.6 A

Convert the following to ordinary notation.

1) 69.5×10^3 2) $.0059 \times 10^{-3}$ 3) 5.64×10^{-5} 4) 1.72×10^{-5}

5) 2.15×10^0 6) 2.456×10^7 7) 3.75×10^{-1} 8) 0.16×10^{-3}

9) 18.1×10^4 10) $.0015 \times 10^3$ 11) 14.5×10^{-3} 12) 342.8×10^2

13) 3.3×10^0 14) 1.14×10^{-1} 15) 8.98×10^5 16) 2.15×10^3

17) 7.69×10^4 18) 8.7×10^{-3} 19) 1.03×10^{-5} 20) 9.704×10^2

21) 3.86×10^{-2}

EXERCISE 2.6 B

Choose the one alternative that best completes the statement or answers the question.

Write the number in ordinary notation.

1) 7.546×10^{-6}
a) 0.0000007546   b) 0.000007546   c) −7,546,000   d) 0.00007546

2) 3×10^5
a) 30,000   b) 3,000,000   c) 30,000,000   d) 300,000

3) 7.537×10^4
a) 753,700   b) 301.48   c) 75,370   d) 7537

4) 4.0841×10^{-7}
a) 0.000000040841   b) 0.00000040841   c) −408,410,000   d) 0.0000040841

5) 6.6×10^7
a) 462   b) 66,000,000   c) 6,600,000   d) 660,000,000

6) 2.6483×10^7
a) 264,830,000   b) 185.381   c) 2,648,300   d) 26,483,000

7) 5×10^{-2}
a) 0.005   b) −500   c) 0.05   d) 0.5

8) 6.1309×10^4
a) 61,309 b) 6130.9 c) 245.236 d) 613,090

9) 4.0725×10^{-7}
a) 0.000000040725 b) −407250,000 c) 0.0000040725 d) 0.00000040725

10) 5.239×10^{-5}
a) 0.00005239 b) -523,900 c) 0.0005239 d) 0.000005239

11) 5.606×10^{-5}
a) 0.0005606 b) -560,600 c) 0.00005606 d) 0.000005606

12) 6.6004×10^6
a) 660,040 b) 66,004,000 c) 6,600,400 d) 396.024

13) 4.0913×10^{-7}
a) 0.00000040913 b) 0.0000040913 c) −409130000 d) 0.000000040913

14) 7.4×10^5
a) 7,400,000 b) 370 c) 74,000 d) 740,000

15) 6.6431×10^4
a) 265.724 b) 66,431 c) 6643.1 d) 664,310

16) 7×10^2
a) 70,000 b) 700 c) 70 d) 7000

17) 5.496×10^{-6}
a) 0.0000005496 b) −5,496,000 c) 0.00005496 d) 0.000005496

18) 2.274×10^7
a) 2,274,000 b) 22,740,000 c) 159.18 d) 227,400,000

19) 9.0421×10^{-7}
a) 0.00000090421 b) −904210,000 c) 0.000000090421 d) 0.0000090421

20) 7×10^{-6}
a) −7,000,000 b) 0.00007 c) 0.000007 d) 0.0000007

ADDITION AND SUBTRACTION OF NUMBERS IN SCIENTIFIC NOTATION

> To add or subtract numbers in scientific notation, first change to ordinary notation, and then do the addition or subtraction.

EXAMPLE 1

Find the difference of $\left(2.85 \times 10^{-2}\right)$ and $\left(5.241 \times 10^{-3}\right)$.

SOLUTION

Step 1. $0.0285 - 0.05241$ Change each number from scientific notation to ordinary notation and indicate the subtraction.

Step 2. $= 0.023259$ Do the subtraction.

EXAMPLE 2

Find the sum of $\left(3.14 \times 10^{2}\right)$ and $\left(2.18 \times 10^{3}\right)$.

SOLUTION

Step 1. $314 + 2180$ Change the number from scientific notation to numbers in ordinary notation and indicate the addition.

Step 2. $= 2494$ Add the result.

 MULTIPLICATION AND DIVISION OF NUMBERS IN SCIENTIFIC NOTATION

With scientific notation, working multiplication and division problems involving extremely large or extremely small numbers is simplified. To multiply or divide two or more numbers in scientific notation, we use the associative and the commutative properties to group similar parts of the number in scientific notation (e.g., the power of 10) together and the rules for multiplying and/or dividing exponents to simplify the power of 10. The rules on the next page show the process for multiplying and dividing numbers written in scientific notation.

> To multiply two or more numbers in scientific notation, rearrange the numbers so that the numbers with significant digits are grouped together and the powers of 10 are grouped together. Multiply the numbers together first, and then multiply the powers of 10 and express the result in scientific notation.

> To divide two numbers in scientific notation, divide the numbers with significant digits first, then divide the powers of 10 for the power of 10 in the quotient and express the result in scientific notation.

EXAMPLE 1

Find the product of $(8 \times 10^{-2}) \times (9 \times 10^4)$.

SOLUTION

$(8 \times 10^{-2}) \times (9 \times 10^4)$

$= (72 \times 10^2)$

$= (7.2 \times 10^1) \times 10^2$ Write 72 in scientific notation.

$= 7.2 \times (10^1 \times 10^2)$ Associative property.

$= 7.2 \times 10^3$

EXAMPLE 2

Find the product of $(2.1 \times 10^2) \times (3 \times 10^3)$.

SOLUTION

Step 1. $(2.1 \times 3) \times (10^2 \times 10^3)$ Group 2.1 and 3 together as a product and group the powers of 10 together as a product.

Step 2. $= 6.3 \times 10^5$ Do the multiplication.

EXAMPLE 3
Find the quotient of $(4.2 \times 10^3) \div (2.1 \times 10^2)$.

SOLUTION

Step 1. $(4.2 \times 10^3) \div (2.1 \times 10^2)$ Divide 4.2 by 2.1 and 10^3 by 10^2.

Step 2. 2×10^1 Do the division.

EXAMPLE 4

$$\frac{2.7 \times 10^2}{3 \times 10^5}$$

SOLUTION

Step 1. $\dfrac{2.7}{3} \times \dfrac{10^2}{10^5}$ Divide 2.7 by 3 and 10^2 by 10^5.

Step 2. 0.9×10^{-3} Do the division.

Step 3. $9 \times 10^{-1} \times 10^{-3}$ Change 0.9 to scientific notation.

Step 4. 9×10^{-4} Multiply powers of 10 by adding the exponents.

USING A CALCULATOR

A scientific calculator may be used to perform operations of multiplication and division with numbers written in scientific notation. If you ever tried to multiply two extremely large numbers, such as $3,506,000,000 \times 67,000,000$, the calculator would either give you an error or give the answer in scientific notation. For this example, the display would read 2.34902 E17. This is the answer written in scientific notation. Using our usual notation we would write it this way: $2.34902\ E17 = 2.34902 \times 10^{17}$.

EXAMPLE 5

Write the following output from a calculator in scientific notation.
a) 3.4567 E 5   b) 6.7890 E−4

SOLUTION

a) The number following E gives the power often for the number, thus,
$3.4567\ E5 = 3.4567 \times 10^5 = 345,670$

b) The exponent is negative, so we count to the left giving,
$6.7890\ E{-}4 = 6.789 \times 10^{-4} = .00067890$

EXAMPLE 6

Use a calculator to multiply 1.73×10^8 by 8.41×10^4.

SOLUTION

Enter each number using the E key, and then perform the indicated operation.

1.73×10^8 is entered as 1.73 E8 and 8.41×10^4 is entered as 8.41 E4. To multiply, press the following keys: (1.73 [E] 8) [x] (8.41 [E] 4) =
Display reads: 1.45493 E13 which becomes 1.45493×10^{13}.

EXAMPLE 7

Simplify: $7.21 \times 10^3 \div 4.0 \times 10^8$

SOLUTION

Use the following keystrokes: (7.21 E 3) ÷ (4.0 E 8)
Display reads: 1.8025 E−5 which becomes 1.8025×10^{-5}.

EXERCISES 2.7 – 2.8 A

Add or subtract as indicated.

1) $\left(3.9 \times 10^{-2}\right) - \left(2.8 \times 10^{-3}\right) =$

2) $\left(9.8 \times 10^{3}\right) - \left(3.7 \times 10^{2}\right) =$

3) $\left(7.9 \times 10^{2}\right) - \left(6.4 \times 10^{1}\right) =$

4) $\left(9.84 \times 10^{-2}\right) + \left(3.21 \times 10^{-2}\right) - \left(9.84 \times 10^{-3}\right) =$

5) $\left(4.73 \times 10^{2}\right) + \left(3.14 \times 10^{3}\right) =$

6) $\left(3.72 \times 10^{-2}\right) + \left(2.1 \times 10^{1}\right) =$

7) $\left(1.429 \times 10^{-2}\right) + \left(3.142 \times 10^{-3}\right) =$

Multiply the following. Leave your answer in scientific notation.

8) $\left(7 \times 10^{-1}\right) \times \left(3 \times 10^{-2}\right) \times \left(4 \times 10^{-3}\right) =$

9) $\left(4 \times 10^{3}\right) \times \left(5 \times 10^{3}\right) \times \left(4.1 \times 10^{0}\right) =$

10) $\left(9 \times 10^{2}\right) \times \left(8.1 \times 10^{3}\right) \times \left(7 \times 10^{-3}\right) =$

11) $\left(8.72 \times 10^{2}\right) \times \left(3.4 \times 10^{-4}\right) =$

12) $\left(2.81 \times 10^{-3}\right) \times \left(1.4 \times 10^{5}\right) =$

13) $\left(9.1 \times 10^{2}\right) \times \left(8.7 \times 10^{3}\right) =$

14) $\left(8.6 \times 10^{-2}\right) \times \left(4 \times 10^{-1}\right) =$

15) $\left(7.2 \times 10^{3}\right) \times \left(3 \times 10^{-1}\right) =$

Divide the following. Leave your answer in scientific notation.

16) $\dfrac{\left(4.2 \times 10^{3}\right) \times \left(6.8 \times 10^{-5}\right)}{\left(6.0 \times 10^{2}\right) \times \left(3.4 \times 10^{-4}\right)} =$

17) $\dfrac{\left(4.2 \times 10^{-5}\right) \times \left(68.2 \times 10^{2}\right)}{\left(14.0 \times 10^{3}\right) \times \left(2 \times 10^{-7}\right)} =$

18) $\dfrac{\left(3.2 \times 10^{2}\right) \times \left(4 \times 10^{4}\right)}{1.6 \times 10^{3}} =$

19) $\dfrac{87.4 \times 10^{2}}{43.7 \times 10^{3}} =$

20) $\dfrac{3.14 \times 10^{2}}{31.4 \times 10^{1}} =$

21) $\dfrac{8.72 \times 10^{2}}{2.00 \times 10^{4}} =$

22) $\dfrac{8.18 \times 10^{2}}{0.02 \times 10^{2}} =$

23) $\dfrac{6.8 \times 10^{-1}}{2.00 \times 10^{4}} =$

24) $\dfrac{7.68 \times 10^{-4}}{4 \times 10^{-4}} =$

25) $\dfrac{3.66 \times 10^{5}}{3.0 \times 10^{3}} =$

26) $\dfrac{5.8 \times 10^{4}}{2 \times 10^{-1}} =$

EXERCISES 2.7 – 2.8 B

Choose the one alternative that best completes the statement or answers the question.

Use scientific notation to perform the following operations. Leave your answer in scientific notation.

1) $(5 \times 10^{-4}) \div (10 \times 10^{-4})$
   a) 5.0×10^{-1} b) 2.0×10^{8} c) 2.0 d) 5.0×10^{7}

2) $(15 \times 10^{-7}) \div (5 \times 10^{-3})$
   a) 6.0×10^{-4} b) 3.0×10^{-4} c) 6.0×10^{-10} d) 3.0×10^{-10}

3) $(4 \times 10^{4}) \times (2 \times 10^{3})$
   a) 8.0×10^{8} b) 8.0×10^{7} c) 8.0×10^{9} d) 8.0×10^{6}

4) $(6 \times 10^{-4}) \div (12 \times 10^{2})$
   a) 5.0×10^{-7} b) -5.0×10^{-7} c) 5.0×10^{5} d) -5.0×10^{5}

5) $(3 \times 10^{8}) \times (7 \times 10^{-5})$
   a) 2.1×10^{4} b) 2.1×10^{-2} c) 2.1×10^{2} d) 2.1×10^{3}

6) $(2.7 \times 10^{2}) + (2 \times 10^{1})$
   a) 4.7×10^{2} b) 2.9×10^{2} c) 2.9×10^{3} d) 4.7×10^{1}

7) $(5.7 \times 10^{13}) \div (1.9 \times 10^{-7})$
   a) 5.7×10^{20} b) 5.7×10^{6} c) 3×10^{6} d) 3×10^{20}

8) $(4 \times 10^{8}) \div (2 \times 10^{4})$
   a) 2.0×10^{4} b) 2.0×10^{-4} c) -2.0×10^{-4} d) -2.0×10^{4}

9) $(5.7 \times 10^{24}) \div (1.9 \times 10^{8})$
   a) 3×10^{3} b) 3×10^{32} c) 5.7×10^{16} d) 3×10^{16}

10) $(7 \times 10^{2}) - (4 \times 10^{1})$
   a) 3×10^{2} b) 3×10^{1} c) 6.6×10^{1} d) 6.6×10^{2}

CHAPTER 2 REVIEW EXERCISES

1) Estimate the sum of $3.210 + .027 + 5 + 6.013$

2) Estimate the difference: $16.40 - 3.71$

3) Estimate the product of 3.731×1.2

4) Estimate the quotient: $15.864 \div 4.2$

Simplify the following:

5) 5^3 6) 15^2 7) 6^3 8) $\left(\dfrac{1}{4}\right)^2$ 9) $(.04)^3$

Which of the symbols, $<, \le, >,$ and $\ge$ make the following statements true? Give all possible correct answers.

10) .94 ___ .904   11) .87 ___ .865   12) $\dfrac{3}{4}$ ___ $\dfrac{4}{5}$ 13) $\dfrac{2}{3}$ ___ .7

Simplify each statement and then decide whether the statement is true or false.

14) $2 \times 3 + 5(4 + 8) \le 68$ 15) $70 < 9(6 + 2) - 4$ 16) $6^2 - 4^2 \ge 5$

17) $\dfrac{2(1 + 3)}{3(2 + 1)} < 1$

Perform the indicated operation.

18) $\dfrac{4 \times 5 \times 2 - 6 \div 2}{13 + 3 \times 1 \times 5}$ 19) $\dfrac{1}{2}(8 \div 2 \times 4) \div (8 \times 4) \div 2$ 20) $3 + 48 - 16 - 35 \div 7$

21) $\dfrac{3(3 + 1) + 3}{4 + 1}$ 22) $\dfrac{9 - 5(3 - 2)}{3 + 5}$ 23) $5(7 - 4) \div 3 + 2$

24) $6(7 + 2) - 15 \div 5$ 25) $64 \div 8 \div 4 \div 2$

Choose the one alternative that best completes the statement or answers the question.
Use scientific notation to perform the following operations. Leave your answer in scientific notation.

26) $(15 \times 10^3) \div (5 \times 10^{-4})$
     a) 3.0×10^{-1} b) 6.0×10^7 c) 6.0×10^{-1} d) 3.0×10^7

27) $(3 \times 10^2) \times (6 \times 10^3)$
   a) 1.8×10^7 b) 1.8×10^5 c) 1.8×10^8 d) 1.8×10^6

28) $(5.6 \times 10^{14}) \div (1.6 \times 10^{-6})$
   a) 5.6×10^8 b) 4×10^{20} c) 4×10^8 d) 5.6×10^{20}

29) $(7 \times 10^{-2}) \div (35 \times 10^3)$
   a) -2.0×10^{-6} b) 2.0×10^4 c) -2.0×10^4 d) 2.0×10^{-6}

30) $(5.5 \times 10^{40}) \div (1.1 \times 10^{10})$
   a) 5×10^{30} b) 5×10^4 c) 5.5×10^{30} d) 5×10^{50}

31) $(9 \times 10^4) - (4 \times 10^3)$
   a) 8.6×10^3 b) 8.6×10^4 c) 5×10^4 d) 5×10^1

32) $(4.6 \times 10^3) + (4.5 \times 10^2)$
   a) 9.1×10^2 b) 5.05×10^5 c) 9.1×10^3 d) 5.05×10^3

33) $(2 \times 10^7) \times (7 \times 10^{-4})$
   a) 1.4×10^4 b) 1.4×10^2 c) 1.4×10^3 d) 1.4×10^{-2}

34) $(5 \times 10^{-3}) \div (10 \times 10^{-4})$
   a) 2.0×10^1 b) 2.0×10^7 c) 5.0×10^6 d) 5.0

35) $(42 \times 10^5) \div (7 \times 10^7)$
   a) 6×10^{-2} b) -6×10^2 c) 6×10^2 d) -6×10^{-2}

CHAPTER 2 TEST

1) Which is the best estimate for the sum of $5.86 + 4.93$?
   a) 10   b) 9   c) 11   d) none of these

2) Which is the best estimate for the sum of $5,326 + 3,109$?
   a) 8,000   b) 10,000   c) 9,000   d) none of these

3) Which is the best estimate of the difference, $6,988 - 1,089$?
   a) 4,000   b) 6,090   c) 5,000   d) none of these

4) Which is the best estimate of the product of 58×727?
   a) 3,900   b) 4,500   c) 4,200   d) none of these

5) Which is the best estimate of the quotient, $2,514 \div 47$?
   a) 60   b) 600   c) 6   d) none of these

6) Which number is the same as twenty-one billion, seven hundred eighty thousand?
   a) 21,780,000,000   b) 21,780,000   c) 21,000,780,000   d) none of these

7) In which place is the 6 in the number 4,685,392,710?
   a) hundred millions   b) millions   c) billions   d) none

8) Which number is represented by $(6 \times 10^5) + (5 \times 10^4) + (3 \times 10^3) + (5 \times 10^2 + (9 \times 10^1)$?
   a) 653,590   b) 32,319   c) 65,359   d) none

9) Which or the following gives the closest estimate of $59.9 + 6.01 + 8.7$?
   a) 65   b) 70   c) 75   d) 80

10) Simplified, $\left(1\dfrac{1}{2}\right)^3$ equals

    a) 3.16   b) 31.6   c) 3.375   d) 33.75

11) Leon receives a monthly salary of $1500 and saves .09 of his earnings. How many months would it take for him to save 600?
   a) 2   b) 3   c) 4   d) 5

12) Simplified, $\left(\dfrac{4}{3}\right)^2$ equals

    a) $\dfrac{8}{6}$ b) $\dfrac{8}{3}$ c) $\dfrac{16}{9}$ d) $\dfrac{6}{8}$

Write the number in scientific notation.

13) 0.09
   a) 9×10^{-1}
   b) 9×10^{-2}
   c) 9×10^{-3}
   d) -9×10^{2}

14) 23,000,000
   a) 2.3×10^{6}
   b) 2.3×10^{-6}
   c) 2.3×10^{7}
   d) 2.3×10^{-7}

15) 0.005×10^{6}
   a) 0.5×10^{4}
   b) 5×10^{9}
   c) 5×10^{-4}
   d) 5×10^{3}

16) 52,000
   a) 5.2×10^{-4}
   b) 5.2×10^{3}
   c) 5.2×10^{4}
   d) 5.2×10^{-3}

17) 0.0000000475
   a) 4.75×10^{-9}
   b) 4.75×10^{-8}
   c) -4.75×10^{8}
   d) 4.75×10^{-7}

18) 172.8
   a) 1.728×10^{-2}
   b) 1.728×10^{-1}
   c) 1.728×10^{1}
   d) 1.728×10^{2}

19) 868
   a) 8.68×10^{3}
   b) 8.68×10^{-2}
   c) 8.68×10^{2}
   d) 8.68×10^{1}

3 RATIONAL NUMBERS

Rational numbers, in a sense, is the technical name for numbers we call fractions. A **rational number** is a number that can be written or expressed in the form $\frac{a}{b}$ where a and b are whole numbers and $b \neq 0$.

3.1 MEANING OF A FRACTION

There are two ways of looking at a fraction. When a unit is divided into equal parts, one or more of the equal parts is a fraction. Also, when a group of things is divided into equal parts, one or more of the equal parts of the group is considered to be a fraction. We could say that fractions are mainly used to represent parts of a single unit or parts of groups of units. The fraction $\frac{5}{8}$ means 5 parts of 8 equal parts. In the fraction $\frac{5}{8}$, the numbers 5 and 8 are the terms of the fraction. The numerator and the denominator are called the terms of the fraction. In the fraction $\frac{5}{8}$, the number 5 is the numerator and the number 8 is the denominator.

---

1) An **equivalent fraction** is a fraction with different names (denominators) but the same value.

---

2) A **proper fraction** is a fraction in which the numerator is less than the denominator. The following are proper fractions:
$$\frac{1}{3}, \frac{6}{7}, \frac{4}{5}, \frac{3}{5}$$

---

3) An **improper fraction** is a fraction in which the numerator is equal to, or greater than, the denominator. The following are improper fractions:
$$\frac{6}{6}, \frac{9}{8}, \frac{7}{3}, \frac{271}{42}$$

---

4) A **mixed number** is a number consisting of a whole number and a fraction. The following are mixed numbers:

$$5\frac{3}{5}, \ 14\frac{2}{7}, \ 49\frac{36}{37}$$

To obtain equivalent fractions we use the **fundamental principle of fractions** which states:

Fundamental Principle of Fractions: Both the numerator and denominator of any fraction may be multiplied or divided by the same nonzero number without changing the value of the fraction.

EXAMPLE 1

Convert 4/5 to an equivalent fraction with a denominator of 15.

SOLUTION

To convert to an equivalent fraction, we multiply by 1. Since $5 \times 3 = 15$, we multiply the fraction by 1 expressed as $\frac{3}{3}$, thus, $\frac{4}{5} = \frac{4 \times 3}{5 \times 3} = \frac{12}{15}$.

EXAMPLE 2

Write an equivalent fraction for $\frac{9}{12}$ with a numerator of 63.

SOLUTION

Since $9 \times 7 = 63$, we multiply by 1 expressed as $\frac{7}{7}$ which gives $\frac{9}{12} = \frac{9 \times 7}{12 \times 7} = \frac{63}{84}$.

Simplest Terms

A fraction is said to be in simplest form if no number other than 1 will divide both the numerator and denominator. Formally defined, we say that a fraction is in simplest form when its numerator and denominator are relatively primes.

To reduce a fraction to lowest terms: we divide both the numerator and denominator by the GCD.

EXAMPLE 3

Reduce $\dfrac{16}{24}$ to the lowest terms.

SOLUTION

Use the technique of the previous section to find the GCD for 16 and 24.
Prime factors of $16 = 2 \times 2 \times 2 \times 2$
Prime factors of $24 = 2 \times 2 \times 2 \times 3$
GCD $= 2 \times 2 \times 2 = 8$

Now divide both numerator and denominator by 8 to reduce the fraction.

$$\frac{16}{24} = \frac{16 \div 8}{24 \div 8} = \frac{2}{3}$$

In practice, however, we do not formally find the GCD but instead express both the numerator and denominator as the product of primes and then apply the **Fundamental Principle of Fractions**.

EXAMPLE 4

Reduce 78/91 to lowest terms.

SOLUTION

Since 78 is even, we know that it is divisible by two. Expressing the numerator as the product of primes gives, $\dfrac{78}{91} = \dfrac{2 \times 13 \times 3}{7 \times 13}$. If this fraction can be reduced, then 91 must be divisible by one of the factors in the numerator. The number 91 is not even, so 2 does not work, the test for 3 also fails. Dividing by 7 gives 13 so the primes factors of 91 are 7 × 13. Expressing the denominator as primes reduces the fractions as follows:

$$\frac{78}{91} = \frac{2 \times 13 \times 3}{7 \times 13} = \frac{6}{7}$$

EXAMPLE 5

Convert $\dfrac{3}{5}$ to an equivalent fraction with a denominator of 15.

SOLUTION

To convert to an equivalent fraction, we multiply by 1. Since 3 x 5 = 15, we multip
by 1 expressed as $\frac{3}{3}$, thus $\frac{3}{5} = \frac{3 \times 3}{3 \times 5} = \frac{9}{15}$.

EXAMPLE 6

Convert $12\frac{2}{3}$ to an improper fraction.

SOLUTION

By the conversion rule, $12\frac{2}{3} = \frac{36+2}{3} = \frac{38}{3}$.

Reducing Fractions to Lowest Terms

A fraction is in its simple form when no number other than 1 will divide both the denominator.

EXAMPLE 7

Reduce $\frac{14}{21}$ to the lowest term.

SOLUTION

Use the technique of Chapter 1 to find the GCD.

$\frac{14}{21}$ can be reduced since 14 and 21 have the common factor 7.

$\frac{14}{21} = \frac{7 \times 2}{7 \times 3} = \frac{2}{3}$

EXAMPLE 8

Reduce $\frac{13}{26}$ to the lowest term.

SOLUTION

$\frac{13}{26}$ can be reduced since 13 and 26 have the common factor 13.

$\frac{13}{26} = \frac{13 \times 1}{13 \times 2} = \frac{1}{2}$.

EXERCISE 3.1 A

Write as an improper fraction:

1) $5\frac{3}{4} =$ 2) $7\frac{1}{2} =$ 3) $11\frac{1}{3} =$ 4) $3\frac{1}{5} =$ 5) $4\frac{3}{5} =$ 6) $8\frac{7}{8} =$

7) $7 =$ 8) $11\frac{1}{11} =$ 9) $12\frac{3}{4} =$ 10) $2\frac{5}{8} =$ 11) $4\frac{4}{3} =$ 12) $11\frac{1}{5} =$

Write as a mixed number:

13) $\frac{5}{2} =$ 14) $\frac{32}{5} =$ 15) $\frac{21}{3} =$ 16) $\frac{25}{4} =$ 17) $\frac{31}{11} =$ 18) $\frac{16}{9} =$

19) $\frac{41}{7} =$ 20) $\frac{27}{7} =$ 21) $\frac{16}{5} =$ 22) $\frac{16}{7} =$ 23) $\frac{14}{5} =$ 24) $\frac{41}{15} =$

Reduce to lowest terms:

25) $\frac{5}{6} =$ 26) $\frac{121}{132} =$ 27) $\frac{72}{168} =$ 28) $\frac{3}{42} =$ 29) $\frac{21}{35} =$ 30) $\frac{9}{54} =$

31) $\frac{24}{30} =$ 32) $\frac{9}{42} =$ 33) $\frac{12}{32} =$ 34) $\frac{15}{35} =$ 35) $\frac{15}{40} =$ 36) $\frac{14}{8} =$

Insert the missing numerator or denominator so that the fraction is equivalent to the given fraction.

37) $\frac{3}{7} = \frac{}{56}$ 38) $2\frac{3}{16} = \frac{}{32}$ 39) $1\frac{3}{4} = \frac{}{16}$ 40) $5\frac{1}{4} = \frac{}{20}$

41) $7\frac{2}{7} = \frac{}{14}$ 42) $4\frac{3}{5} = \frac{}{20}$ 43) $2\frac{1}{3} = \frac{}{12}$ 44) $1\frac{1}{2} = \frac{}{16}$

45) $\frac{5}{8} = \frac{40}{}$ 46) $\frac{1}{7} = \frac{}{35}$ 47) $\frac{4}{5} = \frac{}{35}$ 48) $\frac{3}{4} = \frac{}{12}$

EXERCISE 3.1 B

Reduce each fraction to the lowest terms.

1) $\dfrac{9}{12}$　　　2) $\dfrac{30}{27}$　　　3) $\dfrac{54}{96}$　　　4) $\dfrac{98}{124}$　　　5) $\dfrac{56}{91}$　　　6) $\dfrac{85}{221}$

7) $\dfrac{189}{243}$　　8) $\dfrac{13}{65}$　　9) $\dfrac{20}{21}$　　10) $\dfrac{116}{117}$　　11) $\dfrac{77}{21}$　　12) $\dfrac{29}{203}$

Insert the missing numerator or denominator so that the fraction is equivalent to the given fraction.

13) $\dfrac{5}{9} = \dfrac{}{45}$　　　14) $\dfrac{3}{16} = \dfrac{}{48}$　　　15) $\dfrac{3}{} = \dfrac{12}{32}$　　　16) $\dfrac{3}{} = \dfrac{12}{16}$

17) $\dfrac{3}{9} = \dfrac{27}{}$　　18) $\dfrac{11}{12} = \dfrac{}{36}$　　19) $\dfrac{11}{18} = \dfrac{}{54}$　　20) $\dfrac{}{24} = \dfrac{102}{144}$

21) $\dfrac{9}{16} = \dfrac{}{272}$　　22) $\dfrac{5}{13} = \dfrac{65}{}$　　23) $\dfrac{9}{} = \dfrac{63}{119}$　　24) $\dfrac{11}{21} = \dfrac{}{189}$

Convert the following improper fractions to mixed numbers.

25) 25/6　　　　26) 43/17　　　　27) 20/9　　　　28) 19/19

29) 33/3　　　　30) 246/28

Convert the following mixed numbers to improper fractions.

31) $14\dfrac{3}{7}$　　32) $13\dfrac{1}{6}$　　33) $12\dfrac{6}{7}$　　34) $5\dfrac{3}{8}$　　　35) $18\dfrac{2}{3}$

36) $133\dfrac{1}{4}$　　　37) $18\dfrac{3}{5}$　　　38) $14\dfrac{7}{8}$　　　39) $13\dfrac{5}{7}$

3.2 ORDERING FRACTIONS

Sometimes we want to know which fraction is larger or smaller (e.g., $\frac{4}{5}$ or $\frac{3}{4}$), or we may want to arrange the fractions in some order; largest to smallest or smallest to largest.

In order to compare quantities, they must have the same unit of measure. For fractions, this unit of measure is the denominator--the name.

If the denominators are the same, we may simply compare their numerators. Unlike fractions must be converted to a common unit of measure called the **Least Common Denominator (LCD)** before they can be compared. The function of the LCD may be seen from this illustration. Suppose you have 3 bicycles and 2 cars; how many would you have in all? The answer is 3 bicycles and 2 cars. However, if we use a common name, for example, vehicles, then we would have 5.

The general method for comparing fractions is to rewrite each fraction so they have the same denominators. The fractions are then compared by looking at the size of their numerators.

EXAMPLE 1

Which is larger, $\frac{4}{5}$ or $\frac{3}{4}$?

SOLUTION

The LCD for 4 and 5 is 20, converting each fraction gives,

$$\frac{4}{5} = \frac{4 \times 4}{5 \times 4} = \frac{16}{20} \qquad\qquad \frac{3}{4} = \frac{3 \times 5}{4 \times 5} = \frac{15}{20}$$

Since 16>15, $\frac{4}{5} > \frac{3}{4}$

EXAMPLE 2

Arrange the fractions $\dfrac{5}{8}$, $\dfrac{11}{12}$, and $\dfrac{15}{18}$ in order from the smallest to the largest.

SOLUTION

Find the LCD for 8, 12, and 18.

Prime factors of $18 = 2 \times 3 \times 3$
Prime factors of $12 = 2 \times 3 \times 2$
Prime factors of $8 = 2 \times 2 \times 2$
LCD $= 2 \times 3 \times 3 \times 2 \times 2 = 72$

Write equivalent fractions with a denominator of 72.

$$\dfrac{5}{8} = \dfrac{5 \times 9}{8 \times 9} = \dfrac{45}{72} \qquad \dfrac{11}{12} = \dfrac{11 \times 6}{12 \times 6} = \dfrac{66}{72} \qquad \dfrac{15}{18} = \dfrac{15 \times 4}{18 \times 4} = \dfrac{60}{72}$$

Arranged in order from the smallest to the largest we get, $\dfrac{5}{8}, \dfrac{15}{18}, \dfrac{11}{12}$.

EXAMPLE 3

Find a fraction between $\dfrac{1}{5}$ and $\dfrac{2}{5}$.

SOLUTION

Convert the fractions to an equivalent fraction, for example, fractions with a denominator of 10. In this instance the equivalent fractions would be $\dfrac{1}{5} = \dfrac{2}{10}$ and $\dfrac{2}{5} = \dfrac{4}{10}$. One fraction between $\dfrac{1}{5}$ and $\dfrac{2}{5}$ would be $\dfrac{3}{10}$.

EXERCISE 3.2

Order the following fractions from the smallest to the largest.

1) $\dfrac{2}{3}, \dfrac{1}{2}, \dfrac{3}{8}, \dfrac{5}{6}$

2) $\dfrac{2}{5}, \dfrac{3}{4}, \dfrac{1}{2}, \dfrac{3}{10}$

3) $\dfrac{3}{4}, \dfrac{5}{6}, \dfrac{11}{12}$

4) $\dfrac{1}{5}, \dfrac{2}{3}, \dfrac{7}{15}$

5) $\dfrac{5}{8}, \dfrac{7}{16}, \dfrac{1}{2}$

6) $\dfrac{5}{12}, \dfrac{7}{5}, \dfrac{7}{18}$

7) $\dfrac{4}{3}, \dfrac{4}{5}, \dfrac{3}{2}, \dfrac{3}{4}$

8) $\dfrac{4}{7}, \dfrac{7}{12}, \dfrac{5}{14}$

9) Find a fraction between $\dfrac{7}{8}$ and $\dfrac{9}{11}$.

10) Which of the following fractions is closest to 1: $\dfrac{10}{9}, \dfrac{5}{6}, \dfrac{8}{9}$, or $\dfrac{7}{8}$?

11) Which of the following is less than $\dfrac{1}{2}$: $\dfrac{1}{3}, \dfrac{3}{4}, \dfrac{7}{10}, \dfrac{5}{6}$?

12) Which is larger, $\dfrac{6 \times 12 \times 18}{6 \times 6 \times 6}$ or 7?

13) Which is larger, $\dfrac{1}{3} + \dfrac{1}{4}$ or $\dfrac{1}{4} + \dfrac{1}{5}$?

Determine which of the following fractions are larger.

14) $\dfrac{3}{8}$ or $\dfrac{1}{2}$

15) $\dfrac{3}{4}$ or $\dfrac{2}{3}$

16) $\dfrac{3}{20}$ or $\dfrac{3}{10}$

17) $\dfrac{3}{4}$ or $\dfrac{5}{8}$

18) $\dfrac{15}{16}$ or $\dfrac{7}{8}$

3.3 ADDITION AND SUBTRACTION OF FRACTIONS

In 3.1 we have learned how to reduce fractions to lowest terms and convert a rational number to an equivalent fraction by reducing. Addition and subtraction of fractions are similar to addition and subtraction of whole numbers. In this section we will learn about situations that call for adding and subtracting fractions, and we will learn how to subtract and add certain parts of fractions.

To add or subtract fractions:

Step 1. Find the LCD of the fraction.

Step 2. Write each fraction as an equivalent fraction with denominator equal to the LCD.

Step 3. The numerator will be the indicated sum or difference; the denominator will be the LCD.

Step 4. Reduce to lowest terms.

EXAMPLE 1

$$\frac{2}{3}+\frac{3}{4}=$$

SOLUTION

$$\frac{2}{3}+\frac{3}{4}=$$ Determine the LCD for the denominators 3 and 4.

Step 1. LCD is 12   The LCD has the same value as the LCM, which is 12.

Step 2. $\dfrac{2}{3}+\dfrac{3}{4}=\dfrac{2\times4+3\times3}{12}$ Divide the denominator 3 into 12. Multiply the quotient 4 by the numerator 2. Divide the denominator 4 into 12. Multiply the quotient 3 by the numerator 3. Set the sum of these indicated products over the LCD.

Step 3. $=\dfrac{8+9}{12}$ Do the multiplication.

Step 4. $=\dfrac{17}{12}$ Do the addition.

Step 5. $= 1\dfrac{5}{12}$ Convert to a mixed number and reduce the
fractional part if possible.

Addition and Subtraction of Mixed Number Fractions

Recall that a mixed number fraction consists of a whole number and a fractional part. For example, the mixed number $7\dfrac{1}{8}$, has the whole number 7 and a fraction, $\dfrac{1}{8}$.

> **To add or subtract a mixed number fraction:** we find the sum or difference of their whole parts and fractional parts separately. The fractional part is added or subtracted as before, except it may be necessary to regroup or rename some of the fractions.

EXAMPLE 2

$$4\dfrac{1}{2} + 17\dfrac{3}{5}$$

SOLUTION

Step 1. $4\dfrac{1}{2} + 17\dfrac{3}{5}$

$= \dfrac{4}{1} + \dfrac{1}{2} + \dfrac{17}{1} + \dfrac{3}{5}$ Convert each mixed number to the sum of the integer and the fraction.

Step 2. $= \dfrac{4 \times 10 + 1 \times 5 + 17 \times 10 + 3 \times 2}{10}$ Determine the LCD, and divide each denominator into the LCD. Multiply each numerator by the respective quotient.

Step 3. $= \dfrac{40 + 5 + 170 + 6}{10}$ Do the multiplication in the numerator.

Step 4. $= \dfrac{221}{10}$ Do the addition in the numerator.

Step 5. $= 22\dfrac{1}{10}$ Convert to a mixed number.

EXAMPLE 3

$$\frac{3}{4} - \frac{1}{5} + \frac{1}{2}$$

SOLUTION

$$\frac{3}{4} - \frac{1}{5} + \frac{1}{2} \quad = \frac{(3 \times 5) - (1 \times 4) + (1 \times 10)}{20} \qquad \text{The LCD is 20.}$$

$$= \frac{15 - (4) + 10}{20}$$

$$= \frac{21}{20}$$

$$= 1\frac{1}{20}$$

EXAMPLE 4

$$13\frac{2}{5} - 9\frac{7}{15}$$

SOLUTION

$$13\frac{2}{5} - 9\frac{7}{15} \quad = \frac{13}{1} + \frac{2}{5} - \left(\frac{9}{1} + \frac{7}{15}\right)$$

$$= \frac{13}{1} + \frac{2}{5} - \frac{9}{1} - \frac{7}{15} \qquad \text{Clear the parentheses.}$$

$$= \frac{13 \times 15 + 2 \times 3 - (9 \times 15) - (7 \times 1)}{15}$$

$$= \frac{195 + 6 - (135) - 7}{15}$$

$$= \frac{195 + 6 + (-135) - 7}{15}$$

$$= \frac{59}{15}$$

$$= 3\frac{14}{15}$$

EXERCISE 3.3 A

Add or subtract as indicated.

1) $\dfrac{2}{7}+\dfrac{1}{7}$

2) $\dfrac{9}{13}+\dfrac{7}{13}$

3) $\dfrac{5}{3}+\dfrac{7}{3}$

4) $\dfrac{3}{14}+\dfrac{5}{7}$

5) $\dfrac{3}{20}+\dfrac{7}{30}$

6) $\dfrac{5}{14}+\dfrac{35}{84}$

7) $\dfrac{5}{8}+\dfrac{4}{9}$

8) $\dfrac{2}{3}+\dfrac{5}{5}+\dfrac{7}{12}$

9) $\dfrac{2}{3}+\dfrac{5}{8}+\dfrac{7}{9}$

10) $\dfrac{9}{11}+\dfrac{1}{2}+\dfrac{1}{6}$

11) $4\dfrac{1}{2}+5\dfrac{7}{12}$

12) $4+5\dfrac{2}{7}$

13) $6\dfrac{8}{9}+12$

14) $7\dfrac{5}{12}+2\dfrac{9}{16}$

15) $8\dfrac{21}{40}+6\dfrac{21}{32}$

16) $\dfrac{5}{9}+17\dfrac{2}{9}+18\dfrac{5}{27}$

17) $\dfrac{7}{8}-\dfrac{5}{16}$

18) $\dfrac{5}{12}-\dfrac{5}{16}$

19) $\dfrac{19}{40}-\dfrac{3}{16}$

20) $\dfrac{11}{15}-\dfrac{5}{27}$

21) $25\dfrac{4}{9}-16\dfrac{7}{9}$

22) $6-4\dfrac{3}{5}$

23) $14\dfrac{3}{5}-7\dfrac{8}{9}$

24) $24\dfrac{3}{8}-7\dfrac{8}{9}$

25) $17\dfrac{3}{4}-8\dfrac{10}{11}$

26) $77\dfrac{5}{18}-61$

27) $137\dfrac{3}{5}-69\dfrac{7}{12}$

28) $13\dfrac{17}{24}-8\dfrac{5}{18}$

EXERCISE 3.3 B

Add or subtract.

1) $\dfrac{1}{2}-\dfrac{3}{4}+\dfrac{11}{5}+\dfrac{1}{20}=$

2) $\dfrac{3}{4}-\dfrac{5}{2}+\dfrac{28}{3}=$

3) $-\dfrac{11}{4}+\dfrac{6}{5}+\dfrac{15}{9}=$

4) $\dfrac{11}{4}-\dfrac{3}{4}+\dfrac{84}{20}=$

5) $\dfrac{11}{3}-\dfrac{3}{2}+\dfrac{5}{6}=$

6) $\dfrac{10}{3}-\dfrac{27}{5}=$

7) $\dfrac{10}{3}-\dfrac{6}{5}=$

8) $\dfrac{11}{4}+\dfrac{1}{2}=$

9) $\dfrac{1}{5}+\dfrac{5}{6}-\dfrac{2}{15}=$

10) $\dfrac{1}{4}+\dfrac{3}{8}-\dfrac{5}{16}=$

11) $\dfrac{1}{5}-\dfrac{1}{3}+\dfrac{7}{15}=$

12) $\dfrac{1}{2}-\dfrac{1}{4}=$

13) $\dfrac{7}{8}-\dfrac{4}{5}=$

14) $\dfrac{7}{12}+\dfrac{1}{3}=$

15) $\dfrac{7}{3}+\dfrac{27}{4}-\dfrac{1}{2}=$

16) $\dfrac{27}{2}+\dfrac{9}{4}-\dfrac{3}{4}=$

17) $\dfrac{11}{4}-\dfrac{28}{5}+\dfrac{37}{10}=$

18) $\dfrac{5}{2}+\dfrac{13}{4}-\dfrac{26}{6}=$

19) $\dfrac{5}{2}-\dfrac{3}{4}+\dfrac{5}{8}=$

20) $\dfrac{3}{2}-\dfrac{9}{4}=$

21) $\dfrac{5}{2}-\dfrac{5}{4}=$

22) $\dfrac{13}{4}-\dfrac{1}{5}=$

23) $\dfrac{1}{5}+\dfrac{6}{7}-\dfrac{1}{2}=$

24) $\dfrac{1}{2}+\dfrac{3}{4}-\dfrac{4}{5}=$

25) $\dfrac{1}{2}+\dfrac{2}{3}-\dfrac{1}{4}=$

26) $\dfrac{2}{3}+\dfrac{3}{5}=$

27) $\dfrac{5}{12}-\dfrac{1}{4}=$

28) $\dfrac{5}{6}-\dfrac{1}{6}=$

29) $3\dfrac{1}{4}+5\dfrac{2}{3}=$

30) $19\dfrac{1}{4}-15\dfrac{5}{6}=$

31) $4\dfrac{1}{4}-3\dfrac{2}{3}=$

32) $2\dfrac{2}{3}-1\dfrac{1}{4}=$

33) $4\dfrac{2}{5}+3\dfrac{2}{3}=$

34) $2\dfrac{1}{2}+3\dfrac{1}{5}=$

3.4 MULTIPLICATION AND DIVISION OF FRACTIONS

To multiply two or more rational numbers, convert all mixed numbers to improper fractions and then, if possible, simplify any denominator with any numerator. Then multiply all the numerators together to obtain the numerator of the answer and likewise multiply all the denominators together to obtain the denominator of the answer.

EXAMPLE 1

$$2\frac{1}{2} \times 3\frac{1}{5} =$$

SOLUTION

$$2\frac{1}{2} \times 3\frac{1}{5} = \left(\frac{2}{1} + \frac{1}{2}\right) \times \left(\frac{3}{1} + \frac{1}{5}\right)$$

$$= \frac{(2 \times 2 + 1 \times 1)}{2} \times \frac{(3 \times 5 + 1 \times 1)}{5}$$

$$= \left(\frac{4+1}{2}\right) \times \left(\frac{15+1}{5}\right)$$

$$= {}^1\left(\frac{\cancel{5}}{2}\right)\left(\frac{\cancel{16}}{5}\right)^8$$

$$= 1 \times 8$$

$$= 8$$

> **To divide two rational numbers:** convert the mixed numbers to improper fractions, then invert the divisor and continue with the same procedure as in the multiplication of rational numbers.

EXAMPLE 2

$$\frac{2}{3} \times \frac{9}{14} =$$

SOLUTION

Step 1. $\dfrac{2}{3} \times \dfrac{9}{14} =$

Convert, by dividing both the numerator 2 and the denominator 14 by 2. Then divide both the numerator 9 and the denominator 3 by 3.

Step 2. $\quad = \dfrac{3}{7}$

Multiply the resulting numerators together to obtain the numerator in the answer. Then multiply the resulting denominators together to obtain the denominator of the answer.

EXAMPLE 3

$$2\dfrac{1}{2} \div \dfrac{5}{4} =$$

SOLUTION

$$2\dfrac{1}{2} \div \dfrac{5}{4} \quad = \left(\dfrac{2}{1} + \dfrac{1}{2}\right) \div \dfrac{5}{4}$$

$$= \dfrac{(2 \times 2) + (1 \times 1)}{2} \div \dfrac{5}{4}$$

$$= \dfrac{4+1}{2} \div \dfrac{5}{4}$$

$$= \dfrac{5}{2} \div \dfrac{5}{4}$$

$$= \dfrac{2}{1} \text{ or } 2$$

EXAMPLE 4

$$\dfrac{2}{5} \div \dfrac{8}{15} =$$

SOLUTION

Step 1. $\dfrac{2}{5} \div \dfrac{8}{15} = \dfrac{2}{5} \times \dfrac{15}{8}$

Invert the divisor and then multiply.

Step 2. $= \left(\dfrac{\overset{1}{\cancel{2}}}{\underset{1}{\cancel{5}}}\right) \times \left(\dfrac{\overset{}{15}}{\underset{4}{8}}\right)^{3}$

Do the cancellation.

Step 3. $= \dfrac{3}{4}$

Do the multiplication of the result in step 2.

102

EXERCISE 3.4

Multiply and reduce the answer to lowest terms:

1) $2\dfrac{3}{9} \times \dfrac{3}{7} =$

2) $3\dfrac{4}{7} \times \dfrac{7}{10} =$

3) $2\dfrac{2}{3} \times \dfrac{7}{8} =$

4) $4\dfrac{1}{5} \times \dfrac{5}{7} =$

5) $\dfrac{2}{3} \times \dfrac{3}{4} \times \dfrac{5}{6} =$

6) $4 \times \dfrac{1}{2} \times \dfrac{2}{3} =$

7) $\dfrac{2}{3} \times \dfrac{1}{2} \times \dfrac{3}{4} =$

8) $\dfrac{9}{24} \times \dfrac{2}{3} \times \dfrac{4}{5} =$

9) $\dfrac{3}{5} \times \dfrac{35}{20} =$

10) $5 \times \dfrac{2}{6} =$

11) $2 \times \dfrac{1}{3} =$

12) $\dfrac{2}{3} \times \dfrac{15}{42} =$

13) $\dfrac{7}{16} \times \dfrac{4}{3} =$

14) $\dfrac{15}{7} \times \dfrac{3}{10} =$

15) $4 \times \dfrac{5}{12} =$

16) $\dfrac{3}{10} \times \dfrac{15}{16} =$

17) $\dfrac{2}{5} \times \dfrac{4}{10} =$

18) $\dfrac{1}{2} \times \dfrac{3}{4} =$

19) $\dfrac{7}{8} \times \dfrac{4}{16} =$

20) $\dfrac{1}{4} \times \dfrac{1}{3} =$

Divide and reduce the answer to lowest terms:

21) $2\dfrac{3}{4} \div 3\dfrac{1}{4} =$

22) $2\dfrac{3}{16} \div 1\dfrac{7}{8} =$

23) $1\dfrac{3}{16} \div 1\dfrac{1}{8} =$

24) $2\dfrac{1}{4} \div \dfrac{9}{4} =$

25) $3\dfrac{5}{16} \div \dfrac{3}{16} =$

26) $2\dfrac{4}{5} \div 7\dfrac{7}{10} =$

27) $3\dfrac{1}{5} \div 1\dfrac{3}{10} =$

28) $4\dfrac{1}{5} \div 3\dfrac{1}{2} =$

29) $4\dfrac{1}{2} \div 1\dfrac{1}{2} =$

30) $3\dfrac{1}{4} \div \dfrac{3}{4} =$

31) $\dfrac{1}{2} \div 5 =$

32) $5 \div \dfrac{1}{2} =$

33) $\dfrac{1}{5} \div 2 =$

34) $2 \div \dfrac{1}{5} =$

35) $\dfrac{3}{5} \div \dfrac{1}{5} =$

36) $\dfrac{1}{2} \div 1 =$

37) $4 \div \dfrac{1}{3} =$

38) $\dfrac{1}{3} \div 4 =$

39) $\dfrac{7}{15} \div \dfrac{4}{5} =$

40) $\dfrac{1}{3} \div \dfrac{1}{3} =$

3.5 COMPLEX FRACTIONS

> If the numerator or the denominator or both contain rational numbers that are not whole numbers, then the fraction is called a **complex fraction**.

There are two ways to solve a complex fraction. One method of simplifying a complex fraction is to rewrite the fraction as a division problem.

EXAMPLE 1

Simplify $\dfrac{\dfrac{3}{4}+\dfrac{1}{2}}{1-\dfrac{1}{3}}$.

SOLUTION

Simplifying the numerator, we have $\dfrac{3}{4}+\dfrac{1}{2}=\dfrac{3}{4}+\dfrac{2}{4}=\dfrac{5}{4}$.

Simplifying the denominator, we have $1-\dfrac{1}{3}=\dfrac{3}{3}-\dfrac{1}{3}=\dfrac{2}{3}$.

Therefore, $\dfrac{\dfrac{3}{4}+\dfrac{1}{2}}{1-\dfrac{1}{3}}=\dfrac{\dfrac{5}{4}}{\dfrac{2}{3}}=\dfrac{5}{4}\times\dfrac{3}{2}=\dfrac{15}{8}$.

An alternate method of simplifying a complex fraction is to multiply each term of the fraction, numerator and denominator, by the LCD.

EXAMPLE 2

$\dfrac{2\dfrac{1}{3}}{\dfrac{1}{4}+\dfrac{1}{3}}=$

SOLUTION

The LCD for {4 and 3} is 12

$$\dfrac{2\dfrac{1}{3}}{\dfrac{1}{4}+\dfrac{1}{3}}=\dfrac{\left(\dfrac{7}{3}\right)}{\left(\dfrac{1}{4}+\dfrac{1}{3}\right)}\cdot\dfrac{12}{12}=\dfrac{\dfrac{7}{3}\cdot12}{\dfrac{1}{4}\cdot12+\dfrac{1}{3}\cdot12}=\dfrac{28}{3+4}=\dfrac{28}{7}=\dfrac{7\cdot4}{7\cdot1}=\dfrac{4}{1}=4$$

105

EXERCISE 3.5 A

Express each complex fraction as a simple fraction in the lowest terms.

1) $\dfrac{\dfrac{7}{11}}{\dfrac{5}{13}}$

2) $\dfrac{\dfrac{4}{9}}{\dfrac{5}{21}}$

3) $\dfrac{\dfrac{2}{3}+\dfrac{1}{4}}{\dfrac{2}{3}+\dfrac{1}{6}}$

4) $\dfrac{\dfrac{7}{8}+\dfrac{5}{12}}{\dfrac{3}{4}+\dfrac{2}{3}}$

5) $\dfrac{2-\dfrac{9}{5}}{1+\dfrac{9}{5}}$

6) $\dfrac{2}{2-\dfrac{1}{3}}$

7) $\dfrac{\dfrac{5}{8}+\dfrac{2}{3}}{\dfrac{7}{3}-\dfrac{1}{4}}$

8) $\dfrac{\dfrac{6}{5}-\dfrac{1}{9}}{\dfrac{2}{5}+\dfrac{5}{3}}$

9) $\dfrac{1-\dfrac{3}{8}}{2+\dfrac{1}{4}}$

EXERCISE 3.5 B

Simplify the complex fractions.

1) $\dfrac{2\dfrac{4}{9}+1\dfrac{1}{18}}{1\dfrac{2}{9}-\dfrac{1}{6}}=$

2) $\dfrac{5\dfrac{2}{3}-1\dfrac{1}{6}}{3\dfrac{1}{2}+3\dfrac{1}{6}}=$

3) $\dfrac{7\dfrac{1}{3}+2\dfrac{1}{5}}{6\dfrac{1}{9}+2}=$

4) $\dfrac{3\dfrac{1}{4}+2\dfrac{1}{2}}{5\dfrac{1}{8}+1\dfrac{5}{8}}=$

5) $\dfrac{\dfrac{4}{15}+\dfrac{6}{25}}{\dfrac{3}{5}+\dfrac{3}{10}}=$

6) $\dfrac{\dfrac{3}{5}+\dfrac{4}{7}}{\dfrac{3}{8}+\dfrac{1}{10}}=$

7) $\dfrac{\dfrac{7}{8}-\dfrac{3}{16}}{\dfrac{1}{3}-\dfrac{1}{4}}=$

8) $\dfrac{\dfrac{5}{6}-\dfrac{2}{3}}{\dfrac{5}{8}-\dfrac{1}{16}}=$

9) $\dfrac{\dfrac{2}{3}-\dfrac{1}{4}}{\dfrac{3}{5}-\dfrac{1}{4}}=$

10) $\dfrac{2-\dfrac{1}{3}}{1-\dfrac{1}{3}}=$

11) $\dfrac{7+\dfrac{2}{5}}{2+\dfrac{1}{15}}=$

12) $\dfrac{4+\dfrac{1}{3}}{6+\dfrac{1}{4}}=$

13) $\dfrac{3\dfrac{4}{5}}{11+8}=$

14) $\dfrac{5\dfrac{1}{7}}{2+1}=$

15) $\dfrac{\dfrac{5}{6}-\dfrac{1}{3}}{\dfrac{1}{2}+\dfrac{1}{5}}=$

16) $\dfrac{\dfrac{2}{3}+\dfrac{1}{5}}{4\dfrac{1}{2}}=$

17) $\dfrac{\dfrac{1}{5}+\dfrac{1}{6}}{2\dfrac{1}{3}}=$

18) $\dfrac{2+\dfrac{1}{5}}{1+\dfrac{1}{4}}=$

19) $\dfrac{1+\dfrac{3}{7}}{\dfrac{2}{3}}=$

3.6 APPLICATIONS INVOLVING FRACTIONS

Many practical problems in everyday situations give rise to the need for fractions. For example, if we wanted to cut a 5-foot piece of board into 3 equal pieces, each piece would be $1\frac{2}{3}$ feet long.

In this section, application problems involving fractions are discussed. These problems are solved just like application problems involving whole numbers. As a matter of fact, an approach to solving fractional application problems is to disregard the fractional part of the number, determine what operation would be used if the problem had only whole numbers, and then apply the same operation(s) to the fractional numbers.

The example below shows you how to solve fractional word problems.

EXAMPLE 1

A carpenter used $6\frac{1}{2}$ pounds of roofing nails at a cost of 12 cents a pound, $15\frac{2}{3}$ pounds of eight-penny nails at 15 cents a pound, and $14\frac{1}{4}$ pounds of six-penny nails at 10 cents a pound. What was the total cost of these nails?

SOLUTION

In each case we have more than one pound of nails. Recall that when we have more than one of the same, we multiply. So to find the total cost of each type of nail, we multiply.

$$\text{Roofing nails } 12 \times 6\frac{1}{2} = 12 \times \frac{13}{2} = \frac{2 \times 2 \times 3 \times 13}{2} = 78 \text{ cents}$$

$$\text{8 penny nails } 15\frac{2}{3} \times 15 \text{ cents } = \frac{47}{3} \times 15 = \frac{47 \times 3 \times 5}{3} = 235 \text{ cents}$$

$$\text{6 penny nails } 14\frac{1}{4} \times 10 \text{ cents } = \frac{57 \times 5 \times 2}{2 \times 2} = \frac{285}{2} = 142\frac{1}{2} \text{ cents}$$

Now, to find the total cost, add amounts for

Roofing Nails + 8 Penny Nails + 6 Penny Nails = Total Cost

78 cents   + 235 cents + $142\frac{1}{2}$ cents = $455\frac{1}{2}$ cents

Another problem-solving strategy is drawing a model or diagram to help identify the problem. The example below shows this technique.

EXAMPLE 2

James left his estate to his wife and two sons. If the wife receives $\frac{1}{3}$ of the estate and each son receives $\frac{1}{2}$ of the remainder, find the value of the entire estate if each son receives $4,000 as his share.

SOLUTION

Sketch a model of the problem.

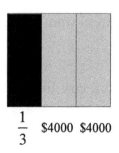

$$\frac{1}{3} \quad \$4000 \quad \$4000$$

When the wife receives $\frac{1}{3}$, the remainder is $\frac{2}{3}$. Half of this amount is $\frac{1}{3}$ with each $\frac{1}{3}$ being equal to $4000. Since there are 3 one-third shares, the total amount of the estate was $12,000.

EXAMPLE 3

It takes $1\frac{1}{6}$ gallons of paint to cover one room, and $2\frac{3}{4}$ gallons to paint another room. How much will it take to paint both rooms?

SOLUTION

We find the total amount of paint needed by adding the amounts needed for each room.

$$1\frac{1}{6} = 1\frac{2}{12}$$
$$2\frac{3}{4} = 2\frac{9}{12}$$
$$\overline{\qquad\qquad}$$
$$3\frac{11}{12} \text{ gallons of paint}$$

EXERCISE 3.6 A

Solve each problem. Drawing a model may help you solve the problems.

1) If 5 gallons of oil are added to 25 gallons of water, what fraction of the new mixture is oil?

2) Three hundred shares of a stock were purchased at a cost of $54\frac{3}{4}$ dollars a share. What is the cost of 300 shares?

3) How many more containers would be needed to package 140 items in cartons of 14 rather than 10?

4) Ron picked two numbers, 6 and 8, out of a hat. Today, every sixth person will get a free sandwich and every eighth person will get a free drink. Suppose 64 customers come into the store, which ones will get both a free drink and a free sandwich?

5) Jasmine made enough money by selling candy bars at 15 cents each to buy several cans of pop at 48 cents each. If she had no money left over, what is the fewest number of candy bars she could have sold?

6) In January Marion spent $\frac{1}{4}$ of the money in his savings account. The following month he spent $\frac{1}{5}$ of the remainder. If he then had $3,000 left, how much was in his savings account originally?

7) The first of four identical bottles is $\frac{1}{2}$ full of water, the second is $\frac{3}{4}$ full, the third is $\frac{7}{8}$ full, and the fourth is $\frac{15}{16}$ full. After water is poured from the fourth bottle to fill each of the first three bottles, what fraction of the fourth bottle will be full?

8) A gasoline tank that is $\frac{1}{2}$ full has 8 gallons removed. The tank is then $\frac{1}{10}$ full. What is the capacity in gallons of the tank?

9) Deloris owned $\frac{5}{8}$ of an interest in a house. She sold $\frac{1}{5}$ of her interest at cost for $1,000. What was the total value of the house?

111

10) Esther does $\frac{1}{3}$ of her homework and then goes to dinner. After dinner she completes $\frac{3}{4}$ of the remainder of her assignments and then decides to go to the football game. What part of her homework will be left incomplete if she spends no additional time on her assignments?

11) Delvin used $\frac{1}{3}$ of his inheritance to pay off the mortgage on his house and $\frac{3}{5}$ of what was left to purchase a new automobile. How much was left from his total $30,000 inheritance after these two expenditures?

EXERCISE 3.6 B

1) A certain bus trip consists of three parts. The first part takes $2\frac{1}{3}$ hours; the second part takes $3\frac{3}{4}$ hours, and the third part takes $2\frac{14}{15}$ hours. How many hours does the trip take?

2) To make a new suit, Jamison bought $2\frac{3}{16}$ yards and $1\frac{7}{8}$ yards of cloth. How many yards of cloth did she buy?

3) On four sales trips, Pat drove $428\frac{3}{8}$ miles, $576\frac{3}{16}$ miles, $498\frac{1}{20}$ miles, and $550\frac{3}{4}$ miles. What was the total distance he drove?

4) The measurements along each side of a house were $42\frac{3}{4}$ feet, $23\frac{1}{2}$ feet, $42\frac{7}{8}$ feet, and $22\frac{9}{10}$ feet. What was the total distance around the house?

5) A construction company received government contracts to build freeway sections of $20\frac{7}{8}$ miles, $3\frac{3}{10}$ miles, $14\frac{3}{16}$ miles, and $11\frac{1}{24}$ miles. What was the total mileage represented by all the contracts?

6) Flying against the wind, an airline pilot, making two stops, had flying times of $2\frac{3}{5}$ hours, $2\frac{5}{12}$ hours, and $3\frac{1}{4}$ hours. On the return trip with the wind at his tail, the times were $2\frac{7}{8}$ hours, $2\frac{1}{2}$ hours, and $2\frac{1}{5}$ hours. How much slower was his total flying time against the wind?

112

7) A certain box was $4\frac{5}{8}$ inches long, $3\frac{3}{16}$ inches wide, and $2\frac{3}{8}$ inches deep. The dimensions of another box were $7\frac{1}{2}$ inches, $5\frac{3}{10}$ inches, and $4\frac{4}{5}$ inches. What was the difference between the total dimensions of the two boxes?

8) A teacher graded two sets of test papers. The first set took $3\frac{3}{4}$ hours to grade and the second set took $2\frac{3}{5}$ hours. How much faster did the teacher grade the second set?

9) A man can paint a room in $3\frac{3}{5}$ hours and his wife can paint the same size room in $4\frac{1}{5}$ hours. How much time is saved by having the man paint the room?

10) In order to meet the demands of a customer, a chef must multiply the ingredients of a recipe by $16\frac{1}{2}$. If the ingredients include $3\frac{3}{5}$ lbs. of sugar and $8\frac{1}{10}$ lbs. of flour, how many pounds of sugar and flour does the chef need?

11) In constructing a small building, $\frac{5}{16}$ of a cement piling must be underground while $\frac{11}{16}$ is above ground. If a certain piling is $32\frac{1}{5}$ feet long, how much of it is above ground and how much below?

12) A length of pipe is $27\frac{5}{8}$ feet. What would be the total length if $36\frac{1}{2}$ of these pipes were laid end to end?

13) A man driving to work drives $17\frac{7}{10}$ miles one way five days a week. How many miles does he drive each week going to and from work?

14) The total distance around a square (its perimeter) is found by multiplying the length of one side by four. Find the perimeter of a square if the length of one side is $5\frac{1}{16}$ inches.

15) The product of $7\frac{2}{3}$ and some other rational number is $4\frac{1}{2}$. What is the other number?

16) The result of multiplying two numbers is $10\frac{1}{3}$. If one of the numbers is $7\frac{1}{6}$, what is the other number?

17) Three different men measured a house for new tile flooring, getting measurements of $430\frac{1}{2}$ sq. ft., $430\frac{1}{4}$ sq. ft., and $430\frac{1}{9}$ sq. ft. The actual measurement was $429\frac{7}{8}$ sq. ft. What was the average error of the three men's measurements?

18) Find the average of the numbers $\frac{7}{8}$, $\frac{9}{10}$, $2\frac{1}{2}$, $1\frac{3}{4}$

19) Find the average of the numbers $\frac{5}{6}$, $\frac{7}{15}$, $\frac{8}{21}$.

ADDING AND SUBTRACTING OF DECIMALS

> A **decimal fraction** is a fraction whose denominator is not written but is some power of 10.

Thus, 0.6 (6 tenths) is a decimal fraction. If it is written in the form $\frac{6}{10}$, it becomes a common fraction. The expression 0.6 means the same as .6. The 0 does not have to be written before the decimal point.

Usually, decimal fractions are called decimals. The period placed at the left of tenths is called the **decimal point**. The decimal fraction is one of the most important discoveries in mathematics. For a long time, no one ever thought of placing digits at the right of the units place.

In a decimal fraction, the digits to the right of the decimal point have the following values.

Tenths	Hundredths	Thousandths	Ten-thousandths	Hundred-thousandths	Millionths
.6	9	5	4	3	2

Adding and Subtracting Decimals

As with whole numbers, we make many mistakes with problems involving decimals. There is little difference in adding or subtracting decimal numbers. Since decimal numbers are used often in many problems, it is important to perform the operations with great accuracy.

> **To add or subtract decimals** we use the following procedure:
>
> 1) Express each decimal to the same number of decimal places; annexing zeros to the right of the decimal point, if needed.
>
> 2) Arrange the numbers in a column so that the decimals are lined up.
>
> 3) Add or subtract as indicated.

EXAMPLE 1

Add 37.09 + 0.86 + 5.2 + 9.0

SOLUTION

Express each decimal to the same number of decimal places, line up the decimal points, and then add.

$$
\begin{array}{rcr}
37.09 = & & 37.09 \\
0.86 = & & 0.86 \\
5.2 \ = & & 5.20 \\
+\,9.0 = & & 9.00 \\
\hline
\text{Sum} = & & 52.15
\end{array}
$$

EXAMPLE 2

Add: 0.237 + 4.9 + 27.32

SOLUTION

Tens	Ones		Tenths	Hundredths	Thousandths
1	1				
	0	.	2	3	7
	4	.	9		
2	7	.	3	2	
3	2	.	4	5	7

Note that by placing the decimal points on a vertical line, digits of the same place value are added.

116

EXAMPLE 3

Subtract $21.532 - 9.875$ and check.

SOLUTION

Tens	Ones		Tenths	Hundredths	Thousandths
1	10		14	12	12
2	1	.	5	3	2
	9	.	8	7	5
1	1	.	6	5	7

Note that by placing the decimal points on a vertical line, digits of the same place value are subtracted.

Check: Subtrahend   9.875
     + Difference   + 11.657

    = Minuend   21.532

EXAMPLE 4

A worker earned a salary of $138.50 for working five days this week. The food server also received $22.92, $15.80, $19.65, $39.20, and $27.70 in tips during the five days. Find the total income for the week.

SOLUTION

To find the total income, add the tips ($22.92, $15.80, $19.65, $39.20, and $27.70) to the salary ($138.50).

$138.50
  22.92
  15.80
  19.65
  39.20
+ 27.70
$263.77

The worker's total income for the week was $263.77.

EXAMPLE 5

You had a balance of $62.41 in your checking account. You then bought a CD for $8.95, film for $3.17, and a skateboard for $39.77. After paying for these items with a check, how much do you have left in your checking account?

SOLUTION

To find the new balance:
Find the total cost of the three items ($8.95 + $3.17 + $39.77).
Subtract the total cost from the old balance ($62.41).

EXAMPLE 6

A student bought 10.5, 8.7, and 9.2 gallons of gasoline in one month. How much gas did the student buy during the month?

SOLUTION

To find the total amount of gasoline purchased, add the three amounts (10.5, 8.7, and 9.2).

$$
\begin{array}{r}
10.5 \\
8.7 \\
+\ 9.2 \\
\hline
28.4
\end{array}
$$

The student bought 28.4 gallons of gasoline during the month.

EXERCISE 3.7 A

Add or subtract as indicated.

1) 16.008 + 2.0385 + 132.06   2) 1.792 + 67 + 27.0526   3) 24.037 – 18.41

4) 214 – 7.143   5) 23.004 – 7.2175

6) 4.307 + 99.82 + 3 + 9.078   7) Subtract .32 from .675   8) 92 – 19.2909

9) 9.06 + 4.976 + 59.6   10) 247.39 + 39.46 + 8.746   11) 8.07 – 5.392

12) 3 – 1.296

13) Twenty-seven hundredths + one and five thousandths.

14) Find the difference of two, and ninety-two hundred-thousandths.

15) Subtract eight from sixty-four and thirty-seven hundredths.

16) Twenty-seven hundredths + two and four thousandths.

EXERCISE 3.7 B

1) 2.051 + .2006 + 5.4 + 37   2) 52.3 + 6 + 21.01 + 4.005

3) 43.766 + 9.33 + 17 + 206   4) 4.0086 + 0.34 + .6 + .05

5) 37.02 + 25 + 6.4 + 3.89   6) 3.488 + 16.593 + 25.002

7) .59 + 6.91 + .05   8) 5 + 6.1 + .4   9) .6 + .4 + 1.3

10) 5.0015   11) 4.128   12) 1.007   13) 47.3
     2.443   .02   20.063   42.03
      .0469   3.   .49   29.003

Find each of the indicated differences:

14) 22.418   15) 4.   16) 31.009   17) 4.8
     17.523   .0026   .534   .0026

18) 22.418   19) 78.015 – 13.068   20) 1.0057 – .03
     17.523

21) 29.5 – 13.61   22) 17.83 – 8.9   23) 5.2 – 3.76

119

24) Find the difference between sixty-five and four hundred twenty-six ten-thousandths, and fifty-eight and two thousand, five hundred seventy-one ten-thousandths.

25) Find the difference between five hundred forty-eight and thirteen hundredths, and three hundred ninety-three and fifty-seven thousandths.

26) Find the sum of the three numbers: seven hundred sixty-three ten-thousandths; fourteen thousandths; one thousand four hundred twenty-nine ten-thousandths.

27) Find the sum of the three numbers: four hundred thirty-six and seventeen thousandths; five hundred and three tenths; one hundred seventy-seven and three hundred forty-eight thousandths.

3.8 MULTIPLICATION AND DIVISION OF DECIMALS

To understand the procedure for multiplying decimals, we can consider the corresponding fractions. For example, the product of 3.2 and 4.16.

$$(3.2)(4.16) = 3\frac{2}{10} \times 4\frac{16}{100} = \frac{32}{10} \times \frac{416}{100} = \frac{32 \times 416}{10^1 \times 10^2} = \frac{13312}{10^3} = 13.312$$

Observe that the answer was obtained by multiplying the whole numbers 32 and 416 and then dividing the result by 10^3. From this observation the following algorithm for multiplying decimals is derived.

> **To multiply decimals:** multiply the numbers as if they were whole numbers. The number of decimal places in the product will equal the sum of the number of decimal places in the factors.

EXAMPLE 1

Multiply $.369 \times .207$

SOLUTION

```
    .369
    .207        3 decimal places
    2583        3 decimal places
   7380
 .076383        3 + 3 = 6 decimal places
```

Note that it was necessary to annex one zero to the left so that the product would have 6 decimal places.

> **To divide decimals:** multiply by a number that will make the divisor a whole number. Then move the decimal point straight up and divide as you would whole numbers.

EXAMPLE 2

Divide: $58.092 \div 82$ (Round to the nearest thousandth).

SOLUTION

$$
\begin{array}{r}
0.7084 \approx 0.078 \\
82 \overline{\smash{)}\ 58.0920} \\
-57\ 4 \quad\ \ \\
\overline{69\ \ } \\
-\ \ 0\ \ \\
\overline{692} \\
-656\ \ \\
\overline{360} \\
-328\ \ \\
\overline{32}
\end{array}
$$

EXAMPLE 3

A retail tire store pays $3.76 excise tax on each tire sold. During one month the store paid $300.80 in excise tax. How many tires were sold during the month?

SOLUTION

To find the number of tires sold, divide the total excise tax paid ($300.80) by the excise tax paid for each tire ($3.76).

$$
\begin{array}{r}
80 \\
\$3.76 \overline{\smash{)}\ \$300.80} \\
-\ \ 300\ 8\ \ \\
\overline{00} \\
-\ \ 0\ \\
\overline{0}
\end{array}
$$

EXERCISE 3.8

Find each of the indicated products.

1) 4(.75)   2) 16(.875)   3) 8(.125)   4) .66(.33)

5) 4.7(.02)   6) .51(.13)   7) 4.15(2.6)   8) 5.9(.25)

9) 6(3.1)   10) .2(.02)   11) .5(.05)   12) .03(.03)

13) (.6) (.7)   14) 3(2.5)   15) 1.4(.2)   16) (3.5) (.6)

Find each of the indicated quotients.

17) $9\overline{)7.6}$ 18) $3.2\overline{).416}$ 19) $7\overline{)6.6}$ 20) $.14\overline{).042}$

21) $1.8\overline{).0036}$ 22) $5.6\overline{)28}$ 23) $2.4\overline{)48}$ 24) $.03\overline{)16.02}$

25) $.04\overline{)82.24}$ 26) $.7\overline{).63}$ 27) $.8\overline{).064}$ 28) $.9\overline{)1.62}$

29) $.5\overline{)4.95}$ 30) $3\overline{)1.71}$ 31) $2\overline{)4.68}$

32) If the postal rates on educational material are 10¢ for the first pound and 5¢ for each additional pound, what will be the cost per pound to mail 8 pounds of books?

33) If a motorcycle can travel 450.6 miles in 7.81 hours, what will be its average speed?

34) A road grader burns 8.4 gallons of fuel oil each hour that it operates. How long will the grader operate on 100.4 gallons of fuel oil?

35) If an automobile averages 17.2 miles per gallon, how far will it go on 18 gallons of gasoline?

36) If a store advertises three cans of beans for $1.00, what will be the charge for one can?

37) Find the average, correct to the nearest hundredth, of the numbers: 44.62, 57.8, 39.49, 76.2, and 61.523.

38) If an automobile dealer makes $150.70 on each used car he sells and $425.30 on each new car he sells, how much did he make the month that he sold 11 used and 6 new cars?

39) If an architect makes a drawing to the scale that 1 inch represents 6.75 feet, what distance is represented by 5.5 inches?

40) Mr. Marshall bought two suits for $65.30 each, one pair of shoes for $22.95, three pairs of socks for $1.25 a pair, four shirts for $5.97 each, and six ties for $2.65 each. What was his total bill?

3.9   COMPARING AND CONVERTING FRACTIONS TO DECIMALS

Sometimes we have expressions involving rational numbers written as decimals and fractions. To work with these expressions, we need to be able to convert fractions to decimals and vice versa. The rules for converting decimals to fractions, and fractions to decimals, follow.

> **To convert a decimal to a fraction:** remove the decimal point and place the decimal part over a denominator equal to the place value of the digits in the decimal part of the number.

EXAMPLE 1

Convert 8.65 to a fraction.

SOLUTION

8.65 means 8 and 65 hundredths written $8 + \dfrac{65}{100}$ or $8\dfrac{65}{100}$ which simplifies to $8\dfrac{13}{20}$.

> **To convert a fraction to a decimal:** divide the numerator of the fraction by the denominator.

EXAMPLE 2

Convert $2\dfrac{3}{4}$ to a decimal.

SOLUTION

To find a decimal equivalent for $2\dfrac{3}{4}$, convert to an improper fraction and divide the numerator by the denominator.

$$2\dfrac{3}{4} = \dfrac{11}{4}$$

$$
\begin{array}{r}
2.75 \\
4\overline{)\ 11} \\
-\ 8 \\
\hline
30 \\
-\ 28 \\
\hline
20 \\
-\ 20 \\
\hline
0
\end{array}
$$

Being able to convert fractions to decimals, and vice versa, allows us to work problems involving both representations of rational numbers as illustrated in the following examples.

EXAMPLE 3

How much is $70.1 - 23\dfrac{1}{4}$?

SOLUTION

We can convert 70.1 to a fraction or $23\dfrac{1}{4}$ to a decimal and then subtract. Let us convert $23\dfrac{1}{4}$ to a decimal which gives us 23.25.

$$\begin{array}{r} 70.10 \\ -\ 23.25 \\ \hline 46.85 \end{array}$$

EXAMPLE 4

Express the product of 4.5 and .5 as a fraction.

SOLUTION

The product of $4.5 \times .5 = 2.25$. As a fraction $2.25 = 2\dfrac{25}{100} = 2\dfrac{1}{4}$.

EXAMPLE 5

Use a calculator to write $\dfrac{8748}{320}$ as a mixed number fraction.

SOLUTION

$\dfrac{8748}{320} = 27.3375$. Converting the decimal part to a fraction, we get $27.3375 = 27\dfrac{3375}{10000} = 27\dfrac{27}{80}$ written in lowest terms.

Ordering Decimals

Decimals are ordered just like fractions. We express each number to the same LCD and then compare their numerators. Consider the decimal fractions .61 and .601. Written as common fractions we get $\dfrac{61}{100}$ and $\dfrac{601}{1000}$. The LCD is 1000, therefore $\dfrac{61}{100} = \dfrac{610}{1000}$ and $\dfrac{601}{1000} = \dfrac{601}{1000}$. Since 610, the numerator, is larger than 601, $.61 > .601$.

These observations lead to the following rule for ordering decimals.

To order decimals

1) Convert each decimal to the same LCD by expressing each decimal to the same place value; annex zeros to the right of the decimal point as necessary.

2) Ignore the decimal point and compare the numbers just as you would whole numbers.

EXAMPLE 6

Order .15, .0051, 1.105, .051 from smallest to largest.

SOLUTION

Express each number to the same place value (LCD).

$$
\begin{aligned}
.15 &= \quad .1500 \\
.0051 &= \quad .0051 \\
1.105 &= \quad 1.1050 \\
.051 &= \quad .0510
\end{aligned}
$$

Ignoring the decimals, the numbers (numerators of the fractions) are 1500, 51, 11, 050 and 510. Ordered by whole numbers from smallest to largest.

Whole number	51	510	1,500	11,050
Related decimal	.0051	.051	.15	1.105

Rounding and Estimating

In rounding, locate the place value to which you are rounding and then look at the digit one place to the right. If the digit to the right is 5 or more, increase the digit to which you are rounding by 1. If not, leave it as it is. In estimating, numbers are rounded so that their approximations are easy to work with. For example, in estimating the product of 4.8 and 27.9, 4.8 would be rounded to 5 and 27.9 would be rounded to 30.

EXAMPLE 7

The closest whole number approximation to $\dfrac{(198.27)(0.5012)}{2.02}$ is?

SOLUTION

Approximating each number gives $\dfrac{(198.27)(0.5012)}{2.02} \approx \dfrac{(200)\left(\dfrac{1}{2}\right)}{2} \approx 50.$

127

EXERCISE 3.9

Write as decimal numbers (round to two decimal digits).

1) $\dfrac{1}{6} =$ 2) $\dfrac{1}{4} =$ 3) $\dfrac{1}{8} =$ 4) $\dfrac{7}{8} =$

5) $\dfrac{5}{6} =$ 6) $\dfrac{3}{4} =$ 7) $\dfrac{5}{9} =$ 8) $\dfrac{3}{11} =$

9) $\dfrac{1}{12} =$ 10) $\dfrac{1}{16} =$ 11) $\dfrac{5}{13} =$ 12) $\dfrac{3}{7} =$

13) $\dfrac{3}{25} =$ 14) $\dfrac{6}{10} =$ 15) $\dfrac{4}{5} =$ 16) $\dfrac{4}{17} =$

17) $\dfrac{3}{50} =$ 18) $\dfrac{4}{100} =$ 19) $\dfrac{5}{12} =$ 20) $\dfrac{7}{20} =$

Write as a fraction in lowest terms.

21) 3.12   22) 6.55   23) 6.5   24) 1.04

25) 0.013   26) 0.0005   27) 0.025   28) 1.0625

29) 1.82   30) 3.875   31) 2.125   32) 1.25

33) 0.56   34) 0.42   35) 0.7

Order the following from smallest to largest.

36) .068, .2356, .7, $\dfrac{1}{2}$ 37) 1.1, 1.01, .011, .101, .1001

38) Round 15.1874 to the nearest hundredths

39) Which is between 1.7 and 1.8?

   (a) 1.078   (b) 1.870   (c) 1.708

40) $\dfrac{3}{10} + \dfrac{1}{1000} + \dfrac{5}{100} =$

41) $\dfrac{4.91 \times 51.7}{.51}$ is closest to what integer?

42) What is the quotient of 156 divided by 35 rounded to the nearest hundredths?

43) $\dfrac{3}{5} + \dfrac{16}{17} + \dfrac{12}{13} + \dfrac{6}{13}$ is closest to what integer?

44) What is the difference between $\dfrac{1}{2}$ and $\dfrac{1}{4}$ expressed as a decimal?

CHAPTER 3 REVIEW EXERCISES A

1) Reduce $\dfrac{42}{147}$ to lowest terms.

2) Write an equivalent fraction for $\dfrac{5}{7} = \dfrac{}{49}$.

3) Convert $7\dfrac{4}{15}$ to an improper fraction.

4) Convert $\dfrac{69}{17}$ to mixed number fraction.

5) Order from the smallest to the largest: $\dfrac{2}{3}, \dfrac{1}{5}, \dfrac{7}{15}$.

6) Which fraction is greater than $\dfrac{2}{5}$?

    a) $\dfrac{1}{3}$ b) $\dfrac{7}{25}$ c) $\dfrac{3}{10}$ d) none

Perform the indicated operation.

7) $27\dfrac{2}{3} - 19\dfrac{4}{5}$ 8) $5\dfrac{3}{7} \times 5\dfrac{1}{4}$ 9) $2\dfrac{7}{12} + 6\dfrac{5}{16}$

10) $9\dfrac{3}{7} \div 4\dfrac{5}{14}$ 11) $\dfrac{6 - \dfrac{3}{4}}{5\dfrac{1}{3} + \dfrac{1}{2}}$ 12) $82.006 + 9.95 + .927$

13) $3 - 1.2096$ 14) $98.5 \times .089$ 15) $141.38 \div 3.9$

16) Convert 7.048 to a fraction.   17) Convert $6\dfrac{7}{8}$ to a decimal.

18) Arrange in order from the largest to the smallest:

    $1\dfrac{27}{100}$, 1.0127, 1.1027, 2

19) The product of 1.5 and .4 expressed as a common fraction is?

20) Find a fraction between $\frac{1}{3}$ and $\frac{3}{7}$.

21) Find a decimal between 1.5 and 1.6.

22) The sum of $\frac{1}{5}$, .015, 69.82, and $2\frac{1}{4}$ is?

23) A company uses two sizes of boxes, 8 inches and 12 inches long. These boxes are to be shipped. What is the shortest length that will accommodate all boxes of either size without any room left over?

24) A board $11\frac{1}{2}$ feet long is cut into 3 equal pieces. How long is each piece?

25) Of the 30,000 tickets for the state football championship game, $\frac{1}{3}$ were sold at $3.00, $\frac{1}{3}$ were sold at $2.50, and the rest were sold at $1.25. How many tickets were sold at $1.25?

26) The sum of $\frac{14}{15} + \frac{12}{13} + \frac{16}{17} + \frac{3}{5}$ is closest to 1, 2, 3, or $3\frac{1}{2}$?

CHAPTER 3 REVIEW EXERCISES B

State whether each of the following rational numbers is a proper fraction or an improper fraction.

1) $\dfrac{15}{14}$

2) $\dfrac{5}{5}$

3) $\dfrac{4}{1}$

4) $\dfrac{25}{27}$

5) $\dfrac{0}{14}$

6) Build an equivalent fraction with the given denominator $\dfrac{5}{12} = \dfrac{\ }{60}$.

7) Write $\dfrac{22}{5}$ as a mixed number.

8) Write $4\dfrac{5}{8}$ as an improper fraction.

Fill in the missing terms so that each equation is true.

9) $\dfrac{0}{9} = \dfrac{\ }{27}$

10) $\dfrac{15}{11} = \dfrac{75}{\ }$

11) $\dfrac{9}{10} = \dfrac{54}{\ }$

12) $\dfrac{1}{6} = \dfrac{\ }{12}$

Reduce each of the following to lowest terms.

13) $\dfrac{150}{120}$

14) $\dfrac{99}{88}$

15) $\dfrac{0}{4}$

16) Add: $5\dfrac{7}{12} + 3\dfrac{7}{18}$

17) Subtract: $9\dfrac{5}{9} - 3\dfrac{11}{12}$

18) Multiply: $\dfrac{9}{16} \times \dfrac{4}{27}$

19) Multiply: $2\dfrac{1}{8} \times 4\dfrac{5}{17}$

20) Divide: $\dfrac{11}{12} \div \dfrac{3}{4}$

21) Divide: $2\dfrac{3}{8} \div 1\dfrac{2}{3}$

22) Evaluate: $\dfrac{5}{8} \cdot \dfrac{3}{10} + \dfrac{1}{4} \div 2$.

23) Find the average of $\dfrac{3}{4}$, $\dfrac{5}{8}$, $\dfrac{9}{10}$, and $\dfrac{2}{3}$.

Simplify.

24) $\dfrac{6\dfrac{1}{2} + 2\dfrac{3}{11}}{3\dfrac{1}{9} - 2\dfrac{5}{6}}$

25) $\dfrac{\dfrac{7}{8} - \dfrac{3}{16}}{\dfrac{1}{3} - \dfrac{1}{4}}$

Write in words.

26) 6.5781

27) 900.5

28) 4.008

133

Write in decimal notation.

29) two hundred and seventeen hundredths

30) eighty-four and seventy-five thousandths

31) three thousand three and three thousandths

Round off as indicated.

32) .0385 (nearest thousandth)

33) 72.997 (nearest hundredth)

Add or subtract as indicated.

34) $16.92 - 7.9$

35) $5 - 1.0377$

36) $34.967 + 40.8 + 9.451 + 8.2$

37) $32.5 - 14.71$

38) Subtract   29.005
    $-$ 7.9286

39) Multiply   9.074
    $\times$ 6.09

40) Divide $8.09\overline{)17.42963}$. Round to the nearest thousandth.

41) Convert $\dfrac{11}{15}$ to a decimal. Round to the nearest thousandth.

42) Convert $0.16\dfrac{2}{3}$ to a fraction.

43) An investor purchased $4\dfrac{3}{4}$ ounces of gold for $1615. What was the price for one ounce?

44)   A machine takes 0.017 inch from a brass bushing that is 1.412 inches thick. Find the resulting thickness of the bushing.

45) A car costs $3876. The down payment is $1500 and the balance is paid in 24 equal monthly installments. Find the monthly payment.

46) The tax on your business is $820 plus 0.08 times your profit. You made a profit of $64,860 last year. Find the amount of tax you paid last year.

47) Joe writes checks of $42.98, $16.30, $42.15, and $101.67. Find the total of the four checks.

48) During a three-week period, a rain gauge collected $\frac{3}{8}$ inches of rain, $3\frac{1}{3}$ inches of rain, and $1\frac{1}{2}$ inches of rain. Find the total rainfall for the three weeks.

CHAPTER 3 TEST

1) Write 0.35 as a fraction in lowest terms.

a) $\dfrac{35}{100}$ b) $\dfrac{7}{20}$ c) $\dfrac{35}{10}$ d) $\dfrac{35}{1000}$ e) $\dfrac{7}{2}$

2) Write 5.080 as a fraction in lowest terms.

a) $5\dfrac{4}{5}$ b) $5\dfrac{8}{10}$ c) $5\dfrac{2}{25}$ d) $5\dfrac{2}{5}$ e) $5\dfrac{8}{20}$

3) Write $\dfrac{3}{16}$ as a decimal.

a) .53   b) .188   c) .18   d) 0.1875   e) .533

4) What part of 1.2 is 5.16?

a) 4.3   b) .43   c) 43   d) 0.23   e) none of them

5) The closest estimate of $1\dfrac{11}{12}$ and $\dfrac{3}{7}$ is?

a) $2+\dfrac{1}{5}$ b) $2+\dfrac{1}{2}$ c) $1+\dfrac{1}{2}$ d) $2+1$

6) The balance owed on a loan is $5,600. After 9 payments of $324, how much is still owed on the loan?

a) $2916   b) $2684   c) $5276   d) none

7) The average of $\dfrac{5}{6}$ and $\dfrac{2}{3}$ is?

a) $\dfrac{7}{9}$ b) $\dfrac{3}{4}$ c) $\dfrac{9}{6}$ d) none

8) Which of the following brands of toothpaste is the least expensive?

a) 5 oz of 89 cents   b) 9 oz for $1.44
c) 7 oz for $1.19   d) 11 oz for $1.65

9) Multiply 0.08×2.69

a) 2.512   b) .2152   c) 21.52   d) none

10) $4.923 \div 0.067$ rounded to the nearest hundredth is

a) 73.47   b) 7.348   c) 73.48   d) 72.47

11) Subtracting 1.887 from 2 gives a difference of

    a) .113   b) 2.887   c) 1.887   d) none

12) Which of the following pairs are not equal?

    a) $\frac{2}{5}$ and .4   b) $\frac{1}{3}$ and $\frac{11}{33}$ c) $\frac{7}{11}$ and $\frac{72}{112}$ d) $\frac{13}{4}$ and $3\frac{1}{4}$

13) Which fraction is smaller than $\frac{1}{2}$?

    a) $\frac{5}{6}$ b) $\frac{3}{8}$ c) $\frac{2}{3}$ d) none

14) How much is $12.2 - 2\frac{1}{4}$?

    a) 9.95   b) $1\frac{9}{20}$ c) 9   d) none

15) The product of $17\frac{5}{8}$ and 128 is?

    a) 2200   b) 2305   c) 2356   d) 2256

16) Converted to a mixed number, $\frac{27}{7}$ equals?

    a) $2\frac{1}{7}$ b) $3\frac{6}{7}$ c) $6\frac{1}{3}$ d) $7\frac{1}{2}$

17) Converted to a decimal, $\frac{3}{8}$ equals?

    a) .0375   b) 37.5   c) .375   d) 375

18) If each bag of cookies weighs $5\frac{3}{4}$ pounds, how many pounds do 3 bags weigh?

    a) $7\frac{1}{4}$ b) $15\frac{3}{4}$ c) $16\frac{1}{2}$ d) $17\frac{1}{4}$

19) Subtract $27\frac{5}{14}$ from $43\frac{1}{6}$.

    a) $5\frac{8}{21}$ b) 15   c) 16   d) $15\frac{17}{21}$

20) The number of half-pound bags of sugar that can be weighed out of a box that holds $10\frac{1}{2}$ pounds of sugar is?

    a) 21   b) $10\frac{1}{2}$ c) $20\frac{1}{2}$ d) 5

21) How much is 19.6 divided by 3.2 rounded to 3 decimal places?

    a) 6.000   b) 612.5   c) 61.25   d) 6.125

22) Reduced to simplest form, $\frac{56}{91}$ equals?

    a) $\frac{2}{3}$ b) $\frac{5}{9}$ c) $\frac{8}{13}$ d) none

CHAPTER 4 RATIOS, PROPORTIONS, AND PERCENTS

In Chapter 3 we discussed several meanings of a rational number. One was a certain number of equal parts of a whole and the other was to indicate division. Ratios can be misleading unless we are careful to state precisely the items that are represented by the numerators and denominators.

4.1 RATIO AND PROPORTION

A **ratio** is a way of comparing two quantities or numbers. We may denote two quantities, say c and d, as a ratio in the following two ways:

$$\frac{c}{d} \text{ or } c:d.$$

In either case the ratio is read, "c to d" or "c parts to d parts." Ratios should always be written in simplest form and the units of measure should be the same.

Express each of the following examples as a ratio reduced to the lowest form.

EXAMPLE 1

3 quarters to 4 dimes

SOLUTION

Step 1. $\dfrac{75 \text{ cents}}{40 \text{ cents}}$ Express as a ratio in the fractional form and change each quantity to cents.

Step 2. $\dfrac{75 \text{ cents}}{40 \text{ cents}} = \dfrac{15 \text{ cents}}{8 \text{ cents}} = \dfrac{15}{8}$ Reduce.

EXAMPLE 2

12 oranges to 8 apples.

SOLUTION

Step 1.	$\dfrac{12 \text{ oranges}}{8 \text{ apples}}$	Express as a ratio, but we cannot change to
		the same units. Thus, we label each part to avoid confusion.

Step 2.	$\dfrac{12 \text{ oranges}}{8 \text{ apples}} = \dfrac{3}{2}$	Reduce.

PROPORTION

A **proportion** is a statement that two ratios are equal. This is, if $a{:}b$ and $c{:}d$ are two ratios, then the equation $\dfrac{a}{b} = \dfrac{c}{d}$ is called a proportion.

A proportion has four terms: the number in the position of a is the first term; the number in the position of b is the second term; the number in the position of c is the third term; and the number in the position of d is the fourth term. The first and fourth terms are called the **extremes of the proportion** and the second and third terms are called the **means of the proportion**.

Not all proportions need to be true statements. We are more interested in true proportions and one way to tell whether a proportion is true is to find out if the product of the extremes is equal to the product of the means.

EXAMPLE 3

Determine whether the following ratios are proportions.

$$\frac{12}{15} = \frac{4}{5}$$

SOLUTION

$\dfrac{12}{15} = \dfrac{4}{5}$ The extremes are 12 and 5. The means are 15 and 4.

$12 \times 5 = 15 \times 4$

    $60 = 60$ Set the product of the extremes equal to the product of the means.

EXAMPLE 4

Determine whether the following is true or false.

SOLUTION

Step 1.	$\dfrac{8}{36} = \dfrac{2}{9}$	Construct a cross through the equal sign
		with arrows.
Step 2.	$8 \times 9 = ?$ 36×2	Cross multiply.
Step 3.	$72 = 72$	A true statement.

EXAMPLE 5

Determine whether the following ratios are proportions.

$$\frac{9}{24} = \frac{3}{7}$$

SOLUTION

Step 1.	$\dfrac{9}{24} = \dfrac{3}{7}$	
Step 2.	$9 \times 7 = ?$ 24×3	
Step 3.	$63 \neq 72$	This is a false equation. The proportion is false.

FINDING THE UNKNOWN TERM IN A PROPORTION

A large number of everyday life problems can be solved by proportions. Basically, a proportion will always have four terms.

The procedure for solving the equation (finding the value of x that makes the proportion true) is to use the fact that the product of the extremes must be equal to the product of the means. We need to know that: if both sides of an equation are multiplied by the same number, then the new equation will have the same solution as the original equation.

EXAMPLE 6

Find the missing term in the proportion.

$$\frac{3}{a} = \frac{3}{4}$$

SOLUTION

Step 1.	$\dfrac{3}{a} = \dfrac{3}{4}$	Set up the proportion

141

Step 2. $\dfrac{3}{a} = \dfrac{3}{4}$ Construct the cross

$3 \times 4 = 3a$ Cross multiply

Step 3. $\begin{array}{c} 12 = 3a \\ 4 = a \end{array}$ Divide by 3

EXAMPLE 7

$$\dfrac{x}{16} = \dfrac{\dfrac{3}{4}}{\dfrac{2}{3}}$$

SOLUTION

$\left(\dfrac{2}{3}\right)(x) = 16\left(\dfrac{3}{4}\right)$ Cross multiply

$\dfrac{2x}{3} = 12$ Divide by $\dfrac{2}{3}$. Recall, dividing by $\dfrac{2}{3}$ is the

same as multiplying by its reciprocal, $\dfrac{3}{2}$

$18 = x$

SOLVING WORD PROBLEMS USING PROPORTIONS

A common use of proportions is to solve word problems when three terms of a proportion are known and the problem involves finding the value of the missing term that will make the proportion true.

Our first step in solving word problems is to learn to find the values of missing or unknown terms of proportions that make the proportions true.

Steps in solving word problems using a proportion:

1) Identify the unknown quantity.
2) Define the ratio to be compared.
3) Make sure the corresponding quantities of the proportion appear in the numerator and the denominator of each ratio.
4) Solve the proportion.

EXAMPLE 8

If 6 carts cost $75, what would be the cost of 10 carts?

SOLUTION

First, we find the quantities to be compared: $\dfrac{\text{carts}}{\text{dollars}}$.

Now label the units to help ensure the proportion is set up correctly.

$$\frac{\text{carts}}{\text{dollars}} \qquad \frac{6 \text{ carts}}{75 \text{ dollars}} = \frac{10 \text{ carts}}{x \text{ dollars}}$$

$$10(75) = 6x \qquad \text{Cross multiply}$$

$$\frac{750}{6} = \frac{6x}{6} \qquad \text{Divide by 6}$$

$$\$125 = x$$

EXAMPLE 9

Carbon combines with hydrogen in a ratio of 3 parts by weight of carbon to 1 part by weight of hydrogen to form a gas called methane. How much hydrogen is needed to combine with 1200 grams of carbon to form methane?

SOLUTION

We are comparing parts by weight of hydrogen to parts by weight of carbon, therefore our proportion is

$$\frac{1 \text{ part of hydrogen}}{3 \text{ parts of carbon}} = \frac{x \text{ parts of hydrogen}}{1200 \text{ grams of carbon}}$$

$$3x = 1(1200) \qquad \text{Cross multiply}$$

$$\frac{3x}{3} = \frac{1200}{3} \qquad \text{Divide both sides by 3}$$

$$x = 400 \text{ grams of hydrogen}$$

EXAMPLE 10

Convert 130 miles to kilometers.

SOLUTION

From a standard table of measures, we find the conversion factor for miles to kilometer is 1 mile = 1.609 kilometers. Let x = number of kilometers, and our proportion is

$$\frac{\text{Miles}}{\text{Kilometers}} \qquad \frac{1 \text{ mile}}{1.609 \text{ kilometers}} = \frac{130 \text{ miles}}{x \text{ kilometers}}$$

$$x = 130(1.609) \qquad \text{Cross multiply}$$

$$x = 209.17 \qquad \text{kilometers}$$

EXERCISE 4.1 A

Express the ratio of the following.

1) 20 gallons of gas to 70 gallons of gas.

2) 16 parts of hydrogen to 38 parts of hydrogen.

3) 2 feet to 36 inches.

4) There were 18 questions on a test. A student got 6 questions incorrect. What is the ratio of the number of questions answered correctly to the total number of questions on the test?

5) In a class, there are 12 men and 15 women. What is the ratio of men to women in the class?

6) 8 months to 8 years.

Determine whether the following ratios are proportions.

7) $\dfrac{8}{11} = \dfrac{24}{33}$

8) $\dfrac{17}{19} = \dfrac{3}{4}$

9) $1\dfrac{1}{3} = \dfrac{12}{9}$

10) $\dfrac{\frac{1}{5}}{\frac{2}{3}} = \dfrac{3}{10}$

11) $\dfrac{2}{3} = \dfrac{7}{8}$

12) $\dfrac{5}{4} = \dfrac{10}{6}$

Use a proportion to write an equivalent fraction for each of the following.

13) $\dfrac{2}{3} = \dfrac{?}{4}$

14) $\dfrac{5}{6} = \dfrac{?}{18}$

15) $\dfrac{8}{12} = \dfrac{64}{?}$

16) $\dfrac{8}{5} = \dfrac{7}{?}$

Find the missing part for each proportion.

17) $\dfrac{x}{6} = \dfrac{10}{15}$

18) $\dfrac{5}{4} = \dfrac{x}{16}$

19) $\dfrac{4}{x} = \dfrac{2}{11}$

20) $\dfrac{1}{8} = \dfrac{\frac{1}{2}}{x}$

Use a proportion to convert the following measurements to the unit indicated.

21) If 1 meter equals 1.1 yards, then 27 meters equal how many yards?

22) If 1 liter=1.06 qt. then 56.3 liters equal how many quarts?

23) If 1 kg=2.2 lbs, then 9.86 pounds equal how many kilograms?

24) If 1 km = .6 miles, then 25.97 kilometers equal how many miles?

25) On a road map, 3 inches represents 8 miles. How many inches would represent a distance of 36 miles?

26) If there should be 3 calculators for every 4 students in an elementary school class, how many calculators are needed for 44 students?

27) How long will it take to go 1500 miles if a plane travels at 550 miles in 2 hours?

28) A recipe calls for 2 cups of sugar for every 5 cups of flour. If a cook is using 4 cups of flour, how many cups of sugar are needed?

29) If Abbie can bake 27 pies in twelve days, how many pies can she bake in 20 days?

30) A recipe for punch which serves 20 people requires $\frac{3}{8}$ cup of fruit punch mix. How many cups of mix would be required if the recipe is increased to serve 75 people?

31) It takes 2 minutes to fill a tank $\frac{3}{8}$ full. How long would it take to fill the rest of the tank?

32) Carbon combines with hydrogen in a ratio of 3 parts by weight of carbon to 1 part by weight of hydrogen to form a gas called methane. How much hydrogen is needed to combine with 1500 grams of carbon to form methane?

33) Nitrous oxide (laughing gas) can be decomposed to give 7 parts by weight of nitrogen and 4 parts by weight of oxygen. How much nitrogen (by weight) would be obtained if enough nitrous oxide was decomposed to give 36 grams of oxygen?

34) The ratio of boys to girls in Akilah's class is 8:7. If the class has 60 students, how many boys are in the class?

35) A library orders fiction and non-fiction books: The library orders 4 fiction books for every non-fiction book. If the library purchases 40 books, how many of each type of book does it purchase?

36) Brand A cereal contains about .43 parts of sugar and Brand B contains about .46 parts of sugar. What proportion of sugar would there be in a mixture of 100 grams of Brand A and 200 grams of Brand B?

37) Ron's high school has 1500 students. The teacher to pupil ratio is 1:30. How many additional teachers would have to be hired to reduce the ratio to 1:20?

38) To estimate the crowd at an event, a survey team distributed 125 identification tags. Later during the day, the team counted the number of people passing an imaginary point with and without tags. Of the 700 people crossing the line, 16 had tags. To the nearest hundred, approximately, how many people attended the event?

EXERCISE 4.1 B

Write the following comparisons as ratios reduced to lowest terms.

1) Two yards to five feet

2) 15 boys to 7 girls

3) One hour to two days

4) 3 pints to 15 pints

5) 8 days to 2 weeks

6) 9 minutes to 2 hours

7) 8 ounces to 2 pounds

8) 3 months to 2 years

Determine if the proportion is true or not true.

9) $\dfrac{9}{7} = \dfrac{6}{5}$

10) $\dfrac{3}{4} = \dfrac{54}{72}$

11) $\dfrac{15}{7} = \dfrac{17}{8}$

12) $\dfrac{7}{40} = \dfrac{7}{8}$

13) $\dfrac{45}{135} = \dfrac{3}{9}$

14) $\dfrac{7}{8} = \dfrac{11}{12}$

15) $\dfrac{15}{5} = \dfrac{3}{1}$

16) $\dfrac{3}{18} = \dfrac{4}{19}$

17) $\dfrac{39}{48} = \dfrac{13}{16}$

18) $\dfrac{16}{3} = \dfrac{48}{9}$

19) $\dfrac{27}{8} = \dfrac{9}{4}$

20) $\dfrac{4}{8} = \dfrac{10}{20}$

Find the missing part of each proportion.

21) $\dfrac{n}{4} = \dfrac{6}{8}$

22) $\dfrac{n}{7} = \dfrac{9}{21}$

23) $\dfrac{12}{18} = \dfrac{n}{9}$

24) $\dfrac{3}{n} = \dfrac{15}{10}$

25) $\dfrac{6}{n} = \dfrac{24}{36}$

26) $\dfrac{10}{12} = \dfrac{n}{24}$

27) $\dfrac{n}{6} = \dfrac{2}{3}$

28) $\dfrac{7}{15} = \dfrac{21}{n}$

29) $\dfrac{9}{4} = \dfrac{18}{n}$

30) $\dfrac{8}{5} = \dfrac{n}{6}$

31) $\dfrac{4}{n} = \dfrac{9}{5}$

32) $\dfrac{n}{5} = \dfrac{7}{8}$

33) $\dfrac{36}{20} = \dfrac{12}{n}$

34) $\dfrac{5}{12} = \dfrac{n}{8}$

35) $\dfrac{3}{4} = \dfrac{8}{n}$

36) $\dfrac{32}{n} = \dfrac{1}{3}$

37) $\dfrac{40}{n} = \dfrac{15}{8}$

38) $\dfrac{n}{15} = \dfrac{21}{12}$

Solve the following word problems.

39) Three units of hydrogen react with one unit of nitrogen. How many units of hydrogen will react with $2\dfrac{1}{3}$ units of nitrogen?

40) A man drew plans for a city park using a scale of $\frac{1}{4}$ inch to represent 25 feet. How many feet would 2 inches represent?

41) If $\frac{1}{2}$ inch represents 10 miles on a map, how many inches would represent 25 miles?

42) A building 14 stories high casts a shadow of 30 feet at a certain time of day. What would be the length of the shadow of a 20-story building at the same time of day in the same city?

43) Three and one-half cups of flour are needed to bake 2 small cakes. How many cups of flour are needed to bake 5 cakes of the same size?

44) The company needs 35 men to build 10 houses in a certain amount of time. How many men would they need to build 18 homes in the same amount of time?

45) Three pizzas sell for $2.91 at a local supermarket. What would be the cost of 5 pizzas?

46) If the price of a certain type of paper is $1.75 per yard, how many yards could a dressmaker buy with $35?

47) Suppose gasoline sells for 37¢ per gallon. How many gallons could be bought for $555?

48) If a dozen oranges costs 70¢, what would 1½ dozen cost?

49) If 6 white shirts sell for $25, what would be the cost of 10 white shirts?

50) A property valued at $70,000 is insured with Company A for $30,000 and with Company B for $40,000. A fire caused damage of $28,000. What amount will be paid by each company if they pay proportional amounts?

51) Driving steadily, a man made a trip of 200 miles in 4½ hours. How long would he take to drive 500 miles at the same rate of speed?

52) Playing the Hillcrest Golf Club course three times, Everne used his driver 42 times, his putter 102 times, and hit 12 shots our of sand traps. How many drives, putts, and sand shots would he hit if he played the course 10 times?

53) An electric fan makes 180 revolutions per minute. How many revolutions will the fan make if it runs for 24 hours?

54) A baseball team bought 15 bats for $37.50. What would they have paid for 10 bats?

55) Stacey thinks he should make $6 for every $100 he invests. How much would he expect to make on an investment of $7000?

56) A mixture of antifreeze and water is made so that ½ gallon of antifreeze is mixed with 3 gallons of water. How many gallons of water would be mixed with 2 gallons of antifreeze? What would be the minimum capacity of the container for this mixture?

57) Kenny opened five stores in one year and showed profits of $175,000 that year. How many stores would be needed to produce a yearly profit of one million dollars for the company?

58) An artist, working hard, figures he can do three paintings every two weeks. How long would he take to do 18 paintings?

59) A little girl left alone in a kitchen for two minutes ate two cookies, three crackers, and four pieces of candy. How many cookies, crackers, and pieces of candy might she have eaten if she had been alone for ten minutes?

60) A store sells 3 cans of beans for $1.00. What would be the charge for 2 cans of beans?

61) A man drove 900 miles in two weeks. How far could he expect to drive in fifty-two weeks?

62) An old bus burns 1½ quarts of oil when it is driven 1500 miles. How many quarts of oil will it burn in 4000 miles?

63) If 2 units of a gas weigh 175 grams, what is the weight of 5 units of the same gas?

64) One unit of carbon may react with two units of oxygen. With how many units of oxygen will cause 3¼ units of carbon to react?

4.2 INTRODUCTION TO PERCENTS

The word "percent" is commonly used in our everyday lives. Most people know what is meant by one hundred percent or twenty-five percent, or some other similar expression. The statement, "you are one hundred percent right" means you are completely right. The word percent comes from the Latin, *per centum*, which means per-hundred. The symbol for percent is %.

The expression 50 percent means 50 hundredths or 50 out of every hundred. When we say 50%, we mean 50 out of every hundred, as a common fraction, or 0.50 as a decimal.

Percents are commonly used to calculate taxes on different items. A sales tax of 5% means that, for each dollar you spend, you must pay 5 cents sales tax. Thus, if you purchase a ten-dollar item, you must pay 5 cents sales tax for each dollar or 50¢ sales tax.

EXAMPLE 1

Convert the following to percents.

a) 1.56 b) .056 c) $\dfrac{2}{5}$ d) $1\dfrac{1}{2}$

SOLUTION

Multiply each number by 100 and affix the percent sign.

a) $1.56 = 1.56 \times 100\% = 156\%$

b) $.056 = .056 \times 100\% = 5.6\%$

c) $\dfrac{2}{5} = \dfrac{2}{5} \times 100\% = \dfrac{200}{5}\% = 40\%$

d) $1\dfrac{1}{2} = \dfrac{3}{2} \times 100\% = \dfrac{300}{2}\% = 150\%$

EXAMPLE 2

Convert the following to percents.

a) .27 b) .0037 c) 1.76 d) $\dfrac{3}{5}$ e) $5\dfrac{3}{8}$

SOLUTION

Multiply each number by 100 and affix the percent sign.

a) $.27 = .27 \times 100\% = 27\%$.

b) $.0037 = .0037 \times 100\% = .37\%$

c) $1.76 = 1.76 \times 100\% = 176\%$

d) $\dfrac{3}{5} = \dfrac{3}{5} \times 100\% = \dfrac{300}{5}\% = 60\%$

e) For mixed numbers, change to improper fractions first, then multiply by 100.

$5\tfrac{3}{8} = \dfrac{43}{8} = \dfrac{43}{8} \times 100\% = \dfrac{4300}{8}\% = 537.5\%$

> **To convert a percent to a decimal:** multiply by 100%, which is the same as multiplying by .01 or moving the decimal two places to the left.

EXAMPLE 3

Convert the following to decimals.

a) 54%   b) 120%   c) $2\tfrac{1}{3}\%$

SOLUTION

a) $54\% = 54\left(\dfrac{1}{100}\right) = .54$ or $54\% = 54 \times .01 = .54$

b) $120\% = 120\left(\dfrac{1}{100}\right) = \dfrac{120}{100} = 1.2$

Within the fraction is a mixed number. First convert it to an improper fraction and simplify.

c) $2\tfrac{1}{3}\% = 2\tfrac{1}{3}\left(\dfrac{1}{100}\right) = \dfrac{7}{3}\left(\dfrac{1}{100}\right) = \dfrac{7}{300} = .0233\ldots$

EXAMPLE 4

Convert the following percents to decimals.

a) 25%   b) $5\tfrac{1}{3}\%$ c) 150%

SOLUTION

Multiplying each percent by $\dfrac{1}{100}$, or .01.

152

a) $25\% = 25\left(\dfrac{1}{100}\right) = \dfrac{25}{100} = .25$ or $25\% = 25 \times .01 = .25$

When the fraction is a mixed number, first convert to an improper fraction and simplify.

b) $5\frac{1}{3}\% = 5\dfrac{1}{3}\left(\dfrac{1}{100}\right) = \dfrac{16}{3}\left(\dfrac{1}{100}\right) = \dfrac{16}{300} = 0.0533\ldots$

c) $150\% = 150\left(\dfrac{1}{100}\right) = \dfrac{150}{100} = 1.5$ or $150\% = 150 \times .01 = 1.5$

> **To convert a percent to a fraction:** multiply by $\dfrac{1}{100}$ and simplify the fraction.

EXAMPLE 5

Convert the following to fractions.

a) 58% b) 130% c) $\dfrac{3}{5}\%$

SOLUTION

Since percent means per-hundred, we simply replace the percent sign with $\dfrac{1}{100}$, multiply, and simplify the fraction.

a) $58\% = 58\left(\dfrac{1}{100}\right) = \dfrac{58}{100} = \dfrac{29}{50}$

b) $130\% = 130\left(\dfrac{1}{100}\right) = \dfrac{130}{100} = \dfrac{13}{10} = 1\dfrac{3}{10}$

c) $\dfrac{3}{5}\% = \dfrac{3}{5}\left(\dfrac{1}{100}\right) = \dfrac{3}{500}$

EXAMPLE 6

Convert the following percents to fractions.

a) 46% b) 150% c) $\dfrac{1}{2}\%$

SOLUTION

Since percent means parts per hundred we simply replace the percent sign with $\dfrac{1}{100}$, multiply, and simplify the fraction.

a) $46\% = 46\left(\dfrac{1}{100}\right) = \dfrac{46}{100} = \dfrac{23}{50}$ Notice that the fraction should always be written in the simplest form.

b) $150\% = 150\left(\dfrac{1}{100}\right) = \dfrac{150}{100} = \dfrac{15}{10} = 1\dfrac{1}{2}$

c) $\dfrac{1}{2}\% = \dfrac{1}{2}\left(\dfrac{1}{100}\right) = \dfrac{1}{200}$

When we have a mixed number percent, we change it to an improper fraction first, and then convert. The example below illustrates this process.

> **To convert a mixed number percent to a fraction:** we change it to an improper fraction, and then convert.

EXAMPLE 7

Convert $3\dfrac{3}{4}\%$ to a fraction.

SOLUTION

First converting $3\dfrac{3}{4}$ to an improper fraction gives $\dfrac{15}{4}$; multiply and simplify,

$$3\dfrac{3}{4}\% = \dfrac{15}{4}\left(\dfrac{1}{100}\right) = \dfrac{15}{400} = \dfrac{3}{80}$$

EXAMPLE 8

Convert $5\dfrac{3}{8}\%$ to a fraction.

SOLUTION

First, convert $5\dfrac{3}{8}$ to an improper fraction, which gives $\frac{43}{8}$; now multiply and simplify as before,

$$5\dfrac{3}{8}\% = \dfrac{43}{8}\left(\dfrac{1}{100}\right) = \dfrac{43}{800}$$

EXERCISE 4.2 A

Convert the following decimal numbers to percents.

1) 0.956   2) 3.942   3) 75.43   4) 80.

5) 5.9   6) 0.001   7) 2.14   8) 0.493

Convert the following fractions to a percent.

9) $\dfrac{1}{3}$ 10) $\dfrac{22}{7}$ 11) $\dfrac{14}{9}$ 12) $2\dfrac{1}{7}$

13) $\dfrac{5}{7}$ 14) $14\dfrac{3}{5}$ 15) $\dfrac{9}{7}$ 16) $\dfrac{17}{12}$

17) $2\dfrac{4}{5}$ 18) $\dfrac{5}{8}$ 19) $\dfrac{4}{7}$ 20) $8\dfrac{1}{7}$

21) $15\dfrac{4}{7}$

Convert the following percents to decimal numbers.

22) 0.0068%   23) 37.5%   24) $14\dfrac{9}{10}$%   25) 2.68%

26) 23.57%   27) $15\dfrac{3}{10}$%   28) 46.1%   29) 100%

30) $13\dfrac{3}{5}$%   31) $27\dfrac{1}{4}$%   32) $3\dfrac{1}{2}$%   33) 0.8%

34) 0.049%   35) 482%   36) 46%

37) $15\dfrac{5}{8}$%   38) $22\dfrac{4}{5}$%   39) $1\dfrac{1}{4}$%   40) $\dfrac{1}{4}$%

41) 0.05%   42) 0.5%   43) 213%   44) 159%

45) 210%   46) 3.7%

Convert the following percents to fractions in reduced form.

47) 3.2%

48) 250%

49) 375%

50) 0.0045%

51) 0.068%

52) 35%

53) $16\frac{3}{10}$%

54) $2\frac{2}{5}$%

55) $\frac{3}{4}$%

56) 2.4%

57) 0.8%

58) 42.5%

59) 400%

60) 210%

EXERCISE 4.2 B

Convert the following percents to fractions.

1) 46%

2) ½%

3) 4.7%

4) 150%

5) 6.12%

6) .05%

7) 15⅝%

8) 9⅕%

Convert the following percents to decimals.

9) 23%

10) 15%

11) 67.25%

12) 2⅗%

13) 5200%

14) 1¼%

15) .009%

16) 542%

17) 44⅘%

18) 11⅑%

19) 4.9%

20) .00075%

Convert the following decimals to percents.

21) .8

22) 5.2

23) 2.6

24) .055

25) 3.678

26) .00015

27) 1.2

28) 14

29) 4.35

30) .01

31) 14.76

32) .33

Convert the following fractions to percents.

33) $\frac{5}{8}$

34) $\frac{2}{5}$

35) $\frac{7}{40}$

36) $\frac{27}{50}$

37) $1\frac{2}{3}$

38) $2\frac{5}{12}$

39) $\frac{30}{25}$

40) $\frac{1}{8}$

4.3 PERCENT PROBLEMS: PROPORTION METHOD

A problem in percents has three quantities. Usually two of these three quantities are given and the third must be found. These three quantities are called the rate, base, percentage.

In any percentage problem the **rate** is the number with the percent sign. The **base** is the number of which a fractional part is taken. The **percentage or amount** is the result of multiplying the rate by the base.

Using the capital letter A = amount, B = base, and R = percent, the following proportion will be the key to our solution of the three types of problems:

$$\frac{A}{B} = \frac{R}{100}$$

Usually, in the proportion problem, the amount A comes after or before the word **is**. The base B comes after the word **of**. The rate R is the number with the percent sign.

EXAMPLE 1

6% of 125 is?

R B A

SOLUTION

We see that R and B are known and A is the unknown.

The proportion $\dfrac{A}{B} = \dfrac{R}{100}$ will take the form: $\dfrac{A}{125} = \dfrac{6}{100}$.

Solve by cross-multiplying: $\dfrac{100A}{100} = \dfrac{750}{100}$

$A = 7.5$ (answer)

EXAMPLE 2

25% of ____ is 13.

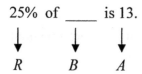

R B A

SOLUTION

We note that R and A are known and B is unknown.

The proportion $\dfrac{A}{B} = \dfrac{R}{100}$ will take the form: $\dfrac{13}{B} = \dfrac{25}{100}$

Solve by cross-multiplying: $\dfrac{1300}{25} = \dfrac{25B}{25}$

$52 = B$

EXAMPLE 3

50 is ____% of 100.

A R B

SOLUTION

Here B and A are known and R is unknown.

The proportion $\dfrac{A}{B} = \dfrac{R}{100}$ takes the form: $\qquad\qquad \dfrac{50}{100} = \dfrac{R}{100}.$

Solve by cross-multiplying: $\dfrac{5000}{100} = \dfrac{100B}{100}$

$50 = R$

PRACTICAL APPLICATIONS

The applications of percents to our everyday lives are very important. For example, a bank charges a percent of its loans as interest; companies operate at a certain loss or profit and we as taxpayers pay a percent for the value of our property.

Most problems involving percents can be solved by the formula. The key is to read the problem carefully to determine which parts are the rate, base, and the percent. Once that is determined, you will see that each problem is one of the types we have discussed in the previous section.

158

EXAMPLE 4

At an Orangeburg department store sale, a perfume originally marked $10 was at a 25% discount. What was the sale price?

SOLUTION

25% of $10.00 is $\dfrac{}{A}$, (where A = amount of discount)

$\downarrow$ $\qquad$ $\downarrow$

R $\qquad\quad$ B

$$\frac{A}{B} = \frac{R}{100}, \qquad \frac{A}{10} = \frac{25}{100}$$

Solve by cross-multiplying: $\quad \dfrac{100A}{100} = \dfrac{250}{100}$

$A = \$2.50$

10.00 (marked price)
$\underline{-2.50}$ (discount)
$\$\ 7.50$

EXAMPLE 5

Bennie was told he would have to pay interest of $30 to borrow $260 for one year. What annual rate of interest was he charged?

SOLUTION

30 is ____ of 260, (where R = the rate).

$\downarrow$ $\quad$ $\downarrow$ $\quad$ $\downarrow$

A $\quad$ R $\quad$ B

30 is ____ % of 260

Solve by cross-multiplying: $\quad \dfrac{30}{260} = \dfrac{R}{100}$

$$\frac{3000}{260} = \frac{260R}{260}$$

(approximately) $11.5 = R$
He was charged 11.5% interest.

EXAMPLE 6

On a recent mathematics test, Marshall got 68% correct out of the total number of questions. He answered 51 questions correct. How many questions were on the test?

SOLUTION

68% of ___ is 51, (where B = the base).

$$\downarrow \qquad \downarrow \qquad \downarrow$$
$$R \qquad B \qquad A$$

Solve by cross-multiplying: $\dfrac{51}{B} = \dfrac{68}{100}$

$$\dfrac{68B}{68} = \dfrac{5100}{68}$$

$$B = 75$$

Therefore, there were 75 questions on the mathematics test.

Problems dealing with investments are similar in usage as percents. The amount is usually called the **principal**; the rate of interest is called the rate; and the money gained or earned is called the **interest**.

The formulas used for these types of problems are:

$$P \cdot R \cdot T = I, \frac{I}{PT} = R, \text{ and } \frac{I}{PR} = T,$$

where P = principal, R = rate, and T = time.

EXAMPLE 7

If $200 is invested at a rate of 7% for 90 days, what would be the interest earned?

SOLUTION

(Please note: The rate is given as a percent, so 90 days is figured as part of a year. Often, in finance, a day-count convention of 360 days = 1 year is used to simplify calculations.).

$$I = P \cdot R \cdot T$$

$$P = 200, R = 7, T = \frac{90}{360}$$

$$I = 200 \times \frac{7}{100} \times \frac{1}{4}$$

$$I = \frac{1400}{400} = \$3.50$$

EXERCISE 4.3 A

Find the following.

1) 10% of 80 is _____

2) _____ is 32.4% of 76

3) 35% of _____ is 25

4) What is 60% of 300?

5) 20 is _____% of 75

6) 17 is _____% of 51

7) 14 is $4\frac{1}{2}$% of _____

8) 156 is 3% of _____

9) $\frac{1}{2}$% of 600 is _____

10) 40 is $33\frac{1}{3}$% of what?

11) 150% of _____ is 48

12) $13\frac{1}{8}$% of 38 is _____

13) 42% of _____ is 12

14) 15 is what percent of 250

15) What percent of 15 is 12?

16) 15% of what number is 9?

EXERCISE 4.3 B

Solve the following percent problems.

1) 10% of 70 is ____

2) ____ is 15% of 60

3) 2% of ____ is 3.

4) 14 is 3% of ____

5) 15 is ____ of 75

6) 17 is ____% of 34

7) 16 is ____% of 48

8) ____ % of 150 is 60

9) 42% of ____ is 10

10) 58% of 120 is ____

11) 50% of 25 is____

12) 16 is $25\frac{1}{2}$% of _____

13) ___ is 11.1% of 19

14) 13.5 is ____% of 16

15) 150% of ___ is 38

16) 48 is 300% of ____

17) ___% of 20 is 180

18) ___% of 70 is 105

19) ____ is $16\frac{2}{3}$% of 36.

20) 40 is $12\frac{1}{2}$% of ___

163

Solve the following word problems.

21) What would be the amount of interest in one year on a savings account of $800 if the bank pays 4% interest?

22) The manufacturer gave a store owner a discount of 3% because he bought $1500 worth of dresses. What was the amount of the discount?

23) The taxes on a house valued at $12,500 were $350. What was the tax rate?

24) At a sale, a department store sold sheets and towels at a discount of 30%. What were the discount prices if sheets were originally $3.50 and towels were originally $2.98?

25) At the Junior High School, the seventh graders had 340 cavities. This was 85% of the cavities of the eighth graders. How many cavities did the eighth graders have?

26) Marshall had a savings account at the local bank of $685. How much interest did his money earn in 6 months if the bank paid 5% interest per year on savings? In 4 months?

27) In one year Mr. Webb, who earned $12,000, spent $2,400 on rent, $4,800 on food, and $1,800 on taxes. What percent of his income went for rent, food, and taxes?

28) Doug played forward on the college basketball team and made 120 of 300 attempted shots. What percent of his shots did he make?

29) Pat missed 3 problems on a math test and received a grade of 85%, how many problems were on the test?

30) If the taxes on a gallon of gasoline are 11¢ per gallon and the service station charges 35.9¢ per gallon, what percent of the cost of a gallon of gas are taxes?

31) A drug store overestimated future sales of toothpaste and was overstocked. To help sell toothpaste marked at 98¢, the toothpaste was put on sale at 3 tubes for $2.49. What was the rate of discount on one tube?

32) On a loan from the bank of $5,000, Mr. Johnson was told he would have to pay 5½% interest per year. If he paid the loan off in 6 months, how much interest did he pay?

33) In baking a large wedding cake, the mother of the bride had to increase the amount of flour in the recipe from 2 cups to 15 cups. What percent of increase was this?

34) By buying a complete set of pots and pans, a young woman received a discount of 10% of the selling price. What was the original selling price of the set if she paid $186?

PERCENT INCREASE-DECREASE PROBLEMS

Percent is often used to state increases or decreases. To find a percent of increase or decrease, find the amount of increase or decrease and then determine what percent this is of the original amount.

To solve these problems we may use either of the following formulas.

$$\text{Percent} = \frac{\text{Part}}{\text{Base}} \quad \text{or} \quad \frac{r}{100} = \frac{\text{Part}}{\text{Base}}$$

where the part is defined as the amount of increase or decrease.

Submitting "the amount of increase or decrease" in the formula, it may be rewritten,

$$\text{Percent Increase/Decrease} = \frac{\text{Amount of Increase or Decrease}}{\text{Base}}.$$

Here are some examples which show you how to compute the percent increase or decrease.

EXAMPLE 1

Theodore received a raise from $4 an hour to $4.25 an hour. What was his percent increase in hourly wages?

SOLUTION

The percent of increase is calculated using the formula, $\text{Percent} = \frac{\text{Amount of increase}}{\text{Base}}$. For this problem the amount of increase is $\$4.25 - \$4 = \$.25$. Substituting in the formula, $Percent = \frac{Amount\ of\ increase}{Base} = \frac{.25}{4.00} \times 100 = 6.25\%$.

Do not become confused, the base is not necessarily the largest number, it is the beginning point.

EXAMPLE 2

In an effort to liquidate his merchandise, George reduced the price of his sweaters from $25 to $20. What was the percent reduction in the price of his sweaters?

SOLUTION

Subtract, $\$25 - \$20 = \$5$, to find the amount of decrease. Then, $Percent = \frac{Amount\ of\ decrease}{Base} = \frac{5}{25} \times 100 = 20\%$.

Sometimes, models are helpful in determining the relationship between the part and the base. In using models, we first determine the base, and then sketch a model to show the relative sizes of the part and base. If the base is larger, we use a bigger box to represent it or vice versa. The related percentages should be placed on the sides of the boxes.

EXAMPLE 3

This year's freshman class of 860 is 8% larger than last year's. What was the enrollment last year?

SOLUTION

First determine the relationship of the given number, 860, to the base. Last year's enrollment was the beginning point—the base. As a percentage, this represented 100%. Which is larger, this year's enrollment of 860 or last year's enrollment? The current enrollment is larger, so our model should reflect this.

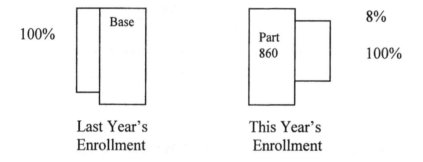

As a percentage, the base is 100% and the part (860) is last year's entire enrollment (100%) plus 8% or 108% of last year's total enrollment.

Now substitute into the formula and solve.

$$\frac{r}{100} = \frac{\text{Part}}{\text{Base}}$$

$$\frac{108}{100} = \frac{860}{\text{Base}}$$

108 (Base) = 100 (860)   Cross multiply
108 (Base) = 86000   Divide both sides by 108

Base = 796 students

166

EXERCISE 4.4 A

Solve the following problems.

1) A service charge of $3.25 is increased to $4.00 per home. Calculate the percent of the increase.

2) Akilah just received a 5% pay increase. Her hourly pay is now $5.65. What was Akilah's hourly wage before the pay increase?

3) The price of milk was increased from $1.80 quart to $1.90 a quart. What was the percent increase in the cost of the milk?

4) Marion decides to build an addition on his house that will increase the floor space in the house by 30%. What is the present floor space if the addition will make a total of 1722.5 square feet?

5) A store advertised, "Save $3 on all shirts. 30% off original price." What was the original price of a shirt and what is the sale price?

6) The population of a town increased 15% in the period 1990 to 1995. In 1990 the population was 26,789. What was the population in 1995?

7) On a diet, a person's weight dropped from 200 pounds to 178 pounds. By what percent did the person's weight decrease?

8) What would be the price of a suit listed at $210 if a discount of 25% is offered?

9) Mary now weighs 130 pounds. If this is an 8% loss in weight, what was her weight before going on a diet?

10) In a local election, 335,000 people cast their votes. If this number of votes represented a gain of 11% over the last election, what was the number of voters in the last election?

11) Shirley's Boutique collects 5% state sales tax on all sales. If total sales including taxes are $25,864.20, find the portion that is tax.

12) Viola's monthly salary of $1200 is 10% more than her monthly salary last year. Find her monthly salary last year.

13) What percent of 5/6 is 3/4?

14) Tony bought a home two years ago for $65,000 and is now selling it for $72,000. What was the percentage gained in the price of the home over this two-year period?

15) A TV was purchased on sale for 30% off the original price. If the sale price was $490, what was the original price?

16) A bookstore keeps 20% of the receipts from the sale of a new book. The remainder goes to the publisher. The authors get 12% of all money received by the publishers. Timmy and Vincent split evenly all authors' money paid. If 800 copies of this book are sold at $24.95 per copy, how much money does each author get?

17) On the average, after one year, a car's value is 75% of its original list price. At the end of the second year, its value is 55% of the original price. If you purchased an automobile for $16,700 two years ago, what is its value now?

18) The U.S. Census Bureau reported the following median prices for homes by region in the United States.

	1980	1992
Northwest	$60,300	$170,000
Midwest	54,400	124,000
South	50,300	118,000
West	54,000	140,900
U.S.	54,750	138,225

a) What was the percent increase in the median prices of homes for each region and the U.S. over the 12 years?

b) In which regions was the increase in the median price of homes lower than that of the U.S. as a whole?

EXERCISE 4.4 B

Solve the following problems.

1) The monthly food budget for Pat increased from $175 per month to $210 per month in one year. What percent increase does this represent?

2) A manager plans to increase its 5000-unit-per-month production schedule by 6.5%.
   a) How many more units will be produced each month?
   b) How many units will be produced each month?

3) Marshall's salary this year is $26,000. This salary will increase by 11% next year.
   a) What is the amount of increase?
   b) What will the salary be next year?

4) Derrick increased his 1250-square-foot home by 20% by adding a family room. How much larger, in square feet, is the home now?

5) A girl's 60-word-per-minute typing speed increased by 20 words per minute. What is the percent increase?

6) A stock that sold for $30 per share increased in price by $1.50 in one day. What percent increase does this represent?

7) A store increased its weekly hours of operation from 60 hours to 70 hours. What percent increase does this represent?

8) A new company increased its number of employees from 500 to 525.
   a) How many employees were added to the company?
   b) What percent increase does this represent?

9) A labor contract called for a 8.5% increase in pay for all employees.
   a) What is the amount of the increase for an employee who makes $224 per week?
   b) What is the weekly wage of this employee after the wage increase?

10) A maker of radios increased its monthly output of 2000 radios by 15%. What is the amount of increase?

11) A student's reading speed increased 70 words per minute from 280 words per minute. What is the percent increase in reading speed?

12) The value of a $2500 investment increased $500. What percent increase does this represent?

13) During a warehouse sale, all merchandise was reduced 15% off the regular price. What is the sale price of a sofa which normally sells for $650?

14) An advertised brand of paint which regularly costs $16 per gallon is on sale for $12 per gallon.
   a) What is the discount?
   b) What is the discount rate?

15) Beef which regularly sells for $1.80 per pound is on sale for 20% off the regular price.
   a) What is the discount?
   b) What is the sale price?

16) An oven which regularly costs $400 is on sale for $100 off the regular price. What is the discount rate?

17) A store is selling its $28 water filter for 15% off the regular price. What is the discount?

18) A store manager offers an automatic coffee maker which regularly sells for $24 at $8 off the regular price. What is the discount rate?

169

CHAPTER 4 REVIEW EXERCISES A

Write the following comparisons as ratios reduced to lowest terms:

1) 20 inches to three feet

2) Earnings of $500 to investments of $5,500

3) 75 students to 80 students

Find the solution set for each of the following proportions:

4) $\dfrac{52}{26} = \dfrac{39}{z}$

5) $\dfrac{7\frac{1}{8}}{1} = \dfrac{x}{8}$

6) $\dfrac{y}{72} = \dfrac{10}{36}$

7) $\dfrac{45}{9} = \dfrac{x}{5}$

Determine whether each of the following proportions is true or false. If a proportion is true, write three other true proportions using the same four numbers:

8) $\dfrac{9}{32} = \dfrac{4\frac{1}{2}}{16}$

9) $\dfrac{4}{5} = \dfrac{80}{100}$

10) $\dfrac{2}{3} = \dfrac{66}{100}$

11) $\dfrac{4}{12} = \dfrac{7}{21}$

12) On a map, $\dfrac{3}{4}$ inch represents 30 miles. What is the actual distance between two cities that are marked 5 inches apart on the map?

13) To buy a house for $26,500, a down payment of $5,300 is needed. What down payment is needed to buy another house for $45,600?

14) If 5 pizzas cost $9.25, what would 3 pizzas cost?

15) If apples sell for 39¢ a pound, what would be the cost of 5 pounds of apples?

16) An automobile was slowing down at the rate of 5 m.p.h. for every 3 seconds. If the automobile was going 65 m.p.h. when it began to slow down, how fast would it be going at the end of 12 seconds?

17) If a machine produces 5,000 hair pins in 2 hours, how many will it produce in 2 8-hour days?

Convert the following rational numbers to percents:

18) $\dfrac{5}{12}$

19) $5\dfrac{3}{4}$

20) $3\dfrac{1}{4}$

21) $\dfrac{6}{5}$

22) $\dfrac{13}{10}$ 23) $\dfrac{6}{10}$ 24) $\dfrac{37}{100}$ 25) $\dfrac{85}{100}$

Convert the following percents to fractions or mixed numbers:

26) $\dfrac{1}{4}\%$ 27) $16\dfrac{2}{3}\%$ 28) 27% 29) 2500%

30) 400% 31) 66% 32) 14%

33) By purchasing 4 tires, a customer is given a discount of 10% on each tire. If he paid $108 for the tires, what was the original price of each tire?

34) What is the percent of mark-up (to the nearest 1%) based on the cost of an automobile which cost $1,500 and sold for $2,000? What is the percent of mark-up based on the selling price?

35) The salesman in Problem 37 earned a commission of $1,524 on the sale of a home. What was the selling price of the home?

36) If a principal of $1000 is invested at a rate of 9% for 30 days, what would be the interest earned?

37) A salesman works on a 6% commission. What would be his commission on the sale of an apartment building for $67,400?

Solve each of the following problems using either the proportion method or the decimal method.

38) 62 is _?_ % of 31.   39) _?_ is $6\dfrac{2}{3}\%$ of 15.   40) 14 is $5\dfrac{1}{2}\%$ of _?_

41) 5 is 10% of _?_   42) _?_ is $\dfrac{1}{4}\%$ of 8.   43) 75 is _?_ % of 300.

44) 7 is _?_ % of 10

45) If 7 miles is equivalent to 11.27 kilometers, then 24 kilometers is equivalent to how many miles?

46) Convert the following numbers to percents.
a) 3.45   b) .056   c) $\dfrac{3}{200}$ d) $\dfrac{4}{5}$ e) $7\dfrac{4}{9}$ f) $\dfrac{9}{25}$

47) Convert to a decimal.
a) $\dfrac{3}{4}\%$ b) 350%   c) 15%

172

48) Convert the following percents to fractions.

   a) 56%   b) $4\frac{1}{2}$%   c) $\frac{1}{2}$%

49) Solve each problem for the unknown.

   a) 12 is what percent of 4?   b) 8 is what percent of 4?

   c) What number is 50% of 39?   d) What is 60% of 300?

   e) What is 125% of 80?   f) What percent of 15 is 12?

   g) What percent of 10 is 40?   h) 15% of what number is 277?

   i) 200% of _____ is 14   j) $\frac{1}{4}$% of __ is 40

   k) $5\frac{3}{8}$% of 760 is _____   l) 72 is 80% of what number?

50) A team won 65% of its games. How many games did it win if the number of games played was 120?

51) If the sales tax is 9%, how much is the sales tax to the nearest cent on a purchase of $40?

52) Isaiah's lunch at a coffee shop cost $3.95. If he tips 15%, how much should the tip be?

53) A color TV with remote control regularly sells for $455. Janet buys one at a 30% discount. What is the sale price of the TV?

CHAPTER 4 REVIEW EXERCISES B

1. Write an equivalent fraction for each of the following.

   a) $\dfrac{3}{9} = \dfrac{12}{?}$ b) $7\dfrac{1}{8} = \dfrac{}{6}$

2) Decide if the following proportions are true or false.

   a) $\dfrac{3}{8} = \dfrac{6}{16}$ b) $\dfrac{\frac{2}{3}}{\frac{4}{5}} = \dfrac{15}{18}$

3) Solve the following proportions.

   a) $\dfrac{3}{8} = \dfrac{x}{24}$ b) $\dfrac{7}{9} = \dfrac{x}{8}$ c) $\dfrac{2}{3} = \dfrac{60}{x}$ d) $\dfrac{4}{x} = \dfrac{44}{55}$

Use proportions to solve the following.

4) If 8 lbs. of fertilizer cover 1600 square feet, how many pounds of fertilizer would you need to cover 200 square feet?

5) Charraine's school orders 2 cartons of chocolate milk for every 7 students. If there are 581 students in the school, how many cartons of chocolate milk should be ordered?

6) The store sells 2 pens for 35 cents. If you buy 6, what will they cost?

7) An item on sale sold for $55 after a 30% reduction. What was the original cost of the item?

8) The price of apples dropped from 70 cents a dozen to 63 cents a dozen. Find the percent decrease in price.

9) After a 6% raise, John's hourly wages increased to $8.64. What was his hourly wage before the raise?

10) A class of 80 is 25% girls. If 10% of the boys and 20% of the girls attended a picnic, what percent of the class attended?

CHAPTER 4 TEST A

1) A bike, regularly priced at $95, is on sale for 30% off. What is the sale price of the bike?
   a) $65   b) $28.50   c) $66.50   d) None

2) If the ratio 6:8 equals the ratio k:24, then $k=$
   a) 2   b) 14   c) 18   d) 22

3) 2/25 expressed as a percent is
   a) .08%   b) 8%   c) 80%   d) 125%   e) 12.5%

4) At a restaurant, your meal cost $13.56. At 15%, how much should you leave for a tip?
   a) 15 cents   b) $2.03   c) $1.15   d) $1.50   e) $1.35

5) Which of the following is not equivalent to 40%?
   a) 40/100   b) 2/5   c) .40   d) 4/5

6) You made purchases amounting to $136 at a store. If the sales tax rate is 6%, what is the total cost of your purchases including taxes?
   a) $8.16   b) $144.16   c) $ 152.32   d) $181.60

7) What percent of 16 is 80?
   a) 20%   b) 5%   c) 500%   d) 12.8%

8) 40% of what quantity is 120?
   a) 48   b) 300   c) 30   d) 4.8

9) After a 5% raise, Tonya's hourly wages increased to $5.67. What was her salary before the raise?
   a) $5.33   b) $5.62   c) $5.40   d) cannot be determined

10) A coat has a price tag on it of $39.95. You can purchase the coat for $32.95 by paying cash. What is the percent of the cash discount?
   a) 3.03%   b) 17.52%   c) 21.2%   d) none

11) Converted to a common fraction, $\frac{3}{8}$% is

   a) $\frac{3}{8}$ b) $\frac{375}{1000}$ c) $\frac{3}{800}$ d) $\frac{15}{40}$

12) Converted to a percent, $4\frac{3}{8}$ equals

   a) 4.375%   b) 437.5%   c) 375%   d) 35/8%

13) 36% of 84 is

a) 30.24 b) 233.33   c) 3024   d) none

14) George is making a 750-mile trip. He drove 430 miles on the first day. What percent of the trip does he have to travel on the second day to complete the trip in two days?
   a) 57.33%   b) 74.4%   c) 42.66%   d) none of these

15) Which of these is not a correct way to find 66% of 30?
   a) $(66 \times 30) \div 100$ b) 66.0×30 c) $\frac{2}{3} \times 30$ d) $.66 \times 30$ e) $\frac{66}{100} \times 30$

16) In the proportion, $\frac{x}{5} = \frac{8}{20}$, $x = ?$
   a) $\frac{8}{5}$ b) 4   c) 2   d) 20

17) Is $\frac{5}{8} = \frac{3}{4}$ a proportion?
   a) yes   b) no   c) cannot be determined

18) On a test of 50 items, 15 were answered incorrectly. What percent of the items were answered correctly?
   a) 30%   b) 70%   c) 15%   d) none of these

19) In a class, $\frac{1}{8}$ of the students were freshmen and $\frac{1}{4}$ of the students were sophomores. What percent of the class were freshmen and sophomores?
   a) 12.5%   b) 25%   c) 37.5%   d) none of these

20) In a referendum, 3,260 people voted. Of this number, $\frac{1}{4}$ of the persons voted against the referendum. What percent voted for the referendum?
   a) 25%   b) 75%   c) $\frac{3}{4}$%   d) $\frac{1}{4}$%

21) If 1 kilogram=2.2 pounds, then 60.7 pounds equals how many kilograms?
   a) 27.59   b) 133.54   c) 2.2   d) 60.7

22) The missing number in the expression $\frac{7}{8} = \frac{?}{24}$ is
   a) .875   b) 3   c) 7   d) 21

23) Written as a percent, .003 equals,
   a) .03%   b) .003%   c) 30%   d) .3%

24) 74 is _____% of 37?

a) 200%   b) 50%   c) 14%   d) 20%

25) 8.67 % in fractional form is

a) $8\dfrac{67}{100}$ b) $\dfrac{867}{1,000}$ c) $\dfrac{867}{10,000}$ d) 867

26) $\dfrac{1}{4}$% of 500 =

a) 250   b) 1.25   c) 125   d) none of these

27) For the purchase of a used car, a 30% down payment was needed. If the purchase price was $6,200, how much down payment was needed?
a) $4,340   b) $1,860   c) $8,060   d) none of these

28) An appliance was marked 20% off the original price. If the sale price was $480, what was the original price?
a) $96   b) $576   c) $600   d) $384

CHAPTER 4 TEST B

Solve the following proportions for x.

1) $6 : x = 12 : 8$

    a) 4   b) 9   c) 16   d) 576   e) none of these

2) $\dfrac{4}{5} = \dfrac{x}{30}$

    a) 20   b) 24   c) $27\dfrac{1}{2}$ d) $\dfrac{2}{3}$ e) none of these

3) Express as a ratio in reduced form: 12 and 16.

    a) $\dfrac{3}{4}$ b) $\dfrac{4}{3}$ c) 192   d) $-1\dfrac{1}{3}$ e) none of these

4) Express as a ratio in reduced form: 46 and 8.

    a) $\dfrac{4}{23}$ b) 368   c) $\dfrac{23}{4}$ d) $\dfrac{21}{2}$ e) none of these

5) If yard fencing costs \$15 per meter installed, how much will it cost to fence a yard with a perimeter of 106 meters?

    a) \$1,090   b) \$1,690   c) \$159   d) \$1,590   e) none of these

6) If 5 men can paint 2 houses in 3 days, how long will it take 2 men to paint the 2 houses, if they all work at the same rate of speed?

    a) 15 days   b) 4 days   c) $3\dfrac{1}{3}$ days   d) $7\dfrac{1}{2}$ days   e) none of these

7) If it takes 4 men 3 days to paint a house, how many houses can 10 men paint in 3 days if they all work at the same rate of speed?

    a) 3 houses   b) $2\dfrac{1}{2}$ houses   c) $1\dfrac{1}{3}$ houses   d) $2\dfrac{1}{3}$ houses   e) none of these

8) If John can drive 300 miles in 10 hours, how many miles can he travel in 12 hours traveling at the same rate of speed?

    a) 360 miles   b) 25 miles   c) 250 miles   d) 380 miles   e) none of these

9) Find the mean proportional to: 2 and 288.

    a) 144   b) 576   c) 25   d) 24   e) none of these

10) What percent of 85 is 120.7?

    a) 0.604%   b) 6.04%   c) 60.4%   d) 14.2%   e) 142%

11) What percent of 64 is 12?

    a) $18\frac{3}{4}\%$ b) 1.875%   c) $187\frac{1}{2}\%$ d) $5\frac{1}{3}\%$ e) $53\frac{1}{3}\%$

12) Find 31.5% of 16.6.
    a) 20   b) 52.29   c) 5.229   d) 0.5229   e) 2

13) Find $7\frac{1}{2}\%$ of 210.

    a) 3   b) 30   c) 1.575   d) 15.75   e) 157.5

14) Find 180% of 60.
    a) 3   b) .108   c) 1.08   d) 10.8   e) 108

15) Find 55% of 11.
    a) 5   b) 605   c) .605   d) 6.05   e) 60.5

16) Find 32% of 125.
    a) 40   b) 4   c) 400   d) 3.9   e) 39

17) Write 44% as a fraction.

    a) $\frac{4}{25}$ b) $\frac{4}{11}$ c) $\frac{11}{25}$ d) $\frac{4}{100}$ e) $\frac{44}{10}$

18) Write 160% as a decimal.
    a) 1.6   b) 16   c) 160   d) .16   e) 1.2

19) Write 35% as a decimal.
    a) .035   b) 350   c) 35   d) 0.35   e) 3.5

20) Write 2.15 as a percent.

    a) $2\frac{3}{20}\%$ b) $2\frac{1}{7}\%$ c) 21.5%   d) 2.15%   e) 215%

21) Write 0.72 as a percent.
    a) 720%   b) 7.2%   c) 72%   d) .72%   e) 36%

22) Write $1\frac{7}{8}$ as a percent.

    a) 186%   b) $186\frac{1}{4}\%$ c) $18\frac{1}{4}\%$ d) $187\frac{1}{2}\%$ e) None

23) Write $\frac{5}{6}$ as a percent.

a) 83%   b) $83\frac{1}{3}$%   c) 84%   d) $8\frac{1}{2}$%   e) 833%

24) If the retail sales tax is 5%, what would be the total cost of a $189.95 camera?
a) $195.45   b) $193.73   c) $191.19   d) $199.44   e) None

25) What is the interest on a personal loan of $650 at $8\frac{1}{2}$% over 27 months?

a) $149   b) $14.92   c) $149.18   d) $12.43   e) $124.32

26) A dress is marked as being "30% off." What is its sale price if the list price was $13.95?
a) $18.60   b) 18.14   c) $4.65   d) $4.19   e) $9.77

27) A TV salesman earns a commission of $12\frac{1}{2}$% of all sales. What will he earn on a $464 sale?
a) $3.71   b) $37.12   c) $58.00   d) $5.80   e) $27

28) After a 20% discount an article cost $14.00. What was the original cost?
a) $17.50   b) $70.00   c) $16.80   d) $42.00   e) $49.00

29) A pound of sugar cost 39¢ before the price was increased 65%. Find the new cost.
a) 26¢   b) 65¢   c) 42¢   d) $1.04   e) None

30) If 160% of a number is 0.12, find the number.
a) 1.33   b) 0.75   c) 0.075   d) 7.5   e) 13.3

31) 23% of what number is 80.5?
a) 3.5   b) 2.8   c) 35   d) 350   e) 28

32) What percent of $\frac{5}{6}$ is $\frac{2}{3}$?

a) 8%   b) 80%   c) $\frac{4}{50}$%   d) 18%   e) 55.5%

33) What percent of 1.32 is 0.858?
a) 15.4%   b) 65%   c) 6.5%   d) 650%   e) 1.54%

34) What percent of 0.38 is 2.16?
a) 450%   b) 45%   c) 4.5%   d) 22%   e) None

CHAPTER 5 FORMULAS, SEQUENCES, AND SERIES

A **formula** is a mathematical statement in which more than one letter is used to express a relationship between quantities. Examples of formulas you have probably used in mathematics courses are $D = RT$, $I = PRT$, and $A = LW$. But the use of formulas is not limited to mathematics only. Formulas are used in business, engineering, the sciences, and medicine to solve problems.

Solving a formula, from an algebraic perspective, is simply evaluating an algebraic expression. In this chapter we will use formulas to solve various application problems.

5.1 SIMPLE AND COMPOUND INTEREST

As consumers, we make purchases ranging from small household appliances to automobiles. For the most part, we do not pay for these purchases in total at the time of acquisition. Instead, we arrange financing over a period of time. The fee we pay for the privilege of borrowing money to acquire these goods and services over time and then repaying it is called interest. One type of interest is simple interest which is found by the following formula.

Simple Interest

Simple-interest Formula
Interest is the charge made for using a certain amount of money (called the principal) for a definite period of time on the basis of a percent (called the rate of interest). The formula for calculating interest is

$$I = P \times R \times T$$

where

I = interest, the amount charged for the use of money
P = principal, the amount loaned or borrowed
R = rate of interest, the percent charged for the use of money on an annual basis
T = time, the time allowed or taken to pay back the principal and the interest, stated in years, months, or days

EXAMPLE 1

Calculate the interest on a loan of $500 at 12% for 1 year and for $\frac{1}{2}$ year.

SOLUTION

For one year:

$I = P \times R \times T$

$I = \$500 \times 12\% \times 1$ or $I = \$500 \times .12 \times 1$

$I = \$500 \times \dfrac{12}{100} \times 1$ $I = \$60 \times 1$

                $I = \$60$

$I = \$5 \times 12 \times 1$

$I = \$60$

For $\frac{1}{2}$ year:

$I = P \times R \times T$

$I = \$500 \times 12\% \times \dfrac{1}{2}$ or $I = \$500 \times .12 \times \dfrac{1}{2}$

$I = \$500 \times \dfrac{12}{100} \times \dfrac{1}{2}$ $I = \$60 \times \dfrac{1}{2}$

                $I = \$30$

$I = \$60 \times \dfrac{1}{2}$

$I = \$30$

EXAMPLE 2

Calculate the amount of interest on a loan of $300 at 12% for 4 months.

SOLUTION

$I = P \times R \times T$

$I = \$300 \times 12\% \times 4$ months   or $I = \$300 \times .12 \times \dfrac{4}{12}$

$I = \$300 \times \dfrac{12}{100} \times \dfrac{4}{12}$ $I = \$300 \times .12 \times \dfrac{1}{3}$

$I = \$3 \times 1 \times 4$

$I = \$12$ $I = \$36 \times \dfrac{1}{3}$

$I = \$12$

EXAMPLE 3

Esther borrowed $400 from the bank. She agrees to pay the bank 12% interest at the end of two years. Find the amount of interest and the total amount due.

SOLUTION

The principal (P) is $400. The rate ($R$) is 12%, converted to a decimal .12, and the time (T) is 2 years. Substitute the values in the formula and simplify.

$I = PRT$
$I = 400 \times .12 \times 2 = \$96$

The amount due = Principal + Interest
= $400 + $96
= $496

EXAMPLE 4

Alex purchases a stereo system for $930. He paid $100 down and finances the balance over 3 years at a rate of 13%. Find the total amount owed and his monthly payments.

SOLUTION

Balance to be Financed = Purchase Price minus Down Payment
= $930 - $100 = $830

Now substitute the following in the formula for interest:

Principal = $830, Rate = 13% or .13, and Time = 3 years.

187

Interest = Principal x Rate x Time
    = $830 x .13 x3 = $323.70

Amount owed = Principal +Interest
    = $830 + $323.70
    = $1,153.70

Monthly payments = Total owed ÷ Number of months
    = $1,153.70 ÷ 36
    = $32.04

EXAMPLE 5

Emanuel bought a compact disc player for $388. He agreed to pay for the player in 9 equal installments at a rate of 9.7%. Find the amount of his monthly payments.

SOLUTION

In the formula, time is expressed in years, so before we can use the formula, 9 months must be converted to years. To change 9 months to years, divide 9 by 12 (9 ÷ 12) to get .75. Substituting,

Interest = Principal × Rate × Time
    = $388 x .097 × .75
    =$28.23

Total owed = Principal +Interest
    = $388 + $28.23
    = $416.23

Monthly payments = Total owed ÷ Number of months
    = $416.23 ÷ 9
    = $46.25

Compound Interest

Financial institutions pay compound interest on savings accounts. **Compound interest** is interest paid on both the principal and the interest already paid. Such interest is usually compounded annually, semi-annually, or quarterly. To see how interest accumulates when it is compounded, let us find the balance in an account with an initial deposit of $3000 at 8% compound annually after three years.

We use the formula for finding simple interest ($I = PRT$) three times.

First Year: $3000
            ×   .08
            ——————————
              $ 240   amount of interest

End of first year total = Original Principal + Interest
$$= \$3000 + \$240 \quad \text{or} \quad \$3240$$

Second Year: $3240
$$\underline{\times \qquad .08}$$
$$\$ \ 259.20 \qquad \text{amount of interest}$$

End of second year total = $3240 + $259.20 = $3499.20

Third Year: $3499.20
$$\underline{\times \qquad .08}$$
$$\$ \ 279.94 \qquad \text{amount of interest}$$

End of third year total = $3499.20 + $279.94 = $3779.14

A simpler method of finding the new principal after compounding is to use the compound interest formula. The derivation of this formula is covered in more advanced mathematics courses. For our purposes in this book, we need to know only how to use the formula.

Compound Interest Formula

$$A = P\left(1 + \frac{r}{t}\right)^{nt} \quad \text{where}$$

P = the initial principal
t = number of compounding periods during the year
r = interest rate
n = number of years

Note: Intermediate calculations should be rounded to four decimal places and the final answer should be rounded to the nearest cents.

EXAMPLE 6

Find the amount on deposit and the interest earned for an initial deposit of $1,250 compounded semi-annually for 6 years at 6%.

SOLUTION

Semi-annually means twice a year, therefore $t = 2$, the number of compounding periods in a year. Substituting, $P = \$1,250$, $t = 2$, $r = 6\%$ or .06, and $n = 6$, we have

$$A = 1250\left(1 + \frac{.06}{2}\right)^{6(2)}$$

The TI-81 automatically performs the order of operation so we simply enter the expression as follows.

$$1250\left(1 + \frac{.06}{2}\right)^{\wedge(6\times2)}$$, then enter, and the display reads \$1,782.20.

For the TI-35, we simplify within parentheses first,

$$.06[\div]2 + 1[=]1.03$$

Then we have $A = 1250(1.03)^{(6\times2)}$, which gives

$$1.03[y^x](6\times2)[=][\times]1250[=]$$

Display reads: \$1,782.20

EXAMPLE 7

Find the amount if \$2,500 is compounded quarterly for five years at 9.6%.

SOLUTION

The following values are substituted in the formula:

Principal $(P) = \$2,500$
$t = 4$, since it is compounded quarterly and there are four quarters in a year.
$r = 9.6\%$ or $.096$
$n = 5$

$$A = P\left(1 + \frac{r}{t}\right)^{nt} = 2,500\left(1 + \frac{.096}{4}\right)^{5(4)}$$

Using the TI-81 we have, $2,500\left(1 + \frac{.096}{4}\right)^{\wedge(5\times4)}$

Display reads \$4,017.35.

EXERCISE 5.1

Find the simple interest owed for the following.

1) $457.80 at 13% for 8 years

2) $5,678 at 6% for 1 year

3) $759 at 7.8% for 1 year

4) $9,188 at 8% for 9 months

5) $2,500 at 13% for 8 months

6) $3,700 at 11.65% for 4 months

7) Michael purchased a stereo system for $856. He agrees to pay for the stereo in 9 equal payments over 9 months. The interest rate charged is 9%. Find his total monthly payments.

8) Josh bought a television set for $499. He paid $50 down and financed the balance at 18% for 2 years. Find the total amount that he must repay and his monthly payments.

9) Jackie owes the Internal Revenue Service $1,530 in underpayment of taxes. If she pays the taxes in 4 months and the IRS charges a penalty of 15% per year, how much will she owe at the end of the four months?

10) If a bank pays interest at the rate of 5% per year payable twice a year, how much interest is earned a year on a principal of $2,980?

11) On January 1,Timmy had $484 in his bank account. The bank pays compound interest on June 30 and December 31 at the rate of 6% per year. If he makes no deposits or withdrawals, how much will he have on deposit at the end of the year?

12) You have $1,000 to invest over a five-year period. Which option would yield the maximum return, investing 7% compounded annually or 6% compounded quarterly?

13) An investment counselor suggests buying a $5,000 bond which pays 8% annual interest, compounded quarterly. Find the amount that would be earned after five years.

14) Determine the interest on a loan of $500 at 9% for 3 years.

15) Solve for interest in the following problems:

Principal	Rate	Time	Interest
a) $ 500	18%	1 year	_____
b) $ 240	9%	$\frac{1}{6}$ years	_____
c) $1500	10%	$\frac{1}{3}$ years	_____

d) $ 50   16% $\frac{1}{2}$ years   _____

e) $ 175   12% $\frac{2}{3}$ years   _____

16) Determine the interest on a loan of $2000 at 8% for 2 years.

17) A store charges 12% on payments for merchandise one-fourth of a year after the bill is due. How much interest does a borrower owe on a bill for $200 at the end of 3 months?

18) Find the principal in the following:

Principal	Rate	Time	Interest
a) _____	10%	90 days	$ 16
b) _____	14%	45 days	$ 14
c) _____	24%	240 days	$ 300
d) _____	6%	270 days	$ 45
e) _____	8%	60 days	$ 4
f) _____	12%	30 days	$ 20
g) _____	18%	180 days	$ 369
h) _____	15%	January 6 – October 13	$ 49
i) _____	17%	March 14 – June 12	$ 102
j) _____	12½%	April 26 – August 4	$ 380

19) Find the time in days, using ordinary interest, in the following. Ordinary interest assumes 360 days in a year.

Principal	Rate	Time	Interest
a) $ 3,000	12%	_____	$ 20.00
b) $ 2,000	14%	_____	$ 35.00
c) $ 6,500	9%	_____	$ 71.50
d) $ 1,750	6%	_____	$ 3.50
e) $ 2,140	12%	_____	$ 32.10
f) $ 9,165	17½%	_____	$ 400.97

Solve: Round to the nearest cent.

20) Jamison borrows $50,000 for 8 months at an annual interest rate of 11.5%. What is the simple interest due on the loan?

21) Karon borrowed $150,000 at a 9.5% annual interest rate for four years. What is the simple interest due on the loan?

22) A credit card company charges a customer 1.6% per month on the customer's unpaid balance. Find the interest owed to the credit card company when the customer's unpaid balance for the month is $1256.

23) A television is purchased and an $800 loan is obtained for two years at a simple annual interest rate of 12.4%.
   a) Find the interest due on the loan.
   b) Find the monthly payment.

$$\left(\text{Monthly payment} = \frac{\text{loan amount} + \text{interest}}{\text{number of months}} \right)$$

24) Patricia purchases a small plane for $57,000 and finances the full amount for five years at a simple annual interest rate of 14%.
   a) Find the interest due on the loan.
   b) Find the monthly payments.

25) In a store, 25 new cash registers were installed for a total cost of $24,000. The entire amount was financed for 2½ years at a simple annual interest rate of 11.2%. Find the monthly payment.

26) To finance the purchase of 15 new cars, a delivery service operator borrows $100,000 for 9 months at an annual interest rate of 12%. What is the simple interest due on the loan?

27) Bennie was offered a $25,000 loan at a 13.5% annual interest rate for four years. Find the simple interest due on the loan.

28) A bank charges its customers an interest rate of 2% per month for transferring money into an account which is overdrawn. Find the interest owed to the bank for one month when $250 was transferred into an overdrawn account.

29) A company purchased a new weaving machine for $225,000 and financed the full amount at 8% simple annual interest for four years.
   a) Find the interest due on the loan.
   b) Find the monthly payment.

$$\left(\text{Monthly payment} = \frac{\text{loan amount} + \text{interest}}{\text{number of months}} \right)$$

Find the balance in each account at the end of the period indicated:

	Principal	Rate	Compounded	Time
30)	$300	4%	Annually	1 year
31)	$864	5%	Semi-annually	2 years
32)	$1,250	6%	Quarterly	4 years
33)	$2,156	7%	Annually	2 years
34)	$5,488	8.0%	June 30 & Dec. 31	1 year

EFFECTIVE INTEREST AND ANNUITIES

With the increase in the assortment of extenders of credit, abusers become common. Hence, in 1969, Congress passed legislation to standardize the calculation of interest. Since 1969, all extenders of credit must reveal to the borrower what the effective interest rate (interest on the unpaid balance) is.

Effective-Interest Formula

$$\text{effective interest rate} = \frac{2 \times \text{number of pay periods in 1 year} \times \text{total interest}}{\text{principal} \times (\text{number of payments} + 1)}$$

Definitions

1) *Number of pay periods in 1 year* does not refer to the length of the loan. It refers to the frequency of pay periods. For example, the number of pay periods for a loan that is to be paid back in five monthly installments is 12, since the payments are monthly and there are 12 months in a year.

2) *Total interest* refers to all charges in excess of the principal. Regardless of the title given to such charges by retailers (service charges, etc.), for calculative purposes, it is interest. To find the amount of interest it is necessary to find the total amount to be paid and subtract the amount of the principal:

   a) amount to be paid = number of payments × amount per payment

   b) principal = amount borrowed

   c) total interest = amount to be paid − principal

Personal Loans

Using the above formulas and information, effective interest can be calculated for personal loans as follows.

EXAMPLE 1

Find the rate of interest on a loan of $375 to be repaid in ten equal monthly installments of $43 each.

SOLUTION

   a) amount to be paid = number of payments × amount per payment

     amount to be paid = 10 × $43 = $430

   b) principal = $375

   c) total interest = amount to be paid − principal

$$I = \$430 - \$375 = \$55$$

$$\text{effective interest rate} = \frac{2 \times \text{number of pay periods in 1 year} \times \text{total interest}}{\text{principal} \times (\text{number of payments} + 1)}$$

$$R = \frac{2 \times 12 \times \$55}{\$375 \times (10 + 1)}$$

$$R = \frac{2 \times 12 \times \overset{\scriptscriptstyle 5}{55}}{\underset{75}{375} \times \underset{1}{11}} \quad \text{or} \quad R = \frac{24 \times 55}{375 \times 11}$$

$$R = \frac{24}{75} \qquad\qquad R = \frac{1320}{4125}$$

$$R = .32 \qquad\qquad R = .32$$

$$R = 32\% \qquad\qquad R = 32\%$$

Annuities

An **annuity** is a sequence of equal payments made at equal time intervals. Unlike compound interest in which the principal is invested all at one time, with an annuity we make equal payments over a given time period. The sum of all payments plus their interest is called the **amount of an annuity** or **future value of an annuity**. To find the amount after making equal payments or deposits in an account over a specified time the following formula is used.

Amount of annuity or future value of annuity formula

$$S = R\left[\frac{(1+i)^n - 1}{i}\right] \text{ where } S = \text{amount of the annuity, } R = \text{periodic payment,}$$

i = rate per period divided by the number of compounding periods per year,
n = number of payments over the period. The payment period and the interest period must be the same.

196

EXAMPLE 3

Vincent deposits $50 per quarter in a savings account that pays 5% interest compounded quarterly. How much will be have in the account after 9 years?

SOLUTION

Since the funds are compounded quarterly, the rate period (i) is 5% or .05 divided by 4. Thus $i = .0125$. The other values are $n = 4(9) = 36$.

Substituting in the formula we have, $S = 50 \left[\dfrac{(1+.0125)^{36} - 1}{.0125} \right]$

Simplifying, $1[+].0125[+][y^x]36[=][-]1[=][\div].0125[\times]50[=]$
Display reads, $2,255.76
Using the TI-81, we can use the same sequence as above, or the following:

$$S = 50 \left(\frac{(1+.0125)^{\wedge 36 - 1}}{.0125} \right) = \$2,255.78.$$

EXERCISE 5.2

1) Calculate the effective interest rates on the following loans (round off to nearest percent):

	Amount of loan	Amount of each payment	Number of payments	Frequency of payments	Total amount to be paid	Total interest	Effective interest rate
a)	$ 375	$ 52.00	8	monthly			
b)	400	32.08	15	monthly			
c)	750	50.00	16	weekly			
d)	460	60.00	9	quarterly			
e)	1,500	50.00	36	monthly			
f)	895	26.00	36	weekly			
g)	9,200	3,000.00	4	semiannually			
h)	12,000	280.00	48	monthly			
i)	698	60.00	12	quarterly			
j)	45	8.00	6	monthly			

2) What is the effective interest rate on a $250 loan to be paid in ten equal monthly installments of $27.50? (Round off to the nearest percent.)

Determine the amount in the account for each of the following given periodic deposits of the amounts indicated. Use the formula for the amount of an annuity.

	Amount of Deposit	Interest Rate	Compounding Period	Time
3)	$60 per month	7%	Monthly	5 years
4)	$200 per quarter	8%	Quarterly	10 years
5)	$125 per month	7.78%	Monthly	6 years
6)	$350 semi-annually	6.5%	Semi-annually	9 years

LOAN PAYMENTS, CREDIT CARDS, AND MORTGAGES

Do you have a credit card? Do you have a loan for your car? Do you have student loans? Do you own a house? Chances are that you owe money for a least one of these purposes. If so, you not only have to pay back the money you borrowed, but you also have to pay interest on the money that you owe.

The two largest single investments that most of us will make in our lives is the purchase of a car or a home. In both instances, we are unlikely to be able to pay for the purchase in total. We therefore must rely on long-term debt and monthly payments to acquire the purchase. Because the interest is calculated on a decreasing balance, the simple interest formula will not work. Most financial institutions have computer programs that will calculate the monthly payments and the total amount of interest paid over the period. The formula below may be used to calculate monthly payments on long term debt.

Loan Payment Formula (Installment Loans)

$$PMT = \frac{P \times \dfrac{APR}{n}}{\left(1 - \left(1 + \dfrac{APR}{n}\right)^{-nY}\right)},$$

where $\begin{cases} P = \text{starting loan principal (amount borrowed)} \\ PMT = \text{regular payment amount} \\ APR = \text{annual percentage rate} \\ n = \text{number of payment periods per year} \\ Y = \text{loan term in years} \end{cases}$

EXAMPLE 1 *(Student Loan)*

Bennie has a student loan of $7500 with an interest rate of $APR = 9\%$ and a loan term of 10 years. What is his monthly payment? How much will he pay over the lifetime of the loan? What is the total interest he will pay on the loan?

SOLUTION

The starting loan principal is $P = \$7500$, the interest rate is $APR = 0.09$, the loan term is $Y = 10$ years, and $n = 12$ for monthly payments. We put these numbers in the loan payment formula.

$$PMT = \frac{P \times \left(\dfrac{APR}{n}\right)}{\left(1 - \left(1 + \dfrac{APR}{n}\right)^{-nY}\right)} = \frac{\$7500 \times \left(\dfrac{0.09}{12}\right)}{\left(1 - \left(1 + \dfrac{0.09}{12}\right)^{-(12 \times 10)}\right)}$$

$$= \frac{\$7500 \times (0.0075)}{\left(1 - (1.0075)^{-120}\right)}$$

$$= \frac{\$56.25}{(1 - 0.407937305)}$$

$$= \$95.01$$

Bennie's monthly payments will be $95.01 on this student loan. Over the 10-year lifetime of the loan he will pay a total of

$$10 \text{ yr} \times 12 \frac{\text{months}}{\text{yr}} \times \frac{\$95.01}{\text{month}} = \$11{,}401.20.$$

Of this amount, $7500 pays off the principal, so the total interest Bennie will pay is

$$\$11{,}401 - \$7500 = \$3901.$$

Credit Cards

Credit card loans differ from other loans in that you are not required to pay off your balance in any set period of time. Instead, you are required to make only a "minimum monthly payment" that depends on your balance and the interest rate. Most credit cards have high interest rates compared to other types of loans, so it is to your advantage to pay off credit card balances as quickly as possible. Once you decide to pay off a balance in a certain period of time, you can use the loan payment formula to calculate the payments you will need to make.

EXAMPLE 2 *(Credit Card Debt)*

Kenneth has a credit card balance of $2300 with an annual interest rate of 21%. He decides he would like to pay off his balance over one year. How much will he need to pay each month, and how much total interest will he pay? Assume that he makes no further purchases with his credit card.

SOLUTION

Kenneth's starting loan principal is $P = \$2300$, the interest rate is $APR = 0.21$, and he will make $n = 12$ payments per year. He plans to pay off his balance in $Y = 1$ year. The loan payment formula gives

$$PMT = \frac{P \times \left(\dfrac{APR}{n}\right)}{\left(1-\left(1+\dfrac{APR}{n}\right)^{-nY}\right)} = \frac{\$2300 \times \left(\dfrac{0.21}{12}\right)}{\left(1-\left(1+\dfrac{0.21}{12}\right)^{-(12\times1)}\right)} = \$214.16.$$

Kenneth will need to make payments of \$214.16 to get his balance paid off in one year—assuming he does not charge anything additional to his credit card!

During the 12 months, he will pay a total of

$$12 \, mo \times \frac{\$214.16}{mo} = \$2569.92.$$

Kenneth loan principal was \$2300, so the remaining \$2569.92 goes to interest.

Mortgages

The interest paid on the purchase of a home is referred to as a **mortgage**. Assuming that you have found your dream home and have negotiated a price, in this section we will explain the steps in computing the monthly cost of the home.

Down Payment

Financial institutions will not usually lend you all of the money needed to purchase a home. Loan amounts typically vary from 80% to 95% of the purchase price, depending upon market conditions, the term of the loan, and the interest rate.

To find the amount of the down payment, multiply the purchase price of the home by the percent that the bank will not lend.

EXAMPLE 3

Find the down payment and loan amount for an 80% loan on a \$62,500 home.

SOLUTION

An 80% loan means that the bank will not pay 20% of the loan.

Down payment = purchase price × percent bank will not lend
= \$62,500 × .20 = \$12,500

Loan amount = purchase price − down payment
= \$62,500 − \$12,500 = \$50,000

EXAMPLE 4

Find the closing costs and the total amount needed to close the loan on the $50,000 loan in Example 3 if the estimated percentage for closing costs is 4%.

SOLUTION

Closing cost = loan amount × estimated percentage closing cost
$$= \$50,000 \times .04 = \$2,000$$

Amount needed to close = down payment + closing costs
$$= \$12,500 + \$2,000 = \$14,500$$

Closing Costs

In addition to the down payment, you must pay closing costs. Closing costs typically include:

- Loan origination fee
- Appraisal
- Cost of credit report
- Mortgage insurance premiums
- Home owner's insurance
- Attorney fees
- Title insurance
- Recording, surveying, and documentary stamps
- Points (prepaid interest compounded at 1% of the loan amount per point)
- Cost of amortization schedule, a table showing the portion of each monthly payment over the loan period that is applied to the principal and the amount applied to the interest.

EXAMPLE 5

Suppose you need a loan of $100,000 to buy your new home. The bank offers a choice of a 30-year loan at an *APR* of 8%, or a 15-year loan at 7.5%. Compare your monthly payments for the two options.

SOLUTION

The starting loan principal is $P = \$100,000$ and $n = 12$. For the 30-year loan, we set $APR = 0.08$ and $Y = 30$. The monthly payments are

$$PMT = \frac{P \times \dfrac{APR}{n}}{\left(1 - \left(1 + \dfrac{APR}{n}\right)^{-nY}\right)} = \frac{\$100,000 \times \left(\dfrac{0.08}{12}\right)}{\left(1 - \left(1 + \dfrac{0.08}{12}\right)^{-(12 \times 30)}\right)} = \$733.76.$$

For the 15-year loan, we set $APR = 0.075$ and $Y = 15$. The monthly payments are

$$PMT = \frac{P \times \dfrac{APR}{n}}{\left(1 - \left(1 + \dfrac{APR}{n}\right)^{-nY}\right)} = \frac{\$100{,}000 \times \left(\dfrac{0.075}{12}\right)}{\left(1 - \left(1 + \dfrac{0.075}{12}\right)^{-(12 \times 15)}\right)} = \$927.01$$

Your payments of \$927.01 on the 15-year loan would be almost \$200 higher than the payments of \$733.76 on the 30-year loan. But let us compare the total amount paid with the two options:

$$30\text{ - year loan} : 30 \text{ yr} \times \frac{12 \text{ mo}}{\text{yr}} \times \frac{\$733.76}{\text{mo}} \approx \$264{,}150$$

$$15\text{ - year loan} : 15 \text{ yr} \times \frac{12 \text{ mo}}{\text{yr}} \times \frac{\$927.01}{\text{mo}} \approx \$166{,}860$$

You would end up paying a total of almost \$100,000 more with the longer-term loan. Your choice is simple, if difficult: the 15-year loan saves you nearly \$100,000 in the long run, but it is a good plan only if you can afford the additional \$200 per month that it will cost you for the next 15 years.

EXAMPLE 6

Find the monthly payments for the \$50,000 loan amount above, if the loan rate is 9.5% for 25 years.

SOLUTION

Identify the value for each variable in the formula and then substitute.

P = amount borrowed = \$50,000
i = interest rate = 9.5% or .095
t = time = 25 years
$-ny = -(12)(25) = -300$

$$\text{Substituting, } PMT = \frac{50{,}000\left(\dfrac{.095}{12}\right)}{\left(1 - \left(1 + \dfrac{.095}{12}\right)^{-300}\right)}$$

Simplifying the formula we get \$436.85.

Escrow

In addition to the monthly payments on the interest and principal, we must pay taxes and insurance on the home. Most banks will require these amounts to be paid into an escrow account.

To find the monthly amount to be paid into the escrow account, sum the yearly costs for homeowner's insurance and taxes and divide by 12. Add this quotient to the monthly payment for principal and interest to get the total monthly payment.

EXAMPLE 7

Find the monthly escrow payment and the total monthly payment on the loan above if annual taxes equal $670 and insurance premiums are $360.

SOLUTION

Total Escrow = Cost of Taxes + Cost of Insurance
$$= \quad \$670 \quad + \quad \$360 \quad = \$1030$$

Monthly Escrow payment = Total escrow $\div 12$
$$= \quad \$1030 \quad \div 12$$

Total Monthly Payments = Principal & Interest Payment + Monthly Escrow
$$= \quad \$436.85 \quad + \quad \$85.83$$

Total Monthly Payments = $522.68

EXERCISE 5.3

1) Calculate the monthly payments on each loan described.
   a) A student loan of $25,000 at a fixed *APR* of 10% for 20 years.
   b) A home mortgage of $150,000 with a fixed *APR* of 9.5% for 30 years.
   c) A home mortgage of $150,000 with a fixed *APR* of 8.75% for 15 years.

2) Loan Payments. Calculate the monthly payments on each loan described.
   a) A student loan of $12,000 at a fixed *APR* of 8% for 15 years.
   b) A home mortgage of $100,000 with a fixed *APR* of 9.5% for 30 years.
   c) A $3000 credit card bill to be paid off in 2 years at an *APR* of 16%.

In Problems 3–6, determine i) your monthly payments; ii) your total payments over the term of the loan; and iii) how much you will pay in interest over the loan term, in dollars.

3) You borrow $5000 over a period of 3 years at an *APR* of 12%.

4) You borrow $10,000 over a period of 5 years at an *APR* of 10%.

5) You borrow $50,000 over a period of 15 years at an *APR* of 8%.

6) You borrow $100,000 over a period of 30 years at an *APR* of 7%.

7) What are the monthly payments required for a 5-year, $13,500 car loan at 9.5%?

8) What are the monthly payments required for a 4-year, $18,200 car loan at 3.9%?

9) The purchase price of a car is $16,000. You make a down payment of 15%. If the percentage rate is 5.6% for 3 years, what are the monthly payments?

10) Priscilla purchases a car for $17,890. She makes a 10% down payment and finances the balance. She is given two options, an 8.9% interest rate for 5 years or 7.6% for 3 years. How much interest does she save by taking the lower interest rate?

Find the (a) amount of down payment, (b) loan amount, (c) closing costs, and (d) monthly payments including escrow (taxes and insurance) for the following.

	Purchase Price	Down Payment	Closing Costs	Interest Rate	No. of Years	Escrow
11)	$63,700	10%	3%	9%	15	none
12)	$55,000	5%	4%	9.25%	30	$800
13)	$47,600	5%	5%	9.5%	25	$2400
14)	$58,650	15%	3.5%	10%	30	$1430

5.4 ARITHMETIC SEQUENCES AND SERIES

A set of numbers arranged in order is called a **sequence**. For example, 2, 4, 6, 8 and 4, 7, 10, 13 are sequences.

Arithmetic Sequences

> When each term in the sequence is obtained from the preceding term by adding the same constant, the sequence is said to be arithmetic.

EXAMPLE 1

Determine which of the following sequences are arithmetic, then find the common difference and list the first ten terms.
- a) 8, 12, 16, 20
- b) 15, 17, 19, 21
- c) 9, 10, 12, 15, 19

SOLUTION

Find the difference between each preceding term and see whether the difference is common. If it is, the sequence is arithmetic.

a) For 8, 12, 16, 20: $12 - 8 = 4, 16 - 12 = 4,$ and $20 - 16 = 4,$
Therefore the common difference is 4.

The first ten terms are 8, 12, 16, 20, 24, 28, 32, 36, 40, 44

b) For 15, 17, 19, 21: $17 - 15 = 2, 19 - 17 = 2,$ and $21 - 19 = 2,$
therefore the common difference is 2.

The first ten terms are 15, 17, 19, 21, 23, 25, 27, 29, 31, 33.

c) For 9, 10, 12, 15, 19: $10 - 9 = 1, 12 - 10 = 2,$
the common difference is not the same, therefore the sequence is not arithmetic.

EXAMPLE 2

List the first six terms of the arithmetic sequence having $a_1 = 15$ and $d = 2$.

SOLUTION

Since, the first term is 15, the second is $15 + 2 = 17$.

Then
$a_3 = 17 + 2 = 19$, $a_4 = 19 + 2 = 21$, $a_5 = 21 + 2 = 23$ and $a_6 = 23 + 2 = 25$.

We can find additional terms of the sequence in similar ways.

$a_4 = 17 + 4 = 21$ or $a_4 = a_1 + 3d = 9 + 12 = 21$

$a_5 = 21 + 4 = 25$ or $a_4 = a_1 + 4d = 9 + 16 = 25$

$a_6 = 25 + 4 = 29$ or $a_6 = a_1 + 5d = 9 + 20 = 29$

EXERCISE 5.4 A

Write the first ten terms for each of the following arithmetic sequences.

1) 8, 12, 16, 20, ...

2) 15, 17, 19, 21, ...

3) 27, 32, 37, 42, ...

4) 52, 59, 66, 73, ...

5) 19, 23, 27, 31, ...

6) 275, 284, 293, 302, ...

7) 80, 73, 66, 59, ...

8) 593, 588, 583, 578, ...

9) 12, $11\frac{1}{3}$, $10\frac{2}{3}$, 10, ...

10) 64, $63\frac{1}{5}$, $62\frac{2}{5}$, $61\frac{3}{5}$, ...

11) $d = 4$, $a_1 = 13$

12) $d = 12$, $a_1 = 2$

13) $d = 3$, $a_1 = 17$

14) $d = 11$, $a_1 = 8$

15) $d = 3$, $a_4 = 16$

16) $d = 8$, $a_5 = 74$

17) $d = 7$, $a_3 = 52$

18) $d = 6$, $a_5 = 43$

Look through the following sequences. Pick out any that are arithmetic. For those that are arithmetic, find the common difference.

19) 11, 15, 19, 23, 27

20) 47, 52, 57, 62, 67

21) 9, 10, 12, 15, 19

22) 17, 20, 24, 29, 36

23) 58, 69, 80, 91, 102, 113

24) 37, 46, 55, 64, 73, 82

25) 2, 4, 8, 16, 32, 64, 128

26) 2, 4, 6, 8, 10, 12, 14, 16

The nth term of an arithmetic sequence is found by using the formula $a_n = a_1 + (n-1)d$, where a_1 is the first term and d is the common difference.

EXAMPLE 1

Find the 8th term for the sequence 13, 17, 21, 25, . . .

SOLUTION

a_1 = first term = 13, d = common difference = 4, and n = the term = 8.

Substituting, $a_n = a_8 = 13 + (8-1)4 = 13 + 7(4) = 13 + 28 = 41$. The 8th term is 41.

EXAMPLE 2

Find the indicated term for each arithmetic sequence.
a) $a_1 = 7$, $d = 3$; find a_{19}.
b) 13, 19, 25, 31, 37, ...; find a_{15}.

SOLUTION

a) Let $n = 19$ in the formula for a_n.

$$a_n = a_1 + (n-1)d$$
$$a_{19} = 7 + (19-1)3$$
$$= 7 + (18)3$$
$$= 7 + 54$$
$$= 61$$

The nineteenth term of the sequence is 61.

b) Here the first term is $a_1 = 13$. To find d, the common difference, subtract any two adjacent terms. For example, if we choose 25 and 31,

$$d = 31 - 25 = 6.$$

Now we can find a_{15}.

$$a_{15} = 13 + (15 - 1)6 = 13 + (14)6 = 13 + 84 = 97.$$

EXERCISE 5.4 B

Find the indicated term for each of the following arithmetic sequences.

1) $a_1 = 15$, $d = 4$; find a_{10}.

2) $a_1 = 42$, $d = 3$; find a_{10}.

3) $a_1 = 25$, $d = 7$; find a_{18}.

4) $a_1 = 6$, $d = 12$; find a_{25}.

5) $a_1 = 156$, $d = 15$; find a_{10}.

6) $a_1 = 209$, $d = 43$; find a_6.

7) $a_1 = 35$, $d = -4$; find a_7.

8) $a_1 = 59$, $d = -5$; find a_8.

9) $a_1 = 12$, $d = 3$; find a_{15}.

10) $a_1 = 17$, $d = 5$; find a_{13}.

Finding the Number of Terms in an Arithmetic Sequence

The formula $a_n = a_1 + (n-1)d$ can also be used to find the number of terms in an arithmetic sequence. For example, let us find the number of terms in the arithmetic sequence

7, 16, 25, 34, 43, 52, ... , 133.

The three dots show that the sequence continues in the same way from 52 to 133, without our having to list each of the terms.

By looking at the sequence, we know that $a_1 = 7$ and $a_n = 133$. We can find the value of d, the common difference, by subtracting any term from the next term. If we subtract 25 from 34, we find
$$d = 34 - 25 = 9.$$

Now substitute the values of a_1, a_n, and d into the formula

$$a_n = a_1 + (n-1)d$$

This gives
$$133 = 7 + (n-1)9.$$

This result is an equation, or statement of equality. The variable in this equation is n, which represents the number of terms of the sequence.

Now we need to solve the equation.
First, add -7 to both sides.
$$133 + (-7) = 7 + (n-1)9 + (-7)$$
$$126 = (n-1)9$$

Now multiply both sides of the equation by $\frac{1}{9}$,

$$\frac{1}{9} \times 126 = \frac{1}{9}(n-1)9$$

$$14 = n - 1$$

Finally, add 1 to both sides.

$$14 + 1 = n - 1 + 1$$
$$15 = n$$

This tells us the arithmetic sequence 7, 16, 25, 34, 43, 52, ... , 133 has 15 terms.

EXAMPLE 3

Find the number of terms in the arithmetic sequence
35, 33, 31, 29, ..., −25.

SOLUTION

Here $a_1 = 35$, $d = -2$, and $a_n = -25$. (How did we find d?)

Substitute into the formula for a_n,

$$a_n = a_1 + (n-1)d$$
$$-25 = 35 + (n-1)(-2)$$
$$-25 + (-35) = 35 + (n-1)(-2) + (-35)$$
$$-60 = (n-1)(-2)$$
$$-60 = -2n + 2$$
$$-62 = -2n$$
$$31 = n$$

EXERCISE 5.4 C

Find the number of terms in the following arithmetic sequences having the given values of a_1, a_n, and d.

1) $a_1 = 2$, $a_n = 102, d = 5$ 2) $a_1 = 4$, $a_n = 43, d = 3$
3) $a_1 = 9$, $a_n = 47, d = 2$ 4) $a_1 = 17$, $a_n = 117, d = 5$

Find the values of a_1, a_n, and d for each of the following sequences. Then substitute these values into the formula

$$a_n = a_1 + (n-1)d$$

to find the value of n.

5) 6, 14, 22, 30, 38, 46, ... , 246   6) 40, 46, 52, 58, 64, ..., 130
7) 5, 8, 11, 14, ... , 62   8) 23, 42, 61, ..., 213

Series

The sum of the terms of a sequence is called a **series**. For example, in the sequence 2, 6, 18, 54 the related series is $2 + 6 + 18 + 54$.

To find the sum of the first n terms of an arithmetic series we use the formula below.

$$S_n = \frac{n}{2}(a_1 + a_n) \text{ when } a_1 \text{ is the first term and } a_n \text{ is the } n^{th} \text{ term.}$$

EXAMPLE 1

Find the sum of the first 30 terms of the arithmetic sequence, 2, 5, 8, 11, ...

SOLUTION

The formula requires us to know the first and last terms, the first term, $a_1 = 2$, and $n = 30$. The 30th term is not given but may be found by using the formula for finding the nth term of a sequence, $a_1 + (n-1)d = 2 + (30-1)3 = 2 + (29)3 = 89$.

EXAMPLE 2

Find the indicated sum for the following arithmetic sequences.
a) $a_1 = 7$, $a_{12} = 95$; find S_{12}.
b) 43, 51, 59, 67, ..., 163.

SOLUTION

a) Use the formula for S_n,

$$S_n = \frac{n}{2}(a_1 + a_n)$$

$$S_{12} = \frac{12}{2}(7 + 95) = \frac{12(102)}{2} = 612$$

The sum of the first 12 terms of the arithmetic sequence having $a_1 = 7$ and $a_{12} = 95$ is 612.

b) For this arithmetic sequence $a_1 = 43$, $d = 8$, and $a_n = 163$. We must use the methods shown at the end of the previous section to find n, the number of terms. We have

$$a_n = a_1 + (n - 1)d$$
$$163 = 43 + (n - 1)8$$
$$120 = (n - 1)8 \qquad \text{Add } -43 \text{ to both sides.)}$$
$$\frac{120}{8} = \frac{(n-1)8}{8} \qquad \text{(Multiply by } \tfrac{1}{8} \text{.)}$$
$$15 = n - 1$$
$$16 = n$$

The sequence has 16 terms and

$$S_{16} = \frac{16}{2}(43 + 163) = \frac{16(206)}{2} = 1648$$

EXERCISE 5.4 D

Find the indicated sum for each of the following arithmetic sequences.

1) $a_1 = 7$, $a_{12} = 106$; find S_{12}. 2) $a_1 = 4$, $a_{17} = 52$; find S_{17}.

3) $a_1 = 29$, $a_{10} = 101$; find S_{10}. 4) $a_1 = 33$, $a_9 = 233$; find S_9.

5) $a_1 = 10$, $d = 6$; find S_{12}. 6) $a_1 = 8$, $d = 5$; find S_{14}.

7) $a_1 = 143$, $d = -2$; find S_7. 8) $a_1 = 96$, $d = -5$; find S_{12}.

9) $a_1 = 154$, $d = 6$; find S_{100}. 10) $a_1 = 375$, $d = 15$; find S_{100}.

For each of the following, make sure that the listed sequence is really an arithmetic sequence, and if it is, use the formula to find the sum of the numbers in the sequence.

11) 3, 6, 9, 12, 15, 18 12) 11, 13, 15, 17, 19

13) 9, 17, 25, 33, 41, 49 14) 14, 28, 32, 36, 40, 44, 48

15) 100, 110, 120, 130, 140, 150   16) 92, 95, 98, 101, 104, 107, 110

Find each of the following sums.

17) $1 + 2 + 3 + 4 + 5 + 6 + + 49 + 50$

18) $1 + 2 + 3 + 4 + 5 + 6 + + 499 + 500$

5.5 GEOMETRIC SEQUENCES AND SERIES

Geometric Sequences

> A **geometric sequence** is a sequence in which each term is obtained from the preceding term by multiplying by the same constant, the **common ratio**.

For example, the sequence 2, 4, 8, 16, 32 and 1, 3, 9, 27, 81 are geometric sequences.

EXAMPLE 1

Determine which of the following sequences are geometric, then find the common ratio and list the first 6 terms.

a) 1, 4, 16

b) $1, \dfrac{1}{2}, \dfrac{1}{6}, \dfrac{1}{8}$

c) $\dfrac{1}{3}, \dfrac{1}{6}, \dfrac{1}{12}$

SOLUTION

The common ratio is found by dividing any term by the preceding term.

a) $4 \div 1 = 4$, therefore, $r = 4$. The sequence is geometric.
    The first 6 terms are 1, 4, 16, 64, 256, and 1024.

b) $\dfrac{1}{2} \div 1 = \dfrac{1}{2}, \dfrac{1}{8} \div \dfrac{1}{2} = \dfrac{1}{4}$. A common ratio does not exist; therefore, it is not a geometric sequence.

c) $\dfrac{1}{6} \div \dfrac{1}{3} = \dfrac{1}{2}, \dfrac{1}{12} \div \dfrac{1}{6} = \dfrac{1}{2}$, therefore $r = \dfrac{1}{2}$. The sequence is geometric.

The first 6 terms are $\dfrac{1}{3}, \dfrac{1}{6}, \dfrac{1}{12}, \dfrac{1}{24}, \dfrac{1}{48}$, and $\dfrac{1}{96}$.

> The nth term of a geometric sequence is found by the formula $a_1 r^{n-1}$, where a_1 is the first term and r is the common ratio.

EXAMPLE 2

What is the 9th term of the sequence 3, 6, 12, 24, 48?

SOLUTION

$n = 9$, $a_1 = 3$, $r = 2$

Substituting, $a_1 r^{n-1} = 3(2)^{9-1} = 3(2)^8 = 768$

EXAMPLE 3

Suppose a geometric sequence has $a_1 = 4$ and $r = -3$. Find each of the following terms of the sequence.

a) a_5 b) a_{10}

SOLUTION

a) Use the formula $a_n = a_1 \cdot r^{n-1}$ where $n = 5$, $a_1 = 4$, and $r = -3$, we get

$a_5 = 4 \cdot (-3)^{5-1} = 4 \cdot 81 = 324$

b) $a_{10} = 4 \cdot (-3)^{10-1} = 4 \cdot (-3)^9 = -78,732$

EXERCISE 5.5 A

Write the first six terms for each of the following geometric sequences.

1) 1, 3, 9, 27   2) 10, 50, 250   3) 8, 24, 72   4) 7, 14, 28

5) 11, 33, 99   6) 100, 200, 400   7) 1024, 256, 64   8) 6, −12, 24

9) −9, 18, −36   10) −20, 10, −5   11) $a_1 = 6$, $r = 3$ 12) $a_1 = 2$, $r = 5$

13) $a_1 = -4$, $r = 2$ 14) $a_1 = -7$, $r = -4$ 15) $a_1 = -11$, $r = -2$ 16) $a_1 = -9$, $r = -3$

Decide whether each of the following number sequences is a geometric sequence. If it is, find the common ratio.

17) 8, 32, 128, 512   18) 12, 24, 48, 96, 192   19) 7, 10, 13, 16, 19, 22

20) 15, 21, 27, 33, 39, 45   21) 100, 50, 25, 12½   22) 88, 44, 22, 11, 5½

Use the formula $a_n = a_1 \cdot r^{n-1}$ to find the indicated term for each of the following geometric sequences.

23) $a_1 = 3, r = -2$; find a_9.   24) $a_1 = -2, r = 3$; find a_8.   25) $a_1 = \dfrac{1}{2}, r = 4$; find a_7.

26) $a_1 = \dfrac{3}{4}, r = -2$; find a_9.   27) $a_1 = -\dfrac{2}{3}, r = -\dfrac{1}{2}$; find a_{10}

Geometric Series

When we add the terms of a geometric sequence we get a **geometric series**. The formula below is used to find the sum of a geometric series.

$$S_n = \frac{a_1\left(1 - r^n\right)}{1 - r} \text{ when } a_1 \text{ is the first term, } r \neq 1 \text{ and } r \text{ is the common ratio.}$$

EXAMPLE 1

Find the sum of the first nine terms of the sequence 3, 6, 12, 24, 48, . . .

SOLUTION

The formula requires us to know the first and last terms, and the common ratio: $a_1 = 3, n = 9$ and $r = 2$.

$$\text{Substituting, } S_n = \frac{a_1\left(1 - r^n\right)}{1 - r} = \frac{3\left(1 - 2^9\right)}{1 - 2} = \frac{3\left(1 - 512\right)}{1 - 2} = \frac{3\left(-511\right)}{-1} = 1533.$$

The formula for finding the sum of a geometric series of numbers can be written as the following,

$$S_n = \frac{a_1\left(r^n - 1\right)}{r - 1}.$$

EXAMPLE 2

Find the sum of the terms of each of the following geometric sequences:

a) 6, 18, 54, 162, 486   b) −3, 15, -75, 375, −1875, 9375

SOLUTION

a) Here $a_1 = 6$ and $r = 3$. There are five terms, so $n = 5$. The formula above can be used to find S_5, the sum of the five terms.

$$S_5 = \frac{6(3^5 - 1)}{3 - 1}$$
$$= \frac{6(243 - 1)}{2}$$
$$= \frac{6(242)}{2}$$
$$= 726$$

b) In this sequence $a_1 = -3$, $r = -5$, and $n = 6$.

$$S_6 = \frac{-3[(-5)^6 - 1]}{-5 - 1}$$
$$= \frac{-3(15625 - 1)}{-6}$$
$$= \frac{-3(15624)}{-6}$$
$$= 7812$$

EXERCISE 5.5 B

Find the sum of the terms, as indicated, for each of the following geometric sequences.

1) $a_1 = 4, r = 2$; find S_6.

2) $a_1 = 3, r = 4$; find S_5.

3) $a_1 = 7, r = 3$; find S_7.

4) $a_1 = 6, r = 2$; find S_6.

5) $a_1 = 4000, r = \frac{1}{2}$; find S_6.

6) $a_1 = 1620, r = \frac{1}{3}$; find S_5.

7) $a_1 = 5, r = -3$; find S_5.

8) $a_1 = 8, r = -2$; find S_7.

9) $a_1 = -3, r = -4$; find S_6.

10) $a_1 = -7, r = -3$; find S_6.

Find the values of a_1, r, and n for each of the following geometric sequences. Then use the formula above to find the sum of the terms.

11) 2, 4, 8, 16, 32

12) 1, 3, 9, 27, 81

13) 7, 14, 28, 56, 112

14) 1, 4, 16, 64, 256

15) 1, 2, 4, 8, 16, 32, 64, 128

16) 10, 100, 1000, 10,000, 100,000

17) 2, 6, 18, 54, 162, 486, 1458

18) 4, 8, 16, 32, 64, 128, 256

CHAPTER REVIEW EXERCISES 5 A

1) Find the simple interest owed on a loan of $2,600 at 9% for 9 months.

2) If $90,350 is deposited at 6% compounded quarterly, what is the balance after 5 years?

3) The purchase price of an item was $2,000. The down payment was $500 and the interest charges were $100. For 24 payments, what would the monthly payments be?

4) On an "average daily balance" of $470, what is the amount of interest if the annual interest rate is 21%?

5) Find the amount of a monthly car payment for the following:
   a) $16,000 at 12% for 4 years
   b) $22,000 at 10% for 5 years

6) Ceiling Street, Inc., purchased $25,000 worth of 3-month, 5 ½ % government securities. How much interest will the company earn?

7) Calculate the interest rate on a loan of $7230 drawing interest of $72.30 for 4 months.

8) Determine the principal of a loan that earns $12 in 30 days at 16%.

9) Determine the net proceeds and the amount of discount on a $1500 note discounted at 18% for 60 days.

10) Determine the effective rate of interest on a $900 note discounted at 12% for 90 days.

11) An investment of $6750 in 12% corporate bonds produced interest of $101.25. For how long were the bonds held?

12) What rate of interest is required to yield $700 if the principal is $16,000 and the time is 72 days?

13) Jose's company bought a motorcycle for $750. He borrowed the money from the Friendly Loan Co., and the loan is to be paid in 22 monthly installments of $53 each. Determine the effective rate of interest Jose will pay.

14) Suzie borrowed $750 from Dave's Pawnshop. She agreed to pay $50 per week for 16 weeks to repay the loan. What is the effective rate of interest Suzie will pay?

15) Determine the amount of time it will take $1500 to earn $45 at a 9% interest rate.

16) Determine the principal on a note that earns $50 in 4 months at 12%.

CHAPTER REVIEW EXERCISES 5 B

Write the first five terms of each of the following sequences.

1) arithmetic, $a_1 = 9, d = 6$ 2) 17, 9, 1   3) arithmetic, $a_2 = 11, d = 8$

4) arithmetic, $a_1 = 42, d = -18$ 5) geometric, $a_1 = 6, r = 2$

6) geometric, $a_3 = 9, r = \dfrac{1}{3}$ 7) $-10, 20, -40$

Find a_{18} for each of the following arithmetic sequences.

8) $a_1 = 9, d = 3$ 9) $a_1 = 8, d = 4$

10) Find a_{40} for the arithmetic sequence 9, 5, 1, -3, ...

Find the number of terms in each of the following arithmetic sequences.

11) $a_1 = 12, d = 3, a_n = 57$ 12) 1, 4, 7, 10, 13, ..., 52

Find the sum of the first ten terms in each of the following arithmetic sequences.

13) $a_1 = 10, d = 8$ 14) $a_1 = 7, a_2 = 19$ 15) $a_2 = 15, d = 5$

16) $a_1 = -9, d = -4$

17) Find the sum $1 + 2 + 3 + 4 + ... + 1000$.

Find the sum of the first five terms in each of the following geometric sequences.

18) $a_1 = 10, r = 8$ 19) $a_1 = 4, r = 3$

CHAPTER 5 TEST

Directions: In working these problems, the formula sheet which follows the chapter test should be used.

Use for questions 1–2.

Mary opened a savings account with an initial deposit of $2,500. She receives 6% compounded annually on her deposit.

1) After five years, how much will she have?
   a) $3345.56   b) $750.65   c) $3250.90   d) $4000.08

2) How much interest will have been earned during the five years?
   a) 750.67   b) $845.56   c) $1500.90   d) $3346

Use for questions 3–4.

John bought an upright frost-free freezer for $1,230. He paid $100 down and financed the balance over a 3-year period at a rate of 8%.

3) What is the total amount owed?
   a) $295.20   b) $1,425.20   c) $1,401.20   d) $271.20

4) How much are the monthly payments?
   a) $467.06   b) $38.92   c) $40   d) $98.40

Use for 5–9.

Tom and Mary Johnson have a gross monthly income of $2,500. They have a car payment of $150 per month and a furniture bill of $54. Neither bill will be paid off within 6 months. They want to purchase a home for $60,000. They have applied for an 80% loan for 30 years at 10%.

5) How much will their down payment be?
   a) $12,000   b) $6,000   c) $4,800   d) $2,500

6) What will the monthly payments be?
   a) $421.23   b) $166.67   c) $526.80   d) $133.33

7) What is the total amount that will be paid in interest and principal?
   a) $103,718.40   b) $151,642.80   c) $216,000   d) cannot be determined

8) If annual taxes are $509 and annual insurance is $287, how much will the monthly payments be?
   a) $66.33   b) $663   c) $487.56   d) $526.80

9) If they financed the house for 20 years versus 30 years, how much would the monthly payments for interest and principal be?

a) $250 b) $282.36 c) $900 d) $463.21

10) If $5,000 is invested in an account at 10% compounded semi-annually, how much would be in the account at the end of 4 years?

a) $10,718 b) $7,387 c) $6092 d) $7,000

11) You purchased an automobile for $15,000 at 10.5% interest rate for 4 years. How much will your monthly payments be to the nearest dollar?

a) $394 b) $444 c) $384 d) $786

12) Deidre can invest $1,000 at 9% compounded monthly or at 9.1% compounded semi-annually. What investment gives the higher yield?

a) 9% compounded monthly b) 9.1% compounded semi-annually
c) The yields are the same

13) The sequence 5, 13, 21, 29 … is

a) arithmetic b) geometric c) neither

14) The 20th term of the sequence 2, 9, 16, 23, … is

a) 220 b) 135 c) cannot be found d) none of these

15) The sum of the first 18 terms of the sequence 2, 4, 8, 16, … is

a) 262,144 b) 180 c) 524,286 d) none of these

16) If $100 per month is invested at 7.78% compounded monthly, how much would be in the account after 6 years to the nearest dollar?

a) $7,200 b) $9,139 c) $7,760 d)none of these

FORMULA SHEET FOR CHAPTER 5

Simple Interest	Interest = Principal × Rate × Time or $I = PRT$ where Principal (P) is the amount borrowed, Rate (R) is the annual interest given in percent and Time (T) is the length of time in years
Compound Interest	$A = P\left(1+\dfrac{r}{t}\right)^{nt}$ where P = the initial principal t = number of compounding periods during the year r = interest rate n = number of years (round calculations to 4 decimal places)
Effective Annual Rate	$EAR = \left(1+\dfrac{r}{k}\right)^{k} - 1$ where k = number of compounding periods per year and r = annual interest rate
Amount of annuity or future value of annuity	$S = R\left[\dfrac{(1+i)^{n}-1}{i}\right]$ where S = amount of the annuity, R = periodic payment i = rate per period divided by the number of compounding periods per year n = number of payments over the period
Monthly payments and mortgages	$PMT = \dfrac{P \times \left(\dfrac{APR}{N}\right)}{\left(1-\left(1+\dfrac{APR}{N}\right)^{-NY}\right)}$
n^{th} term of an arithmetic sequence	$a_n = a_1 + (n-1)d$, where a_1 is the first term and d is the common difference.
n^{th} term of a geometric sequence	$a_1 r^{n-1}$, where a_1 is the first term and r is a common ratio.
Sum of the first n terms of an arithmetic series	$S_n = \dfrac{n}{2}(a_1 + a_n)$ where a_1 is the first term and a_n is the nth term.
Sum of the first n terms of a geometric series	$S_n = \dfrac{a_1(1-r^n)}{1-r}$ where a_1 is the first term, $r \neq 1$ and r is the common ratio.

CHAPTER 6

INTEGERS AND PROPERTIES OF EQUATIONS AND INEQUALITIES

In this chapter we will learn some basic algebraic terminology that we will be using throughout our study of algebra. We will learn how to perform the operations addition, subtraction, multiplication, and division on algebraic expressions.

We shall also study another basic algebraic concept, which will be used throughout the rest of our study in algebra. We shall use some of the reasoning we learned in solving arithmetic problems, and we shall use algebraic expressions to write a mathematical statement which we shall call an algebraic equation.

Algebraic equations are very useful in solving practical problems stated in words.

6.1 USES OF INTEGERS

In everyday life we use integers. We use integers when we keep an hourly temperature variation throughout each day. The thermometer scale has both negative and positive integers on it. Watching a series of plays made by a team in a football game involves signed or directed numbers when we consider the yardage lost and gained during the series of plays. We can easily see the importance that signed numbers play in our everyday lives.

In Chapter 1 we discussed and defined positive integers.

A **positive integer** is any whole number greater than zero.

A positive integer is written or denoted by a plus sign or by writing the integer without a sign preceding it. Examples of positive integers are +3, +20, +90, 8, 9, and 19. Remember that a positive sign is understood if no sign precedes the integer.

A **negative integer** is any whole number less than zero.

> The set of **integers** is composed of positive whole numbers, zero, and the negative whole numbers.

Absolute Value of a Number

Two numbers that are the same distance from zero on the number line but on opposite sides of zero are called **opposites**.

−5 is the opposite of 5
            and
5 is the opposite of −5.

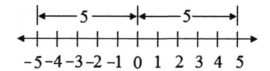

Note that a negative sign can be read as "the opposite of."

−(5) = −5   The opposite of 5 is negative 5.
−(−5) = 5   The opposite of negative 5 is 5.

The **absolute value** of a number, denoted by the symbol | |, is the distance between zero and the number on the number line. Therefore, the absolute value of a number is a positive number or zero.

The distance from 0 to 5 is 5.
Therefore, $|5| = 5$
(the absolute value of −5 is 5).

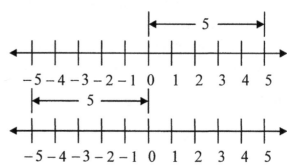

EXAMPLE 1

What is the absolute value of −4?

SOLUTION

On the number line, the distance between zero and −4 is 4 units, therefore, the absolute value of −4 is 4, denoted $|-4| = 4$.

EXAMPLE 2

What is the absolute value of 4?

SOLUTION

On the number line, the distance between zero and 4 is 4 units; therefore, the absolute value of 4 is 4, denoted $|4| = 4$.

EXAMPLE 3

Evaluate $-\left|4\frac{1}{2}\right|$

SOLUTION

$$-\left|4\frac{1}{2}\right| = -4\frac{1}{2}$$

EXERCISE 6.1

Find the value of the following.

1) $-\left|-1\frac{1}{2}\right|$

2) $\left|-2.5\right|$

3) $\left|-2\frac{1}{2}\right|$

4) $\left|-\frac{3}{8}\right|$

5) $-\left|-2\right|$

6) $\left|0\right|$

7) $\left|-1\right|$

8) $\left|7\right|$

9) $\left|-12\right|$

10) $\left|-0.1\right|$

11) $-\left|-6\right|$

12) $\left|\frac{1}{2}\right|$

13) $-\left|-1.7\right|$

14) $\left|3.4\right|$

15) $\left|-6\frac{7}{8}\right|$

16) $\left|-2\right|$

17) $\left|22\right|$

18) $\left|-1\right|$

19) $-\left|4\right|$

20) $\left|-40.6\right|$

21) $-\left|-3\frac{4}{5}\right|$

22) $-\left|-5\right|$

23) $\left|-13.08\right|$

RULES FOR ADDITION OF INTEGERS

> **To add integers with like signs:** combine the absolute value of the numbers and use the common sign.
>
> **To add integers with unlike signs:** get the difference of their absolute values, and use the sign of the integer with the larger absolute value.

The rules for adding integers may be illustrated by adding numbers on the number line. To add numbers on the number line, if the sign of the number is "+" go to the right the number of units given, and if the sign is negative, go to the left the number of units indicated.

EXAMPLE 1

Add 5 + 3

SOLUTION

Suppose that we use the number line to find the sum of (+5) and (+3). Construct a number line and label the points. Beginning at the origin, move 5 units to the right (the positive direction) and place a dot at the end point. From this dot move 3 units to the right and make another dot. Read the value on the number line directly under this dot. From the figure below, the sum is (+8).

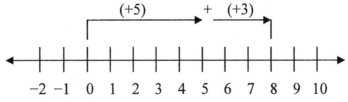

Let us use the number line again to find the sum of (−4) and (−6). Construct a number line and label the points. Starting at the origin, move 4 units to the left (the negative direction) and place a dot. From this point move 6 units to the left and make another dot. Read the value on the number line directly under the dot. From the same figure, note that the sum is (−10).

EXAMPLE 2

Add $+5+(-2)$

SOLUTION

On the number line, go 5 units to the right. From this point, for -2, go two units to the left. We end up at 3. So $+5+(-2)=+3$.

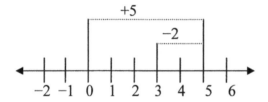

By the addition rule, since the signs are unlike, get the difference of the absolute values of 5 and 2, which is 3, and use the sign of the number with the larger absolute value, in this case 5.

EXAMPLE 3

Add: $-4+(-3)+(-10)+6$

SOLUTION

$-4+(-3)+(-10)+6$
$=(-7)+(-10)+6$
$=(-17)+6$
$=-11$

EXAMPLE 4

Add: $-4\frac{1}{2}+\left(-12\frac{1}{3}\right)$

SOLUTION

$-4\frac{1}{2}+\left(-12\frac{1}{3}\right)$

$=\left(-4\frac{3}{6}\right)+\left(-12\frac{2}{6}\right)$

$=-16\frac{5}{6}$

EXERCISE 6.2

Find the sum of the following.

1) $-\dfrac{3}{8}+\dfrac{3}{4}+\left(-\dfrac{3}{16}\right)$

2) $\dfrac{1}{2}+\left(-\dfrac{3}{8}\right)+\dfrac{5}{12}$

3) $-3\dfrac{1}{3}+\left(-4\dfrac{1}{2}\right)$

4) $2\dfrac{1}{4}+\left(-1\dfrac{1}{2}\right)$

5) $-\dfrac{5}{6}+\left(-\dfrac{7}{9}\right)$

6) $\dfrac{2}{5}+\left(-\dfrac{3}{10}\right)$

7) $-\dfrac{7}{12}+\dfrac{3}{8}$

8) $\dfrac{1}{2}+\left(-\dfrac{1}{3}\right)$

9) $4+56+(-25)$

10) $3+(-7)+4+(-9)$

11) $-57+(-1)+6$

12) $-12+(-8)+9$

13) $-9+(-4)+(-8)$

14) $-21+(-45)+16$

15) $-4+34+(-13)$

16) $16+(-23)+(-4)$

17) $-3+2+(-5)$

18) $3+7+(-8)$

19) $4+8+(-16)$

20) $6+(-12)+3$

6.3 SUBTRACTION OF INTEGERS

> **To subtract two integers**, $a - b$, change the subtraction symbol to addition, add the additive inverse of b to a by changing the sign of the number being subtracted, and apply the rule for addition of integers.

The number line may be used to illustrate the subtraction rule for integers.

EXAMPLE 1

Use the number line to find the difference of $6 - 4$ and compare your answer by applying the subtraction rule for integers.

SOLUTION

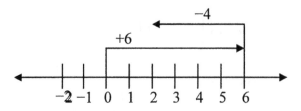

To find the difference of $6 - 4$ on the number line, begin at 0 and move 6 units to the right. From this point, 6, since the number is negative, move 4 units to the left. We end up at 2, showing that $6 - 4 = 2$.

Applying the subtraction rule, we change the subtraction sign to plus, and add the additive inverse of 4. This gives us $6 - 4 = 6 + (-4)$. Now, applying the addition rule for integers, $6 + (-4) = 2$.

EXAMPLE 2

Subtract: $-2 - (-3) - 6 - (-3)$

SOLUTION

$$
\begin{aligned}
-2 - (-3) - 6 - (-3) &= -2 + 3 + (-6) + 3 \\
&= 1 + (-6) + 3 \\
&= -5 + 3 \\
&= -2
\end{aligned}
$$

EXAMPLE 3

Subtract: $-3.9 - 16.78 - (-3.2)$

SOLUTION

$$-3.9 - 16.78 - (-3.2) = -3.9 + (-16.78) + 3.2$$
$$= -20.68 + 3.2$$
$$= -17.48$$

EXAMPLE 4

Subtract: $12\frac{1}{2} - \left(-4\frac{3}{8}\right)$

SOLUTION

$$12\frac{1}{2} - \left(-4\frac{3}{8}\right) = 12\frac{1}{2} + 4\frac{3}{8}$$
$$= 12\frac{4}{8} + 4\frac{3}{8}$$
$$= 16\frac{7}{8}$$

EXERCISE 6.3

Solve the following.

1) $6-(-1)-(-2)$

2) $-9-(-8)-(-5)$

3) $12-(-7)-(-3)$

4) $1-(-3)-(-5)$

5) $-7-(-3)-(-7)$

6) $-6-(-2)+(-9)$

7) $6-(-3)-7$

8) $4-(-12)-3$

9) $-8-7-9$

10) $4\dfrac{1}{2}-2\dfrac{1}{4}-7\dfrac{5}{8}$

11) $2\dfrac{1}{3}-\left(-1\dfrac{1}{2}\right)-\dfrac{3}{8}$

12) $-\dfrac{5}{8}-\left(-\dfrac{2}{3}\right)-\left(-\dfrac{5}{6}\right)$

13) $-2\dfrac{1}{2}-\left(-3\dfrac{1}{4}\right)$

14) $-2\dfrac{11}{15}-\left(-3\dfrac{7}{20}\right)$

15) $\dfrac{3}{4}-\dfrac{3}{7}$

16) $-\dfrac{3}{4}-\left(-\dfrac{2}{3}\right)$

17) $-\dfrac{2}{3}-\dfrac{5}{8}$

18) $\dfrac{2}{5}-\dfrac{14}{15}$

19) $\dfrac{5}{12}-\dfrac{11}{15}$

20) $-\dfrac{7}{12}-\left(-\dfrac{7}{8}\right)$

21) $-\dfrac{3}{4}-\dfrac{5}{8}$

22) $-7.82-1.65$

23) $-6.05-(-3.09)$

24) $4.2-(-3.4)-2.9$

25) $75.9-93.87$

26) $-16.92-(-19.58)$

27) $16.23-9.45$

6.4 MULTIPLICATION AND DIVISION OF INTEGERS

> The product of two negative or two positive integers is positive.
>
> The product of a negative integer and a positive integer is negative.

EXAMPLE 1

Complete the sequence for the following product of numbers.

$$(+4)\ (+4)\ =\ +16$$
$$(+4)\ (+3)\ =\ +12$$
$$(+4)\ (+2)\ =\ +\ 8$$
$$(+4)\ (+1)\ =\ +\ 4$$
$$(+4)\ (\ 0)\ =\ \ \ \ 0$$
$$(+4)\ (-1)\ =\ \ \ \ ?$$
$$(+4)\ (-2)\ =\ \ \ \ ?$$
$$(+4)\ (-3)\ =\ \ \ \ ?$$

SOLUTION

Note that the integers in column 2 are decreasing by 1 and the integers in column 3 are decreasing by 4. To continue the pattern, $(+4)\ (-1)$ must equal -4; $(+4)\ (-2) = -8$; and $(+4)\ (-3) = -12$; thus showing that the product of a negative and a positive integer is negative.

EXAMPLE 2

Complete the sequence for the following product of numbers.

$$(-3)\ (+5)\ =\ -15$$
$$(-3)\ (+4)\ =\ -12$$
$$(-3)\ (+3)\ =\ -\ 9$$
$$(-3)\ (+2)\ =\ -\ 6$$
$$(-3)\ (+1)\ =\ -\ 3$$
$$(-3)\ (\ 0)\ =\ \ \ \ 0$$
$$(-3)\ (-1)\ =\ \ \ \ ?$$
$$(-3)\ (-2)\ =\ \ \ \ ?$$
$$(-3)\ (-3)\ =\ \ \ \ ?$$

SOLUTION

Note that the integers in column 2 are decreasing by 1 unit while the integers in column 3 are increasing by 3 units. To continue the pattern, $(-3)(-1)$ must equal 3; $(-3)(-2) = 6$; and $(-3)(-3) = 9$; showing that the product of two integers with like signs is positive.

Since division and multiplication are inverse operations, the same rules of sign for multiplication hold true for division.

For every division problem there is a related multiplication problem. Therefore, we may think of division as being the inverse operation of multiplication. We can indicate division of two integers in two ways, as

$$5\overline{)-25} \qquad \text{or} \qquad -20 \div 4$$

Note that we cannot divide by zero. Division by zero is undefined. When zero is divided by any other integer other than itself, the result will always be zero. Any number, other than zero, divided by itself has a quotient equal to one.

1) $\quad 4 \div 0 = \text{undefined}$

2) $\quad 0 \div 4 = 0$

Since we may think of division as the inverse operation of multiplication, we may now state a general rule for division of signed numbers.

> The quotient of integers with like signs is positive.

EXAMPLE 3

Find the quotient of $-\dfrac{15}{3}$.

SOLUTION

The quotient is -5. The signs are unlike; therefore, the answer is -5.

EXAMPLE 4

Find the quotient of $\dfrac{(-18)}{(-9)}$.

SOLUTION

The quotient of 18 and 9 is 2. Since the signs are alike, the answer is $+2$.

244

EXERCISE 6.4

Perform the indicated operations.

1) $4\frac{1}{2} \times \left(-1\frac{1}{2}\right)$

2) $-1\frac{3}{5} \times \left(-2\frac{7}{9}\right)$

3) $-2\frac{1}{2} \times \frac{1}{3}$

4) $\frac{4}{15} \times \left(-\frac{2}{22}\right)$

5) $-\frac{3}{5} \times \left(-\frac{5}{20}\right)$

6) $-\frac{1}{3} \times \frac{5}{8}$

7) -14.3×7.9

8) 7.8×9.6

9) -1.6×4.9

10) $-8.9 \times (-3.5)$

11) $-6.7 \times (-4.2)$

12) 4.2×0.9

13) $4 \times (-7) \times 6$

14) $18 \times 0 \times (-14)$

15) $0 \times 6 \times (-2)$

16) $-8 \times (-6) \times (-8)$

17) $-6 \times (-7) \times (-3)$

18) $8 \times (-3) \times (-7)$

19) $-2 \times (-3) \times 6$

20) $-2 \times 7 \times 4$

21) $3 \times 6 \times (-4)$

Divide. Round to the nearest hundredth.

22) $-1.42 \div 6$

23) $0 \div (-8)$

24) $1.8 \div 4$

25) $-882 \div (-65)$

26) $-621 \div (-14)$

27) $814 \div 34$

28) $-275 \div 21$

29) $184 \div (-16)$

30) $-162 \div 9$

31) $105 \div (-4)$

32) $95 \div 9$

33) $-82 \div 0$

34) $16 \div (-9)$

35) $-12 \div 5$

36) $-49 \div (-7)$

37) $-64 \div (-8)$

38) $-24 \div 8$

39) $18 \div (-2)$

6.5 USES OF BRACKETS, PARENTHESES, AND ORDER OF OPERATIONS

In mathematics, brackets and parentheses are often referred to as **grouping symbols**. Both of these mathematical symbols are used in the same manner, either to collect one or more numbers together to be considered as one quantity or to indicate multiplication of quantities.

In the examples below, the parentheses are used to indicate multiplication and the brackets are used to indicate that the product is one quantity.

$$2(9)$$
$$4(-7)$$
$$4 + [(3)(-2)]$$

Note that, if there is no plus or minus sign between the two sets of parentheses, the operation of multiplication should be understood. When both brackets and parentheses are used to group one quantity, simplify the innermost grouping symbols first.

EXAMPLE 1

$-3[(3)(4)]$ Perform the innermost operation.

SOLUTION

Step 1. $-3[(3)(4)] = -3[12]$ Do multiplication and drop the parentheses.

Step 2. $= -36$ The product of (-3) and (12) is (-36).

EXAMPLE 2

$3[(3) + (8-9)]$ Clear the parentheses.

SOLUTION

Step 1. $3[(3) + (8-9)]$
         $= 3[(3) + (8 + (-9))]$ The additive inverse of 9 is (-9).

Step 2. $3[(3) + (-1)]$ Add 8 and -9. The result is (-1).

Step 3. $= 3[2]$ Add (3) and (-1) and we have cleared the parentheses.

Step 4. $= 6$ Multiply.

EXAMPLE 3

$$\left[7-(4)+(9+3)\right]$$

SOLUTION

Step 1. $\quad 7-(4)+(9+3)$ $\qquad$ The sum of 9 and 3 is 12.

$\qquad\quad = 7-(4)+(12)$

Step 2. $\quad 7+(-4)+(12)$ $\qquad$ The additive inverse of 4 is (-4).

Step 3. $\quad = 15$ $\qquad$ Add the signed numbers.

Order of Operations

Many times we are faced with solving problems involving three or four of the basic operations: addition, subtraction, multiplication, and division. The method used to solve these problems may be somewhat confusing to us when only arithmetic numbers are involved. The rules listed below will help us in performing calculations with these types of problems.

---

1) Do all operations inside parentheses.
2) Simplify any expression containing exponents.
3) Multiply or divide as they occur starting from left to right.
4) Add or subtract as they occur from left to right.

EXAMPLE 1

Simplify: $\quad 5-8\div 4+3\times 5$

SOLUTION

Step 1. $\quad = 5-2+15$ $\qquad$ Multiply and divide as they occur from left to right.

Step 2. $\quad = 18$ $\qquad$ Add and subtract as they occur from left to right.

EXAMPLE 2

Simplify: $\quad (-3)^2 - 2\times(8-3)+(-4)$

SOLUTION

Step 1. $(-3)^2 - 2 \times (8-3) + (-4)$ Perform operations inside parentheses.

Step 2. $(-3)^2 - 2 \times 5 + (-4)$ Simplify expressions with exponents.

Step 3. $9 - 2 \times 5 + (-4)$ Multiply and divide as they occur from left to right.

Step 4. $9 - 10 + (-4)$ Add and subtract as they occur from left to right.

$$9 + (-10) + (-4)$$
$$-1 + (-4)$$
$$-5$$

EXAMPLE 3

Simplify: $5 \times 4 - 12 \div 6 + 5 - 3$

SOLUTION

Step 1. $= 20 - 2 + 5 - 3$ Multiply and divide as they occur from left to right.

Step 2. $= 20$ Add and subtract as they occur from left to right.

EXAMPLE 4

Simplify: $(-3)^2 \times (5-7)^2 - (-9) \div 3$

SOLUTION

Step 1. $(-3)^2 \times (-2)^2 - (-9) \div 3$ Perform operations inside parentheses.

Step 2. $9 \times 4 - (-9) \div 3$ Simplify expressions with exponents.

Step 3. $36 - (-9) \div (3)$ Multiply.

Step 4. $36 - (-3)$ Divide.

Step 5. $36 + 3 = 39$ Add and subtract as they occur from left to right.

EXERCISE 6.5

Perform the indicated operations.

1) $-2 \times 4^2 - 3 \times (2-8) - 3$

2) $(-3)^2 \times (5-7)^2 - (-9) \div 3$

3) $16 - 4 \times 8 + 4^2 - (-18) \div (-9)$

4) $3^2 \times (4-7) \div 9 + 6 - 3 - 4 \times 2$

5) $-27 - (-3) - 2 - 7 + 6 \times 3$

6) $10 \times 9 - (8+7) \div 5 + 6 - 7 + 8$

7) $-3 \times (-2)^2 \times 4 \div 8 - (-12)$

8) $-12 \times (6-8) + 1^2 \times 3^2 \times 2 - 6 \times 2$

9) $3 \times 4^2 - 16 - 4 + 3 - (1-2)^2$

10) $-4 \times 3 \times (-2) + 12 \times (3-4) + (-12)$

11) $3 \times 2^2 + 5 \times (3+2) - 17$

12) $7 \times 6 - 5 \times 6 + 3 \times 2 - 2 + 1$

13) $4 \times 2 \times (3-6)$

14) $(-2)^2 - (-3)^2 + 1$

15) $4^2 - 3^2 - 4$

16) $6 - 2 \times (1-5)$

17) $3 \times (8-5) + 4$

18) $2^2 - (-3)^2 + 2$

19) $4 \times (2-7) \div 5$

20) $3 \times (6-2) \div 6$

21) $9 \div 3 - (-3)^2$

22) $3^2 - 4 \times 2$

23) $-3 + (-6) - 1$

24) $4 - (-2)^2 + (-3)$

25) $6 - 2 \times (1-3)$

26) $4 \times (2-4) - 4$

27) $(-3) - (-2)^2 - 5$

6.6 VARIABLES AND ALGEBRAIC EXPRESSIONS

It is important that we understand exactly what is meant by the words *terms* and *coefficients* in algebraic expressions.

A **term** is a single number, or product of a number and one or more variables, raised to powers. The following expressions are terms:

$-2xy$, $9x^2y^4$, $8abc$, $-2x^2$.

The word **coefficient** refers to a numerical coefficient whenever it is expressed in this chapter. By *numerical coefficient* we mean the arithmetic factor. If a literal number is shown without the numerical coefficient, its coefficient is understood to be one. In the product $9x^3y^4$, the number "9" is the numerical coefficient of x^3y^4.

Like terms are terms that are exactly alike in their letter parts. The following are like terms:

$4x^2y^2$, $-7x^2y^2$, $-5x^2y^2$, $-15x^2y^2$, $3x^2y^2$.

Unlike terms are terms that are not exactly alike in their letter parts. The following are unlike terms:

$4x^2y^2$, $8xy$, $-8x^3y^2$, $4xy^2$, $3x$, $-6y^2$.

A **variable** is a symbol used to represent an unknown. Typically, variables are represented by lower case alphabets such as: *a, b, c,* or *x*.

A **monomial** is an algebraic expression consisting of only one term. A **polynomial** is an algebraic expression consisting of more than one term. $4x^3yz$, xy, $5x$, $5x^2yz$ are monomials. The following expressions are polynomials:

$3x^2 + 5x^2y$, $2xy + 7x - 8x^2y$, $4x^3 + 4x^2 + 7x + 9$

A polynomial with two terms is called a **binomial**. A polynomial with three terms is called a **trinomial**. The expression $x^2 + 2xy$ is a binomial. The expression $3x^3 + 5y + 9x^2$ is a trinomial.

An **exponent** is a mathematical notation. Its meaning must be thoroughly understood by a student of algebra. In 7^2, 2 here indicates that two sevens are to be multiplied. In this case 2 is called an exponent. The number 7 on which the exponent is placed is called the **base**.

The **degree** of a term with one variable is the exponent on the variable. For example, $5x^3$ has degree 3, while $8x^{12}$ has degree 12. The term $5x$ has degree 1, and -5 has degree 0 (since -5 is $-5x^0$). The degree of a polynomial in one variable is the highest exponent found in any nonzero term of the polynomial. For example, $5x^6 + 7x^4 + 8$ is of degree 6, the polynomial $7x + 8$ is of degree 1 and 3 (or $3x^0$) is of degree 0.

To combine similar terms we add or subtract their numerical coefficients according to the rules for addition or subtraction of integers and keep the common variable name.

EXAMPLE 1

Add $6xy$ and $2xy$.

SOLUTION

Since the terms are alike, we can combine the numerical coefficients and keep the common variable name.

$6xy + 2xy = 8xy$

EXAMPLE 2

Add $-6a$ and $2a$.

SOLUTION

The terms are similar, so we can combine the numerical coefficients. Thus, $-6a + 2a = -4a$, according to the rule for addition of integers.

Algebraic Expressions

When variables are combined with numbers by any of the fundamental operations of addition, subtraction, multiplication, or division, we have an **algebraic expression**. Examples of algebraic expressions are $x + 4$, $3x + 6$, and $5x + 9$. Recall that a variable is an unknown; therefore, if we knew what value to assign to the variable, we could find a value for the algebraic expression.

254

EXAMPLE 3

Simplify $6xy - 8x + 5x - 9xy$.

SOLUTION

$$6xy - 8x + 5x - 9xy$$
$$= 6xy + (-8)x + 5x + (-9)xy$$
$$= 6xy + (-9)xy + (-8)x + 5x$$
$$= -3xy + (-3)x$$
$$= -3xy - 3x$$

EXAMPLE 4

Simplify $-4z^2 + 8 + 5z^2 - 3$.

SOLUTION

$$-4z^2 + 8 + 5z^2 - 3$$
$$= -4z^2 + 8 + 5z^2 + (-3)$$
$$= -4z^2 + 5z^2 + 8 + (-3)$$
$$= z^2 + 5$$

When an expression involves parentheses, the distributive property is used.

EXAMPLE 5

Combine like terms in the following expressions

a) $14y + 2(6 + 3y)$ 　　　　b) $9k - 6 - 3(2 - 5k)$ 　　　　c) $-(2 - r) + 10r$

SOLUTION

a) $14y + 2(6 + 3y)$ 　$= 14y + 2(6) + 2(3y)$ 　　　Distributive property

　　　　　　　　　$= 14y + 12 + 6y$ 　　　　Multiply

　　　　　　　　　$= 20y + 12$ 　　　　Combine like terms

b) $9k - 6 - 3(2 - 5k)$ 　$= 9k - 6 - 3(2) - 3(-5k)$ 　　　Distributive property

　　　　　　　　　$= 9k - 6 - 6 + 15k$ 　　　　Multiply

　　　　　　　　　$= 24k - 12$ 　　　　Combine like terms

255

c) $-(2-r)+10r$ $= -1(2-r)+10r$ $-(2-r)=-1(2-r)$

$= -1(2)-1(-r)+10r$ Distributive property

$= -2+r+10r$ Multiply

$= -2+11r$ Combine like terms

Evaluating Algebraic Expressions

To evaluate (compute a value) an algebraic expression, assign a given value to each variable, substitute the value into the algebraic expression for each corresponding variable, and then perform the indicated operation.

EXAMPLE 6

Evaluate $3x + 4y$ when $x = -2$ and $y = 3$.

SOLUTION

$3x + 4y$
$3(-2) + 4(3)$
$-6 + 12$
$-6 + 12$
6

EXAMPLE 7

Evaluate $-\dfrac{1}{2}y^2 - \dfrac{3}{4}z$ when $y = 2$ and $z = -4$.

SOLUTION

$-\dfrac{1}{2}y^2 - \dfrac{3}{4}z$

$= -\dfrac{1}{2}(2)^2 - \dfrac{3}{4}(-4)$

$= -\dfrac{1}{2} \cdot 4 - \dfrac{3}{4}(-4)$

$= -2 - (-3)$

$= -2 + (3)$

$= 1$

EXAMPLE 8

Evaluate $-x^2 - 6 \div y$ when $x = -3$ and $y = 2$.

SOLUTION

$$-x^2 - 6 \div y$$
$$= -(-3)^2 - 6 \div 2$$
$$= -9 - 6 \div 2$$
$$= -9 - 3$$
$$= -9 + (-3)$$
$$= -12$$

EXERCISE 6.6

Give the numerical coefficient of the following terms.

1) y^2 2) x^4 3) $-r$ 4) $-z$

5) $35a^4b^2$ 6) $13m^5n^4$ 7) -9 8) 21

Write like or unlike for the following groups of terms.

9) $2, 5, -2$ 10) $-11x, 5x, 7x$ 11) $25y, -14y, 8y$ 12) $7z^3, 7z^2$

Simplify the following expressions.

13) $2p^2 + 3p^2 - 8p^3 - 6p^3$ 14) $5y^3 + 6y^3 - 3y^2 - 4y^2$

15) $6y^2 + 11y^2 - 8y^2$ 16) $-9m^3 + 3m^3 - 7m^3$

17) $1 + 7x + 11x - 1 + 5x$ 18) $-r + 2 - 5r + 3 + 4r$

19) $-10 + x + 4x - 7 - 4x$ 20) $-p + 10p - 3p - 4 - 5p$

21) $16 + 5m - 4m - 2 + 2m$ 22) $6 - 3z - 2z - 5 + z - 3z$

23) $-2x + 3 + 4x - 17 + 20$ 24) $r - 6 - 12r - 4 + 6r$

25) $-5y + 3 - 1 + 5 + y - 7$ 26) $2k - 7 - 5k + 7k - 3 - k$

27) $2k + 9 + 5k + 6$ 28) $2 + 17z + 1 + 2z$

29) $12b + b$ 30) $30x + x$

31) $-4a - 2a$ 32) $-3z - 9z$

33) $9y + 8y$ 34) $15m + 12m$

35) $5x + 2(x + 7)$ 36) $6z - 3(z + 4)$

37) $7w - 2(w - 3)$ 38) $9x - 4(x - 6)$

39) $-2y + 3(y - 2)$ 40) $5m + 3(m + 4) - 6$

41) $8z - 2(z - 3) + 8$ 42) $9y - 3(y - 4) + 8$

43) $3x + 2(x + 2) + 5x$ 44) $7x + 4(x + 1) + 3x$

259

45) $-3y + 2(y - 4) - y$

46) $z - 2(1 - z) - 2z$

47) $3(y - 2) - 2(y - 6)$

48) $7(x + 2) + 3(x - 4)$

Evaluate the variable expression when $a = -3$, $b = 6$, and $c = -2$.

49) $-\frac{2}{3}b - \left(\frac{1}{2}c + a\right)$

50) $\frac{1}{6}b + \frac{1}{3}(c + a)$

51) $\frac{1}{2}c + \left(\frac{1}{3}b - a\right)$

52) $\frac{1}{3}a + \left(\frac{1}{2}b - \frac{2}{3}\right)$

53) $\frac{2}{3}b + \left(\frac{1}{2}c - a\right)$

54) $2b - \left(3c + a^2\right)$

55) $b^2 + c^2 - bc$

56) $a^2 + b^2 - ab$

57) $ac - bc - cd$

58) $ac + bc + ab$

59) $a^2 - b^2 - c^2$

60) $a^2 + b^2 + c^2$

61) $a^2 - (b \div a)$

62) $c^2 - (b \div c)$

63) $3c^2 \div (ab)$

64) $b^2 \div (ac)$

65) $b^2 - a^2$

66) $b^2 - c^2$

67) $4ac \div (b \div a)$

68) $2ac - (b \div a)$

69) $c - (b \div c)$

6.7 SOLVING LINEAR EQUATIONS AND RATIONAL EQUATIONS

The equation is perhaps the most important idea in all of mathematics. Many relationships are concisely described by equations. Different types of equations enable us to solve many applied problems.

LINEAR EQUATIONS

An **equation** is a statement that two expressions are equal. Examples of equations are $x + 2y = 9$, $11y = 5x + 5$, $x^2 - 2x - 1 = 0$.
To solve an equation means to find all numbers that make the equation a true statement. Such numbers are called the **solutions** of the equation.

Equations with the same solution set are called **equivalent equations**. For example, $x + 1 = 5$ and $6x + 3 = 27$ are equivalent equations because they have the same solution set, 4. One way to solve an equation is to rewrite it as a series of simpler equations. We use the Properties of Equality to obtain a series of simpler equivalent equations.

PROPERTIES OF EQUALITY

Equivalent equations may be obtained by

1)   Adding the same numbers to both sides of an equation.
   If $a = b$, then $a + c = c + b$.   Example: If $3 = 3$, then $3 + 4 = 3 + 4$.

2)   Subtracting the same number from both sides of an equation.
   If $a = b$, then $a - c = b - c$.   Example: If $6 = 6$, then $6 - 3 = 6 - 3$.

3)   Multiplying both sides of an equation by the same nonzero number.
   If $a = b$, then $ac = bc$.   Example: If $6 = 6$, then $2(6) = 2(6)$.

4)   Dividing both sides of an equation by the same nonzero number.
   If $a = b$, then $\dfrac{a}{c} = \dfrac{b}{c}$.   Example: If $8 = 8$, then $\dfrac{8}{4} = \dfrac{8}{4}$.

We use the Properties of Equality to rewrite the equation in simplest form by isolating the constant terms in one side of the equation and the variable in the other. Ultimately, we reduce the equation to the form,

variable = constant

EXAMPLE 1

Solve: $3y - 7 = -5$

SOLUTION

$3y - 7 = -5$

$3y - 7 + 7 = -5 + 7$ Isolate y by adding 7 to both sides of the equation.

$3y = 2$ Divide by 3 to make 1 the coefficient of y.

$\dfrac{3y}{3} = \dfrac{2}{3}$

$y = \dfrac{2}{3}$

Check: $3\left(\dfrac{2}{3}\right) - 7 = -5$

$ 2 - 7 = -5$

$ -5 = -5$

EXAMPLE 2

Solve: $2x + 3 = 5x - 9$

SOLUTION

$2x + 3 = 5x - 9$

$2x - 5x + 3 = 5x - 5x - 9$ Subtract $5x$ from each side of the equation.

$-3x + 3 = -9$ Simplify.

$-3x + 3 - 3 = -9 - 3$ Subtract 3 from each side of the equation.

$-3x = -12$ Simplify.

$\dfrac{-3x}{-3} = \dfrac{-12}{-3}$ Divide each side of the equation by -3.

$x = 4$

EXAMPLE 3

Solve: $4 + 5(2x - 3) = 3(4x - 1)$

SOLUTION

$4 + 5(2x - 3) = 3(4x - 1)$

$4 + 10x - 15 = 12x - 3$ Use the distributive property. Then simplify.

$10x - 11 = 12x - 3$

$10x - 12x - 11 = 12x - 12x - 3$ Subtract $12x$ from each side of the equation.

$-2x - 11 = -3$ Simplify.

$-2x - 11 + 11 = -3 + 11$ Add 11 to each side of the equation.

$-2x = 8$ Simplify.

$\dfrac{-2x}{-2} = \dfrac{8}{-2}$ Divide each side of the equation by −2.

$x = -4$

When an equation contains parentheses, use the distributive property to remove the parentheses.

EXAMPLE 4

Solve: $3[2 - 4(2x - 1)] = 4x - 10$

SOLUTION

$3[2 - 4(2x - 1)] = 4x - 10$

$3[2 - 8x + 4] = 4x - 10$

$3[6 - 8x] = 4x - 10$

$18 - 24x = 4x - 10$

$18 - 24x - 4x = 4x - 4x - 10$

$18 - 28x = -10$

$18 - 18 - 28x = -10 - 18$

$-28x = -28$

$\dfrac{-28x}{-28} = \dfrac{-28}{-28}$

$x = 1$

EXAMPLE 5

Solve: $6x - 4(3 - 2x) = 5(x - 4) - 10$

SOLUTION

$6x - 4(3 - 2x) = 5(x - 4) - 10$ Remove parentheses, Watch the signs.

$6x - 12 + 8x = 5x - 20 - 10$ Combine like terms.

$14x - 12 = 5x - 30$ Isolate x by subtracting $5x$.

$14x - 5x - 12 = 5x - 5x - 30$ Combine like terms.

$9x - 12 = -30$

263

$$9x = -18$$

$$\frac{9x}{9} = \frac{-18}{9}$$ Divide by 9 to make 1 the coefficient of x.

$$x = -2$$

Check: $6(-2) - 4[3 - 2(-2)] = 5(-2 - 4) - 10$

$$-12 - 4(3 + 4) = 5(-6) - 10$$

$$-12 - 4(7) = -30 - 10$$

$$-12 - 28 = -40$$

$$-40 = -40$$

RATIONAL EQUATIONS

A rational (or fractional) equation is an equation containing a fraction or decimal. The denominator may or may not contain variables. Examples of fractional equations are

$$\frac{5x}{7} + 7 = 4x - \frac{5}{8} \quad or \quad 46x - 15 = 1.3x + 2.4$$

To solve an equation containing fractions, we multiply all the terms of the equation by the LCD to clear the fractions.

EXAMPLE 6

Solve: $\dfrac{3x - 2}{3} + \dfrac{x - 3}{2} = \dfrac{5}{6}$

SOLUTION

The LCD is 6. Multiply all terms of the equation by 6.

$$\frac{3x - 2}{3} + \frac{x - 3}{2} = \frac{5}{6}$$

$$6\left(\frac{3x - 2}{3}\right) + 6\left(\frac{x - 3}{2}\right) = 6\left(\frac{5}{6}\right)$$ Multiply by 6.

$$2(3x - 2) + 3(x - 3) = 5$$ Remove parentheses.

$$6x - 4 + 3x - 9 = 5$$

$$9x - 13 = 5$$ Combine like terms.

$$9x - 13 + 13 = 5 + 13$$ Isolate the constant, add 13.

$$9x = 18$$

$$\frac{9x}{9} = \frac{18}{9}$$

Divide by 9 to make 1 the coefficient of x.

$$x = 2$$

Check: $\dfrac{3(2)-2}{3}+\dfrac{2-3}{2}=\dfrac{5}{6}$

$$\dfrac{6-2}{3}+\dfrac{-1}{2}=\dfrac{5}{6}$$

$$\dfrac{4}{3}-\dfrac{1}{2}=\dfrac{5}{6}$$

$$\dfrac{8-3}{6}=\dfrac{5}{6}$$

$$\dfrac{5}{6}=\dfrac{5}{6}$$

EXAMPLE 7

Solve: $\dfrac{2x}{x-2}=1+\dfrac{4}{x-2}$

SOLUTION

$$\dfrac{2x}{x-2}=1+\dfrac{4}{x-2}$$

$$(x-2)\dfrac{2x}{x-2}=(x-2)\left(1+\dfrac{4}{x-2}\right)$$

The LCM is $x-2$. Multiply each side of the equation by the LCM.

$$(x-2)\dfrac{2x}{x-2}=(x-2)\cdot1+(x-2)\dfrac{4}{x-2}$$

Simplify using the distributive property and the properties of fractions.

$$\dfrac{\overset{1}{(x-2)}}{1}\cdot\dfrac{2x}{\underset{1}{x-2}}=(x-2)\cdot1+\dfrac{\overset{1}{(x-2)}}{1}\cdot\dfrac{4}{\underset{1}{x-2}}$$

$$2x=x-2+4$$

Solve for x.

$$2x=x+2$$

$$x=2$$

When x is replaced by 2, the denominators of $\dfrac{2x}{x-2}$ and $\dfrac{4}{x-2}$ are zero.

Division by zero is undefined. Hence, there is no solution and the root is said to be extraneous.

EXERCISE 6.7 A

Solve each of the following equations.

1) $-7x + 2 = 3x - 8$ 2) $-6x - 2 = -8x - 4$ 3) $-8 + 5x = 8 + 6x$

4) $-3 - 4x = 7 - 2x$ 5) $3x - 2 = 7 - 5x$ 6) $3 - 7x = -2 + 5x$

7) $4 - 3x = 6x - 8$ 8) $5x + 8 = 4 - 2x$ 9) $4x + 13 = -6x + 9$

10) $12x - 9 = 3x + 12$ 11) $\frac{4}{5}x - 1 = \frac{1}{5}x + 5$ 12) $\frac{5}{7}x - 3 = \frac{2}{7}x + 6$

13) $\frac{3}{4}x + 2 = \frac{1}{4}x - 9$ 14) $\frac{3}{7}x + 5 = \frac{5}{7}x - 1$ 15) $2x - 7(x - 2) = 3(x - 4)$

16) $12 + 2(x - 9) = 3(x - 12)$ 17) $3x + 2(x - 7) = 7(x - 1)$ 18) $2x - 2(x - 1) = 3(x - 2) + 7$

19) $3x - 4(x - 2) = 15 - 3(x - 2)$ 20) $9x - 3(x - 4) = 13 + 2(x - 3)$

21) $x + 5(x - 4) = 3(x - 8) - 5$ 22) $2x - 3(x + 4) = 2(x - 5)$

23) $7 - (x + 1) = 3(x + 3)$ 24) $6x - 3(x + 1) = 5(5 + 2)$ 25) $3x + 2(x + 4) = 13$

26) $6x + 2(x - 1) = 14$ 27) $8x - 3(x - 5) = 30$ 28) $-3 + 4(x + 3) = 5$

29) $5 - 3(x + 2) = 8$ 30) $6 - 2(x + 4) = 6$ 31) $6 - 3(x - 4) = 12$

32) $5 + 7(x + 3) = 20$ 33) $3x - 4(x + 3) = 9$ 34) $2x + 3(x - 5) = 10$

35) $4 + 3(x - 9) = -12$ 36) $3(x - 4) + 2x = 3$ 37) $3x + 2(x + 4) = 13$

38) $6x + 2(x - 1) = 14$ 39) $8x - 3(x - 5) = 30$ 40) $6 - 2(x + 4) = 10$

41) $5 - 3(x + 2) = 8$ 42) $6 - 2(x + 4) = 6$ 43) $6 - 3(x - 4) = 12$

44) $5 + 7(x + 3) = 20$ 45) $3x - 4(x + 3) = 9$ 46) $2x + 3(x - 5) = 10$

47) $2(x - 6) + 7x = 5 - 3(x - 2)$ 48) $3(x - 4) + 3x = 7 - 2(x - 1)$

49) $4.06x + 4.7(x + 3.22) = 1.774$ 50) $3.67x - 5.3(x - 1.932) = 6.99$

EXERCISE 6.7 B

Solve each of the following equations.

1) $\dfrac{1}{4} = \dfrac{x}{2}$

2) $\dfrac{2}{m} = \dfrac{5}{12}$

3) $\dfrac{9}{k} = \dfrac{3}{4}$

4) $\dfrac{15}{f} = \dfrac{30}{8}$

5) $\dfrac{7}{x} = \dfrac{8}{3}$

6) $\dfrac{2}{z} = \dfrac{11}{5}$

7) $\dfrac{6}{x} - \dfrac{4}{x} = 5$

8) $\dfrac{3}{x} + \dfrac{2}{x} = 5$

9) $\dfrac{x}{2} - \dfrac{x}{4} = 6$

10) $\dfrac{4}{x} - \dfrac{2}{3} = 1$

11) $\dfrac{9}{m} = 5 - \dfrac{1}{m}$

12) $\dfrac{3x}{5} + 2 = \dfrac{1}{4}$

13) $\dfrac{2t}{7} - 5 = t$

14) $\dfrac{1}{2} + \dfrac{2}{m} = 1$

15) $\dfrac{x+1}{2} = \dfrac{x+2}{3}$

16) $\dfrac{t-4}{3} = t + 2$

17) $\dfrac{3m}{2} + m = 5$

18) $\dfrac{2k+3}{k} = \dfrac{2}{3}$

19) $\dfrac{5-y}{y} + \dfrac{3}{4} = \dfrac{7}{y}$

20) $\dfrac{x}{x-4} = \dfrac{2}{x-4} + 5$

21) $\dfrac{m-2}{5} = \dfrac{m+8}{10}$

22) $\dfrac{2p+8}{9} = \dfrac{10p+4}{27}$

23) $\dfrac{5r-3}{7} = \dfrac{15r-2}{28}$

24) $\dfrac{2y-1}{y} + 2 = \dfrac{1}{2}$

25) $\dfrac{a}{2} - \dfrac{17+a}{5} = 2a$

26) $\dfrac{m-2}{4} + \dfrac{m+1}{3} = \dfrac{10}{3}$

27) $\dfrac{y+2}{5} + \dfrac{y-5}{3} = \dfrac{7}{5}$

28) $\dfrac{a+7}{8} + \dfrac{a-2}{3} = \dfrac{4}{3}$

29) $\dfrac{m+2}{5} - \dfrac{m-6}{7} = 2$

30) $\dfrac{p}{2} - \dfrac{p-1}{4} = \dfrac{5}{4}$

31) $\dfrac{r}{6} - \dfrac{r-2}{3} = \dfrac{-4}{3}$

32) $\dfrac{5y}{3} - \dfrac{2y-1}{4} = \dfrac{1}{4}$

33) $\dfrac{8k}{5} - \dfrac{3k-4}{2} = \dfrac{5}{2}$

6.8 TRANSLATING WORD PROBLEMS INTO ALGEBRAIC EXPRESSIONS

Algebraic equations are used to solve many kinds of problems that arise in business, industry, and in the scientific fields. One of the first steps in solving word problems is converting the phrases into mathematical equations.

To translate "word problems" into algebraic expressions, we let a variable represent the unknown quantity and then use the fundamental operations of addition, subtraction, division, and multiplication to describe the relationship that exists. In determining the arithmetic operation to use, watch for the following key words or phrases and their mathematical equivalents.

VERBAL EXPRESSION	RELATED ARITHMETIC OPERATION
add, sum of, more than, increased by	Addition
subtract, difference of, decreased by, less than	Subtraction
multiply, of, times, the product of, twice, three times, four times, etc.	Multiplication
divide, divided by, quotient of	Division

Given below are translations of verbal phrases into mathematical statements using the key words above and the arithmetic operations.

VERBAL EXPRESSION	MATHEMATICAL EXPRESSION
a) 7 plus a number	$x + 7$
b) Add 35 to a number	$x + 35$
c) The sum of a number and 1	$x + 1$
d) 2 more than a number	$x + 2$
e) 7 less than a number	$x - 7$
f) A number decreased by 13	$x - 13$
g) Five fewer than x	$x - 5$
h) The product of a number and 7	$(7)(x)$ or $7x$
i) Five times a number	$5x$
j) The quotient of a number and 3	$\dfrac{x}{3}$
k) The reciprocal of a number	$\dfrac{1}{x}$
l) 3 subtracted from 2 times a number	$2x - 3$

m) A number plus its reciprocal $\qquad$ $x + \dfrac{1}{x}$

n) The sum of a number and 5 multiplied by 7 $\qquad$ $(x+5)7 \text{ or } 7(x+5)$

o) The quotient of a number plus 7 and the number $\qquad$ $\dfrac{x+7}{x}$

EXERCISE 6.8

Translate into a mathematical expression.

1) The product of three less than a number and the number

2) The quotient of three times a number and the number

3) The sum of three more than a number and one half of the number

4) The square of a number plus the product of three and the number

5) Seven times the total of a number and eight

6) The product of a number and two more than the number

7) The difference between five times a number and the number

8) Four increased by some number

9) Three-fourths of a number

10) Four times some number

11) A number divided by twenty

12) The square of a number

13) A number increased by eleven more than twice the number

14) The square of a number divided by the sum of the number and twelve

15) Eight more than twice a number and seven

16) A number decreased by the product of five and the number

17) The difference between ten and the quotient of a number and two

18) The quotient of six and the sum of nine and a number

19) Six less than the total of three and a number

20) The ratio of a number and nine

21) The sum of a number and seven

22) The quotient of five and a number

23) The difference between a number and twelve

24) Five less than some number

6.9 TRANSLATING SENTENCES INTO EQUATIONS AND SOLVING

An equation states that two mathematical expressions are equal. Therefore, to translate a sentence into an equation requires recognition of the words or phrases which mean "equals." Some of these phrases are listed below.

$$\left.\begin{array}{l} \text{equals} \\ \text{is} \\ \text{is equal to} \\ \text{amounts to} \\ \text{represents} \end{array}\right\} \text{translate to } =$$

EXAMPLE 1

The quotient of a number and six is five. Find the number.

SOLUTION

The unknown number: x

The quotient of a number and six	five

$$\frac{x}{6} = 5$$

$$\frac{1}{6}x = 5$$

$$6 \cdot \frac{1}{6}x = 6 \cdot 5$$

$$x = 30$$

The number is 30.

EXAMPLE 2

Eight decreased by twice a number is four. Find the number.

SOLUTION

The unknown number: z

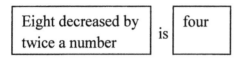

$$8 - 2z = 4$$
$$8 + (-2)z = 4$$
$$8 + (-8) + (-2)z = 4 + (-8)$$
$$(-2)z = -4$$
$$-\frac{1}{2}(-2)z = -\frac{1}{2}(-4)$$
$$z = 2$$

The number is 2.

EXAMPLE 3

Translate "a number decreased by six equals fifteen" into an equation and solve.

SOLUTION

The unknown number: x

$$x - 6 = 15$$
$$x + (-6) = 15$$
$$x + (-6) + 6 = 15 + 6$$
$$x = 21$$

The number is 21.

EXERCISE 6.9

Translate into an equation and solve.

1) Seven less than four times a number is nine. Find the number.

2) The total of a number divided by four and nine is two. Find the number.

3) One third of a number and seven is twelve. Find the number.

4) The sum of three times a number and four is eight. Find the number.

5) The total of twenty and a number is five. Find the number.

6) Five-sixths of a number is fifteen. Find the number.

7) Seven less than a number is three. Find the number.

8) Six times a number is fourteen. Find the number.

9) A number divided by four is six. Find the number.

10) Five more than a number is three. Find the number.

11) The quotient of a number and three is one. Find the number.

12) The product of three and a number is eighteen. Find the number.

13) A number decreased by seven is five. Find the number.

14) The sum of a number and seven is twelve. Find the number.

6.10 SOLVING WORD PROBLEMS

Now that we can translate verbal expressions into mathematical expressions, we can solve real-life problems or "word problems." To solve word problems, read each problem very carefully and determine what is given and what is to be found. Then, follow the steps below.

1) Let some letter, such as x, represent the unknown number.

2) Translate the problem into an equation. Write a true equation from the information given in the problem. Make sure the equation says in symbols exactly what the problem says in words.

EXAMPLE 1

General Problem

A chair and a desk together cost $60. The desk costs $3 less than twice the cost of the chair. Find the cost of each.

SOLUTION

1) Let x represent the cost of the chair. The cost of the desk is given in relationship to the cost of the chair. Specifically, the cost of the desk is "$3 less than twice the cost of the chair." Translated mathematically, the cost of the desk is $2x - 3$.

2) The chair and the desk cost $60 together. Therefore,

$$\begin{array}{ccccc} \text{Cost of Chair} & + & \text{Cost of Desk} & = & \$60 \\ x & + & 2x - 3 & = & \$60 \end{array}$$

3) Solve the equation.

$$x + 2x - 3 = 60$$
$$3x - 3 = 60 \qquad \text{Combine like terms}$$
$$3x - 3 + 3 = 60 + 3 \qquad \text{Add 3 to both sides}$$
$$3x = 63 \qquad \text{Divide by 3}$$
$$x = 21 \qquad \text{Cost of the chair}$$

To find the cost of the desk, evaluate $2x - 3$, when $x = 21$, the cost of the chair.

$$2x - 3 = 2(21) - 3 = 42 - 3 = 39$$

Therefore, the desk costs $39.

The cost of the chair was $21 and the cost of the desk was $39, which is $3 less than twice the cost of the chair. The sum of $21 and $39 is $60, which was the total cost of the two items.

EXAMPLE 2

Age Problems

A man is five years older than three times his son's age. Ten years from now the sum of their ages will be 65 years. What are their ages now?

SOLUTION

1) Let x represent the son's age now. Then, $3x + 5$ is the father's age now. Ten years from now the son will be $x + 10$ years, and the father will be $3x + 5 + 10$ or $3x + 15$. Summarize this information in a chart.

	Son's Age	Father's Age
Now	x	$3x + 5$
In ten years	$x + 10$	$3x + 15$

2) Ten years from now, the sum of their ages will be 65, therefore,

Son's age in 10 years +   Father's age in ten years   =   65

$x + 10$ + $3x + 15$ =   65

3) Solve the equation.

$x + 10 + 3x + 15$	$=$	65	Combine like terms
$4x + 25$	$=$	65	Subtract 25 from both sides
$4x + 25 - 25$	$=$	$65 - 25$	Simplify
$4x$	$=$	40	Divide by 4
x	$=$	10	The son's age

To find the father's age, evaluate the expression for the father's age when $x = 10$, the son's present age.

$3x + 5 = 3(10) + 5 = 30 + 5 = 35$

Therefore, the father is 35.

4) Check.

The son is 10 and the father is 35. We can see that the father is five years older than three times his son's age. Also, in 10 years, the father will be 45 years old and the son will be 20 years old. The sum of their ages will be 65 as stated in the problem.

Consecutive integers are integers that follow one another. For example, 15, 16, 17, 18. If we let x represent the first consecutive integer, then $x + 1$ will represent the second, $x + 2$ the third, $x + 3$ the fourth, and so forth. If we let x represent the first consecutive even or odd integer, then $x + 2$ would be second, $x + 4$ the third, $x + 6$ the fourth, and so forth.

EXAMPLE 3

Find three consecutive odd integers whose sum is 63.

SOLUTION

1) Let x = the first consecutive odd integer. Then $x + 2$ = the second consecutive odd integer, and $x + 4$ = the third consecutive odd integer.

2) Since the sum of the three consecutive integers is 63, we will get the equation

$$x + (x + 2) + (x + 4) = 63.$$

3) Solve the equation.

$x + (x + 2) + (x + 4) = 63$	Clear the parentheses
$x + x + 2 + x + 4 = 63$	Combine like terms
$3x + 6 = 63$	Subtract 6 from both sides
$3x + 6 - 6 = 63 - 6$	Combine like terms
$3x = 57$	Divide by 3
$x = 19$	First consecutive odd integer

To find the other consecutive odd integers, evaluate the expression for each.

Second consecutive odd integer = $x + 2$
$19 + 2 = 21$

Third consecutive odd integer = $x + 4$
$19 + 4 = 23$

4) Check.

The sum of these three consecutive odd integers, 19, 21, and 23, is 63.

EXAMPLE 4

A television is marked with a sale price of $220, which is 20% off the regular price. What is the regular price?

SOLUTION

1) Let r = regular price. The sale price is $100\% - 20\%$, or 80%, of the regular price.

2) Write an equation for the problem.

$$80\% \text{ of the regular price} = \text{the sale price}$$
$$80\% \text{ of } r = 220$$

3) Solve the equation.

$80\%(r) = 220$	Convert percent to decimal
$.80r = 220$	Divide by .80
$r = \$275$	Regular price

4) Check.

80% of $r = .80(275) = \$220$, the sale price.

EXERCISE 6.10

Write an equation and solve.

1) A store sells tennis rackets for $120 each. This price includes the store's cost for the rackets plus a markup of 35%. Find the store's cost for a tennis racket.

2) The selling price of a pair of skates is $150. This price includes the store's cost for purchasing the skates plus a markup of 25%. Find the store's cost for the skates.

3) A toy company has increased their monthly output by 20 bicycles. This represents a 4% increase over last year's production. Find the monthly output a year ago.

4) A town's population has declined by 18,000 within the last three years. This represents a 2% decrease. Find the town's population three years ago.

5) A man pays a monthly rent of $350. This amounts to one fifth of the man's monthly salary. Find the man's monthly salary.

6) The value of Marshall's stock has tripled in the last five years. The value this year is $36. Find the value of the stock five years ago.

7) During a sale, a dress is discounted $40.00. This is 20% off the regular price. Find the regular price.

8) A worker for a company is now making $36,000 a year. This is four times the amount the worker was making 15 years ago. Find the worker's salary 15 years ago.

9) The value of a truck this year is $3000. This is four-fifths of its value last year. Find the value of the truck last year.

10) The value of an apartment this year is $86,000. This is twice its value three years ago. Find the value three years ago.

11) A store is selling hamburger for $2.17 per pound. This is $.30 less than the price at the competitor's store. Find the price at the competitor's store.

12) James paid $1025 in state income tax this year. This is $55 more than last year. Find the state income tax last year.

13) A store paid $27,462 in state income tax. This amounted to $5000 more than 9% of the store's total income. Find the store's total income.

14) An executive receives a base monthly salary of $800 plus a 12.2% commission on total sales per month. During one month the sales executive received $2437.24. Find the total sales for the month.

15) Travil harvested 42,240 bushels of wheat. This is 110% of last year's crop. How many bushels of wheat did Travil harvest last year?

16) Kennie pays a tax which amounts to 6% of the price of a gallon of gas plus $.10. Find the cost of a gallon of gas when the tax paid is $.175.

17) The Fahrenheit temperature equals the sum of 32 and nine-fifths of the Celsius temperature. Find the Celsius temperature when the Fahrenheit temperature is 41 degrees.

18) Kirk charges $65 plus $15 for each yard of cement. How many yards of cement can be purchased for $185?

19) A radio is purchased for $480. A down payment of $30 is made. The remainder is paid in 18 equal monthly installments. Find the monthly payment.

20) Dubose charged $85 to replace faulty wiring in a house. This charge included $35 for materials and $20 per hour for labor. How long did it take Dubose to replace the wiring?

6.11 **PROPERTIES OF INEQUALITIES AND INTERVAL NOTATION**

An inequality is a mathematical statement of comparison that one quantity is less than, greater than, or equal to another quantity. Therefore, it is not an equation.

This relationship among mathematical statements can be expressed by using any of these symbols.

SYMBOLS	MEANING
$<$	IS LESS THAN
$\leq$	IS LESS THAN OR EQUAL TO
$>$	IS GREATER THAN
$\geq$	IS GREATER THAN OR EQUAL TO

The standard number line can help you to visualize how these inequality relationships can be written with respect to the relative position of the real numbers a and b on the real number line.

If a number is to the right of a number on the number line, then it is greater than that number.

ILLUSTRATIONS

a) $3 < 5$ because 3 is to the left of 5 on the number line.

b) $-8 \leq -5$ because -8 is to the left of -5 on the number line and is not equal to -5.

c) $5 > 2$ because 5 is to the right of 2 on the number line.

d) $-9 \geq -10$ because -10 is to the left of -9 on the number line and is not equal to -10.

PROPERTIES OF INEQUALITIES

The solution to an inequality consists of all replacement values that make the inequality true. For the inequality statement, $x \leq 4$, replacement values -6 and -8 make it true while 5 and 6 make the statement false.

To find the solutions of any inequality, we use the properties of inequalities to rewrite the inequality as a series of simpler inequalities.

Using the properties of inequalities, equivalent inequalities may be obtained by

(1) **Adding** the same numbers to both sides of equality.

If $a < b$ then $a + c < b + c$. Example: If $3 < 7$, then $3 + 4 < 7 + 4$.

(2) **Subtracting** the same number from both sides of an inequality.

If $a < b$ then $a - c < b - c$. Example: If $7 < 10$, then $7 - 4 < 10 - 4$.

(3) **Multiplying or dividing** both sides of the inequality by the same nonzero **positive** number.

If $a < b$ and $c > 0$, then $ac < bc$ or $\dfrac{a}{c} < \dfrac{b}{c}$.

Example: If $4 < 8$, then $2(4) < 8(2)$ or $\dfrac{4}{2} < \dfrac{8}{2}$.

(4) **Multiplying or dividing** both sides of an inequality by the same nonzero **negative** number and **reversing** the inequality sign.

If $a < b$ and $c < 0$, then $ac > bc$ or $\dfrac{a}{c} > \dfrac{b}{c}$.

Example: If $9 < 12$, then $(-3)(9) > (-3)(12)$ or $\dfrac{9}{-3} > \dfrac{12}{-3}$.

These properties are used to solve inequalities in much the same manner as equations. Before we solve inequalities, however, we will introduce notations for expressing the solution sets of inequalities.

Interval Notation

A solution of an inequality is a value of the variable that makes the inequality a true statement. The solution set of an inequality is the set of all solutions. The solution set of the inequality $x > 3$, for example, contains all numbers greater than 3. Its graph is an interval on the number line since an infinite number of values satisfy the variable. If we use open/closed-circle notation, the graph of $\{x|x > 3\}$ looks like the following.

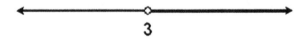

3

In this book, a notation, called **interval notation**, will be used to write solution sets of inequalities. To help us understand this notation, a different graphing notation will be used. Instead of an open circle, we use a parenthesis or a bracket. With this new notation, the graph of $\{x|x > 3\}$ now looks like

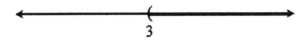

and can be represented in interval notation as $(3, \infty)$. The symbol ∞ is read "infinity" and indicates that the interval includes all numbers greater than 3. The left parenthesis indicates that 3 is not included in the interval. Using a left bracket, [, would indicate that 3 is included in the interval. The following table shows three equivalent ways to describe an interval: in set notation, as a graph, and in interval notation.

Type of Interval	Set	Interval Notation	Graph
Open interval	$\{x|a < x\}$	(a, ∞)	
	$\{x|a < x < b\}$	(a, b)	
	$\{x|x < b\}$	$(-\infty, b)$	
	$\{x|x$ is a real number$\}$	$(-\infty, \infty)$	
Half–open interval	$\{x|a \le x\}$	$[a, \infty)$	
	$\{x|a < x \le b\}$	$(a, b]$	
	$\{x|a \le x < b\}$	$[a, b)$	
	$\{x|x \le b\}$	$(-\infty, b]$	
Closed interval	$\{x|a \le x \le b\}$	$[a, b]$	

285

EXAMPLE 1

Graph each set on a number line and then write in interval notation.

a) $\{x|x \geq 2\}$ 　　　　b) $\{x|x < -1\}$ 　　　　c) $\{x|0.5 < x \leq 3\}$

SOLUTION

a)
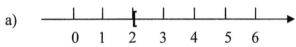

b)

c)

Interval notation can be used to write solutions of linear inequalities. To solve a linear inequality, we use a process similar to the one used to solve a linear equation. We use properties of inequalities to write equivalent inequalities until the variable is isolated.

EXAMPLE 2

Use a number line to graph the solution set indicated.

a) $[3, 6)$ 　　　　b) $(-\infty, 2)$ 　　　　c. $-3 < x \leq 5$

SOLUTION

a)
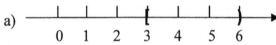

b)

c)

EXAMPLE 3

Solve $x - 2 < 5$.   Graph the solution set.

SOLUTION

$$x - 2 < 5$$
$$x - 2 + 2 < 5 + 2 \qquad \text{Add 2 to both sides.}$$
$$x < 7 \qquad \text{Simplify.}$$

The solution set is $\{x|x < 7\}$, which in interval notation is $(-\infty, 7)$. The graph of the solution set is

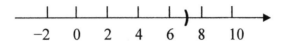

EXAMPLE 4

Solve $3x + 4 \geq 2x - 6$. Graph the solution set.

SOLUTION

$$3x + 4 \geq 2x - 6$$
$$3x + 4 - 2x \geq 2x - 6 - 2x \qquad \text{Subtract } 2x \text{ from both sides.}$$
$$x + 4 \geq -6 \qquad \text{Combine like terms.}$$
$$x + 4 - 4 \geq -6 - 4 \qquad \text{Subtract 4 from both sides.}$$
$$x \geq -10 \qquad \text{Simplify.}$$

The solution set is $\{x|x \geq -10\}$, which in interval notation is $[-10, \infty)$. The graph of the solution set is

EXAMPLE 5

Solve and graph the solution set. $\qquad \dfrac{1}{4}x \leq \dfrac{3}{8}$

SOLUTION

$$\frac{1}{4}x \leq \frac{3}{8}$$
$$4 \cdot \frac{1}{4}x \leq 4 \cdot \frac{3}{8} \qquad \text{Multiply both sides by 4.}$$
$$x \leq \frac{3}{2} \qquad \text{Simplify.}$$

The solution set is $\left\{x \mid x \leq \dfrac{3}{2}\right\}$, which in interval notation is $\left(-\infty, \dfrac{3}{2}\right]$. The graph of the solution set is

287

EXAMPLE 6

Solve: $5(x-2) \geq 9x - 3(2x-4)$.

SOLUTION

$5(x-2) \geq 9x - 3(2x-4)$

$5x - 10 \geq 9x - 6x + 12$ Use the distributive property to remove parentheses.

$5x - 10 \geq 3x + 12$ Combine like terms.

$2x - 10 \geq 12$ Subtract $3x$ from each side of the inequality.

$2x \geq 22$ Add 10 to each side of the inequality.

$\dfrac{2x}{2} \geq \dfrac{22}{2}$ Divide each side of the inequality by the coefficient 2.

$x \geq 11$

The solution set is $\{ x \mid x \geq 11 \}$.

EXAMPLE 7

Solve $-1 \leq \dfrac{2x}{3} + 5 \leq 2$.

SOLUTION

First, clear the inequality of fractions by multiplying all three parts by the LCD of 3.

$-1 \leq \dfrac{2x}{3} + 5 \leq 2$

$3(-1) \leq 3\left(\dfrac{2x}{3} + 5 \right) \leq 3(2)$ Multiply all three parts by the LCD of 3.

$-3 \leq 2x + 15 \leq 6$ Use the distributive property and multiply.

$-3 - 15 \leq 2x + 15 - 15 \leq 6 - 15$ Subtract 15 from all three parts.

$-18 \leq 2x \leq -9$ Simplify.

$\dfrac{-18}{2} \leq \dfrac{2x}{2} \leq \dfrac{-9}{2}$ Divide all three parts by 2.

$-9 \leq x \leq -\dfrac{9}{2}$ Simplify.

EXAMPLE 8

Solve $2 \le 4 - x \le 7$.

SOLUTION

To get x alone, we first subtract 4 from all three parts.

Solve $2 \le 4 - x \le 7$

$2 - 4 < 4 - x - 4 < 7 - 4$ Subtract 4 from all three parts.

$-2 < -x < 3$ Simplify.

$\dfrac{-2}{-1} > \dfrac{-x}{-1} > \dfrac{3}{-1}$ Divide all three parts by -1 and reverse the inequality symbols.

$2 > x > -3$

This is equivalent to $-3 < x < 2$. The solution set in interval notation is $(-3, 2)$, and its graph is shown.

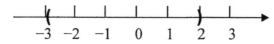

EXAMPLE 9

Solve: $-3 < 2x + 1 < 5$

SOLUTION

$-3 < 2x + 1 < 5$

$-3 - 1 < 2x + 1 - 1 < 5 - 1$ Subtract 1 from each of the three parts of the inequality.

$-4 < 2x < 4$

$\dfrac{-4}{2} < \dfrac{2x}{2} < \dfrac{4}{2}$ Divide each of the three parts of the inequality by the coefficient 2.

$-2 < x < 2$

The solution set is $\{x \mid -2 < x < 2\}$.

289

EXAMPLE 10

Solve: $1 < 3x - 5 < 4$

SOLUTION

$1 < 3x - 5 < 4$

$1 + 5 < 3x - 5 + 5 < 4 + 5$

$6 < 3x < 9$

$\dfrac{6}{3} < \dfrac{3x}{3} < \dfrac{9}{3}$

$2 < x < 3$

$\{x \mid 2 < x < 3\}$

EXAMPLE 11

Solve: $2x + 3 > 11$ or $4x - 1 < 3$

SOLUTION

To solve a compound inequality joined with "or" solves each inequality separately. The solution consists of the numbers in the solution set of either or both inequalities. In set notation, this represents the elements found in the union of the two sets. A graph is helpful in identifying the solution set.

Solving each inequality separately, we get.

$2x + 3 > 11$ $4x - 1 < 3$

$2x > 8$ $4x < 4$

$x > 4$ $x < 1$

The solution set is shown on the graph below.

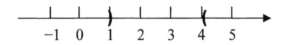

In interval notation the solution is $(-\infty, 1)$ or $(4, \infty)$.

290

EXERCISE 6.11

Solve and graph the following inequalities.

1) $6x > 18$

2) $5k \le -15$

3) $2m - 5 \ge 15$

4) $\dfrac{5z - 6}{8} < 3$

5) $-\dfrac{2x}{3} < -10$

6) $-\dfrac{2x}{5} \ge -4$

7) $\dfrac{3z - 2}{-5} \le 4$

8) $-6p - 2 \ge 16$

9) $-6 < x + 1 < 8$

10) $-15 < 3p + 6 < -9$

11) $2(3k - 5) + 7 > k + 12$

12) $-(9 + k) - 5 + 4k \ge 1$

13) $-2(m - 4) \le -3(m + 1)$

14) $-8 \le \dfrac{3m + 1}{4} \le 3$

15) $-16 < 3t + 2 < -11$

16) $3x - 7 < 6x + 2$

17) $3x - 5 > 7x + 3$

18) $-15 < 3p + 6 < -12$

19) $-19 \le 3x - 5 \le 1$

20) $- < 3t + 2 < -10$

21) $-1 \le \dfrac{2x - 5}{6} \le 5$

22) $-3 \le \dfrac{3m + 1}{4} \le 3$

23) $\dfrac{3}{5}(k - 2) - \dfrac{1}{4}(2k - 7) \le 3$

24) $3(2x - 4) < 2x + 3$

25) $7(4 - x) + 5x < 2(16 - x)$

26) $-1 < x + 1 < 8$

27) $-9 \le k + 5 \le 15$

28) $-4 \le m + 3 \le 10$

29) $-6 \le 2x + 4 \le 16$

30) $-3(x - 6) > 2x - 2$

CHAPTER 6 REVIEW EXERCISES A

Identify each of the following as a term, expression, variable, constant, coefficient, or equation.

1) the 3 in $3p^2$ _____

2) the 6 in $b + 6$ _____

3) the c in $4c^2 + 5$ _____

4) $3n - 20 = 10$ _____

5) $4t + 17$ _____

6) $52y$ _____

7) Simplify by adding like terms
   $7n + 18 - 3p - 4n - 6p$

Write as an algebraic expression:

8) five less than one-fifth of a number

9) the difference between four times a number and sixteen

10) two times the square of a number

11) six more than a certain number

12) Simplify $3y - 2x - 7y - 9x$.

13) Evaluate $c^2 - (2a + b^2)$ when $a = 3$, $b = -6$, and $c = -2$.

14) Simplify $3y + 5(y - 3) + 8$.

15) A loan of $6600 is to be paid in 48 equal monthly installments. Find the monthly payment. Use the formula $L = P \cdot N$, where L is the loan amount, P is the monthly payment, and N is the number of months.

16) Solve $\dfrac{x}{5} - 12 = 7$

17) Solve $3x - 12 = -18$

18) Find: $\pi r^2 h$ for $\pi = 3.14$, $r = 10$, $h = 7$

19) Find: $5a^2 + 9a - 4c$ if $a = 3$, $c = 4$

Multiply and add like terms:

20) $3y - 2x(y - 7) + y(8 - 4x)$

21) Solve $x - 12 = 14$

22) Solve $8 - 3x = 2x - 8$

23) Solve $3x - 4(x - 2) = 8$

24) Solve $\dfrac{5}{8}x = -10$

25) The total of five and three times a number is the number minus two. Find the number.

26) The sum of twice a number and 8 is 10. Find the number.

27) Translate "five times the sum of a number and three" into a mathematical expression.

28) Translate "the sum of x and one third of x" into a mathematical expression.

29) Twenty-one more than two-thirds of a number equals twenty-seven. What is the number?

30) The ratio of the height of the building to its shadow is 5 to 2. If the shadow is 22 ft. long, how high is the building?

CHAPTER 6 REVIEW EXERCISES B

Combine terms whenever possible.

1) $2m + 9m$

2) $15p^2 - 7p^2 + 8p^2$

3) $5p^2 - 4p + 6p + 11p^2$

4) $-2(3k - 5) + 2(k + 1)$

5) $7(2m + 3) - 2(8m - 4)$

6) $-(2k + 8) - (3k - 7)$

Solve each equation.

7) $m - 5 = 1$

8) $y + 8 = -4$

9) $3k + 1 = 2k + 8$

10) $5k = 4k + \dfrac{2}{3}$

11) $(4r - 2) - (3r + 1) = 8$

12) $3(2y - 5) = 2 + 5y$

13) $7k = 35$

14) $12r = -48$

15) $2p - 7p + 8p = 15$

16) $\dfrac{m}{12} = -1$

17) $\dfrac{5}{8}k = 8$

18) $12m + 11 = 59$

19) $-(m + 2) = -3m - 11$

Solve the word problems.

20) If 7 is added to six times a number, the result is 22. Find the number.

21) If 4 is subtracted from twice a number, the result is 8. Find the number.

22) The sum of a number and 5 is multiplied by 6, giving 72 as a result. Find the number.

23) In a marathon, Susan ran $\frac{2}{3}$ as far as Linda. In all, the two people ran 30 miles. How many miles did Susan run?

24. Charles has 15 more college units than Tom. Altogether, the two men have 95 units. How many units does Tom have?

25) The perimeter of a rectangle is ten times the short side. The length is 9 meters more than the width. Find the width of the rectangle.

26) The longest side of a triangle is 11 meters longer than the shortest side. The medium side is 15 meters long. The perimeter of the triangle is 46 meters. Find the length of the shortest side of the triangle.

27) A person has $250 in fives and tens. He has twice as many tens as fives. How many fives does he have?

28) For a party, Joan bought some 15¢ candy and some 30¢ candy, paying $15 in all. If there are 25 more pieces of 15¢ candy, how many pieces of 30¢ candy did she buy?

29) How many liters of 40% salt solution should be added to 72 liters of 80% salt solution to get a 70% mixture?

30) A nurse must mix 15 liters of a 10% solution of a drug with some 60% solution to get a 20% mixture. How many liters of the 60% solution would be needed?

Graph each inequality on a number line.

31) $m \geq -2$ 32) $a < -3$ 33) $-5 \leq p < 6$ 34) $1 \leq p < 4$

Solve each inequality.

35) $y + 5 \geq -2$ 36) $5y > 4y + 8$ 37) $9(k - 5) - (3 + 8k) \geq 5$

38) $3(2z + 5) + 4(8 + 3z) \leq 5(3z + 2) + 2z$ 39) $-6 \leq x + 2 \leq 0$

40) $3 < y - 4 < 5$ 41) $6k \geq -18$ 42) $-11y < 22$

43) $2 - 4p + 7p + 8 < 6p - 5p$ 44) $-(y + 2) + 3(2y - 7) \leq 4 - 5y$

45) $-3 \leq 2m + 1 \leq 4$ 46) $9 < 3m + 5 \leq 20$

Solve each of the following.

47) $\dfrac{y}{y - 5} = \dfrac{7}{2}$ 48) $I = prt$. Solve for r.   49) $-z > -4$

50) $\dfrac{3}{5} = \dfrac{k}{12}$ 51) $2k - 5 = 4k + 7$ 52) $7m + 3 \leq 5m - 9$

CHAPTER 6 TEST A

1) Simplify $3a + 4b + 7a - 2 - 4b$

   a) $10a + 8b - 2$ b) $10a - 2$ c) $10a + 2$ d) $3a - 2 - 4b$ e) $10a - 4b$

Write as an algebraic expression:

2) 7 more than one–third of a number.

   a) $\frac{1}{3}n + 7$ b) $7 + 3n$ c) $\frac{1}{3}(n + 7)$ d) $7n + \frac{1}{3}$ e) $\frac{7}{3}n$

3) The sum of 3 times a number and another number

   a) $3n = n$ b) $3 + n + n$ c) $3x + x$ d) $3(a + b)$ e) $3n + m$

4) Eight less than a certain number

   a) $8 + n$ b) $n + 8$ c) $8 - n$ d) $n - 8$ e) $8n$

5) 5 times a number

   a) $n - 5$ b) $5n$ c) $5 + n$ d) 5×2 e) $5 - n$

6) The 3 in $4y + 3$ is a ?

   a) term   b) expression   c) constant   d) coefficient   e) variable

7) $7t$ is a ?

   a) term   b) expression   c) constant   d) coefficient   e) variable

8) The p in $3 + 4p^2$ is a?

   a) term   b) expression   c) equation   d) coefficient   e) variable

9) $4c - 2b = 10$ is a?

   a) term   b) expression   c) equation   d) coefficient   e) variable

10) The 5 in $5x^2$ is a?

   a) term   b) coefficient   c) variable   d) expression   e) equation

11) $b^2 + 3a$ is a?

   a) equation   b) variable   c) term   d) expression   e) constant

12) If $2\frac{3}{4}$ lbs. of grapes cost 99¢, what will 6 lbs. cost?

a) 46¢   b) $1.75   c) $2.54   d) $2.12   e) $2.16

13) The ratio of my age to his age is $\dfrac{5}{8}$. I am 20. How old is he?

a) 20   b) 16   c) 2   d) 12   e) 32

14) Solve: Three less than twice a number is equal to 13. Find the number.

a) 5   b) 32   c) 16   d) 8   e) -5

15) Solve: $2g - 5 = g - 10$

a) 10   b) 5   c) -5 d) 15   e) -15

16) Solve: $6E + 7 = 43$. $E = ?$

a) 8   b) $8\dfrac{1}{3}$ c) 6   d) 36   e) 7

17) Solve: $4R = 15$ $R = ?$

a) 4   b) $3\dfrac{3}{4}$ c) $4\dfrac{3}{4}$ d) $4\dfrac{1}{4}$ e) $\dfrac{4}{15}$

18) Solve: $\dfrac{x}{3} - 5 = -12$

a) 41   b) $-\dfrac{2}{3}$ c) -21 d) -51

19) Solve: $2x - 4 = -6$

a) -1 b) -6 c) -4 d) 24

20) A car travels 403.1 mi on 12.8 gal of gas. Find the car's mileage per gallon. Use the formula $D = M \cdot G$, where D is the distance, M is miles per gallon, and G is the number of gallons.

a) 32   b) 30.5   c) 31.5   d) 29

21) Solve: $-\dfrac{2}{3}x = 5$

a) $7\dfrac{1}{2}$ b) $-7\dfrac{1}{2}$ c) $-\dfrac{5}{2}$ d) $\dfrac{3}{5}$

22) Solve: $-3x = 9$

a) 3   b) 27   c) $\dfrac{1}{3}$ d) -3

23) Solve: $x - 4 = -5$

a) -1 b) -9 c) 9   d) 1

24) Which number is a solution to $2x + 1 = x - 1$?
    a) -2 b) 2 c) -1 d) 4

25) Simplify $6y - 3(y - 5) + 8$.
    a) $3y + 3$ b) $3y - 7$ c) $3y + 23$ d) $5y + 23$

26) Simplify $3z - 2x + 5z - 8x$.
    a) $6z \cdot 8x$ b) $8z - 10x$ c) $3z - 5x$ d) $z - 3x$

27) Evaluate $3ab - 2ac$ when $a = -2$, $b = 6$, and $c = -3$.
    a) -24 b) -48 c) 24 d) 48

28) Translate "two less than a number is five" into an equation and solve.
    a) 3 b) -3 c) -7 d) 7

29) Three less than three-fourths of a number is two less than one-fourth of the number. Find the number.

    a) 1 b) -1 c) 2 d) $\dfrac{1}{2}$

30) Translate "the product of x and the sum of x and 5" into a mathematical expression.

    a) $x + 5x$ b) $x + x + 5$ c) $x(x + 5)$ d) $2x + 5$

31) Translate "the sum of one-third of n and n" into a mathematical expression.
    a) $\dfrac{1}{3} + n$ b) $\dfrac{1}{3}n + n$ c) $\dfrac{1}{3} + n + n$ d) $3n + n$

32) Solve: $7x - 3(x - 5) = -10$
    a) $-\dfrac{5}{4}$ b) $-2\dfrac{1}{2}$ c) $-6\dfrac{1}{4}$ d) $-3\dfrac{3}{4}$

CHAPTER 6 TEST B

Pick the best choice for each of the following. Simplify each of the following expressions by combining terms.

1) $-9r - 8r + 2r - 11r + 6$
   a) $26r + 6$ b) $6 - 26r$ c) $26r$ d) $-6 - 26r$

2) $12a - 7a + 11a - 14a$
   a) a b) $-2a$ c) $2a$ d) $-a$

3) $-9(2m - 3) - 4m - 12 - m$
   a) $-23m + 15$ b) $23m + 15$ c) $23m - 15$ d) $-23m - 15$

4) $-5(6 - z) - (2 - 11z) + 3z$
   a) $32 - 19z$ b) $-23 - 19z$ c) $32 + 19z$ d) $-32 + 19z$

Solve each of the following equations.

5) $2y - 9 = 7$
   a) -8 b) 10 c) -10 d) 8

6) $7r - 1 = -22$
   a) $-\dfrac{23}{7}$ b) $\dfrac{23}{7}$ c) 3 d) -3

7) $12a - 7 = 11a - 9$
   a) -2 b) 2 c) 5 d) 7

8) $-2(r - 7) = -r - (-1 - 2)$
   a) $-\dfrac{2}{3}$ b) $\dfrac{3}{4}$ c) 11 d) 8

9) $3(2r + 1) + 1 = 27 - 5(r - 2)$
   a) 3 b) $\dfrac{23}{11}$ c) 33 d) -3

10) $\dfrac{2z}{5} = -4$
    a) -10 b) 10 c) $\dfrac{1}{10}$ d) $-\dfrac{1}{10}$

11) $\dfrac{3}{8}p = 2$

a) $-\dfrac{3}{16}$ b) $\dfrac{3}{16}$ c) $-\dfrac{16}{3}$ d) $\dfrac{16}{3}$

12) $-(8-q)=-4-(2-3q)$

    a) -1 b) 1 c) $\dfrac{1}{2}$ d) $-\dfrac{1}{2}$

Solve each of the following word problems.

13) The sum of twice a number and 11 is tripled. The result is -12. Find the number.

    a) $-\dfrac{15}{2}$ b) $\dfrac{15}{2}$ c) -2 d) None of these

14) The perimeter of a triangle is 36. One side is 6. The other two sides are equal in length. Find the length of one of the two equal sides.

    a) 30 b) 9 c) 15 d) None of these

Solve each proportion.

15) $\dfrac{2}{r}=\dfrac{7}{3}$

    a) $\dfrac{6}{7}$ b) $\dfrac{7}{6}$ c) 6 d) -1

16) $\dfrac{r-2}{r}=\dfrac{5}{8}$

    a) 2 b) -1 c) $\dfrac{3}{16}$ d) $\dfrac{16}{3}$

17) $\dfrac{p+1}{p-2}=\dfrac{5}{3}$

    a) $\dfrac{13}{2}$ b) $-\dfrac{1}{4}$ c) $\dfrac{2}{13}$ d) 13

Solve each of the following inequalities.

18) $r+1\ge -7$

    a) $r\ge -8$ b) $r\le 8$ c) $r>8$ d) $r<8$

19) $\dfrac{3}{4}a<-9$

    a) $a\le -12$ b) $a>-12$ c) $a\ge -12$ d) $a<-12$

20) $-q<5$

a) $q > -5$ b) $q < -5$ c) $q \geq -5$ d) $q \leq -5$

21) $-3r \leq -2r + 9$
    a) $r > 9$ b) $r \geq 9$ c) $r \geq -9$ d) $r < 9$

22) $-2(4a - 1) - 3(a - 4) \geq 3$
    a) $a \geq -1$ b) $a < 1$ c) $a \leq 1$ d) $a > -1$ a > -1 (b)

7 GRAPHING LINEAR EQUATIONS

7.1 **LINEAR EQUATIONS IN TWO VARIABLES**

> A **linear equation in two variables** is an equation that can be put in the form
> $$Ax + By = C,$$
> where A, B, and C are real numbers and A and B are not both 0.

A solution of a linear equation requires two numbers, one for each variable. For example, the equation $y = 4x + 5$ is satisfied if x is replaced with 2 and y is replaced with 13, since

$13 = 4(2) + 5$.

Thus, $x = 2$ and $y = 13$ is a solution of the equation $y = 4x + 5$. The phrase "$x = 2$ and $y = 13$" is abbreviated as

(2, 13).

This abbreviation gives the x-value, 2, and the y-value, 13, as a pair of numbers, written inside parentheses. The x-value is always given first. A pair of numbers written in this order is called an ordered pair.

EXAMPLE 1

Decide whether the given ordered pair is a solution of the given equation.
a) (3, 2); $2x + 3y = 12$ b) (−2, −7); $2m + 3n = 12$

SOLUTION

a) Decide whether (3, 2) is a solution of the equation $2x + 3y = 12$ by substituting
3 for x and 2 for y in the given equation,

$$2x + 3y = 12$$
$$2(3) + 3(2) = 12 \qquad \text{Let.}x = 3; \text{ let } y = 2.$$
$$6 + 6 = 12$$
$$12 = 12 \qquad \text{True}$$

This result is true, so (3, 2) is a solution of $2x + 3y = 12$.

b) $2(-2) + 3(-7) = 12 \qquad \text{Let } m = -2; \text{ let } n = -7.$
$\quad -4 + (-21) = 12$
$\quad -25 = 12 \qquad \text{False}$

This result is false, so (−2, −7) is not a solution of $2m + 3n = 12$.

By choosing a number for one variable in a linear equation, the value of the other variable can be found, as shown in the next example.

EXAMPLE 2

Complete the given ordered pairs for the equation $y = 4x + 5$: a) (7,)   b) (, 13)

SOLUTION

a) In this ordered pair, $x = 7$. (Remember that x always comes first.) Find the corresponding value of y by replacing x with 7 in the equation $y = 4x + 5$.

$$y = 4(7) + 5 = 28 + 5 = 33$$

This gives the ordered pair (7, 33).

b) In this ordered pair, $y = 13$. Find the value of x by replacing y with 13 in the equation and then solving for x.

$y = 4x + 5$
$13 = 4x + 5$
$8 = 4x$
$2 = x$

The ordered pair is (2, 13).

EXAMPLE 3

Complete the given ordered pairs for the equation $5x - y = 24$.

Equation   Ordered Pairs
$5x - y = 24$ (5,), (−3,), (0,)

SOLUTION

Find the y-value for the ordered pair (5,) by replacing x with 5 in the given equation and solving for y.

$$5x - y = 24$$
$$5(5) - y = 24 \qquad \text{Let } x = 5.$$
$$25 - y = 24$$
$$-y = -1 \qquad \text{Subtract 25 from both sides.}$$
$$y = 1$$

This gives the ordered pair (5, 1).

Complete the ordered pair (−3,) by letting $x = -3$ in the given equation.
Also, complete (0,) by letting $x = 0$.

If $x = -3$,	If $x = 0$,
then $5x - y = 24$	then $5x - y = 24$
becomes $5(-3) - y = 24$	becomes $5(0) - y = 24$
$-15 - y = 24$	$0 - y = 24$
$-y = 39$	$-y = 24$
$y = -39$	$y = -24$

The completed ordered pairs are as follows,

Equation   Ordered Pairs
$5x - y = 24$ (5,1), (−3, −39), (0, −24)

EXERCISE 7.1

Decide whether the given ordered pair is a solution of the given equation.

1) $x + y = 9$; $(2, 7)$

2) $3x + y = 8$; $(0, 8)$.

3) $2x - y = 6$, $(2, -2)$

4) $2x + y = 5$; $(2, 1)$

5) $4x - 3y = 6$; $(1, 2)$

6) $5x - 3y = 1$, $(0, 1)$

7) $y = 3x$; $(1, 3)$

8) $x = -4y$; $(8, -2)$

9) $x = -6$; $(-6, 8)$

10) $y = 2$; $(9, 2)$

11) $x + 4 = 0$; $(-5, 1)$

12) $x - 6 = 0$; $(5, -1)$

Complete the given ordered pairs for the equation $y = 3x + 5$.

13) $(2,)$ 14) $(5,)$ 15) $(8,)$ 16) $(0,)$ 17) $(-3,)$

18) $(-4,)$.   19) $(, 14)$ 20) $(, -10)$

Complete the ordered pairs, using the given equations.

Equation	Ordered Pairs
21) $y = 2x + 1$	$(3,)$ $(0,)$ $(-1,)$
22) $y = 3x - 5$	$(2,)$ $(0,)$ $(-3,)$
23) $y = 8 - 3x$	$(2,)$ $(0,)$ $(-3,)$
24) $y = -2 - 5x$	$(4,)$ $(0,)$ $(-4,)$
25) $2x + y = 9$	$(0,)$ $(3,)$ $(12,)$
26) $-3x + y = 4$	$(1,)$ $(0,)$ $(-2,)$
27) $x = -4$	$(, 6)$ $(, 2)$ $(, -3)$
28) $y = -8$	$(4,)$ $(0,)$ $(-4,)$
29) $x + 9 = 0$	$(, 8)$ $(, 3)$ $(, 0)$
30) $y - 4 = 0$	$(9,)$ $(-5,)$ $(0,)$
31) $y = -3x$	$(-2,)$ $(0,)$ $(, -6)$
32) $x = 4y$	$(0,)$ $(, 3)$ $(-8,)$

7.2 GRAPHING LINEAR EQUATIONS WITH TWO VARIABLES

The graph of an equation in two variables is a graph of the ordered-pair solutions of the equation. In general, a **linear equation** is an equation that can be written in the form $f(x) = mx + b$, where m is the coefficient of x and b is a constant. Consider $y = 3x - 12$. Choosing $x = 0, 1, 2, 3, 4$, and 5 and determining the corresponding values of y produces ordered pairs of numbers satisfying the equation. These are recorded in the table below

x	$y=3x-12$		y	(x, y)
0	$3(0)-12$		-12	$(0, -12)$
1	$3(1)-12$		-9	$(1, -9)$
2	$3(2)-12$		-6	$(2, -6)$
3	$3(3)-12$		-3	$(3, -3)$
4	$3(4)-12$		0	$(4, 0)$
5	$3(5)-12$		3	$(5, 3)$

The equation $y = 3x - 12$ is an example of a linear equation because its graph is a straight line. Since the graph of a linear equation in two variables is a straight line and a straight line is determined by two points, it is necessary to find only two solutions. However, it is recommended that at least three points be used to ensure accuracy.

EXAMPLE 1

Graph the equation $y = -2x + 3$.

Find three ordered pair solutions, and plot the ordered pairs. The line through the plotted points is the graph. Since the equation is solved for y, let us choose three x-values. Let x be 0, 2, and then -1 to find our three ordered-pair solutions.

SOLUTION

Let $x = 0$

$y = -2x + 3$

$y = -2 \cdot 0 + 3$

$y = 3$ *Simplify*

Let $x = 2$

$y = -2x + 3$

$y = -2 \cdot 2 + 3$

$y = -1$ *Simplify*

Let $x = -1$

$y = -2x + 3$

$y = -2(-1) + 3$

$y = 5$ *Simplify*

The three ordered pairs (0, 3), (2, −1), and (−1, 5) are listed in the table and the graph is shown.

x	y
0	3
−1	5
2	−1

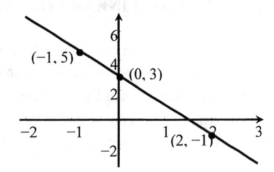

Notice that the graph crosses the y-axis at the point (0,3). This point is called the y-intercept.

EXAMPLE 2

Graph $y = 3x - 2$

SOLUTION

x	y
0	−2
−1	−5
2	4

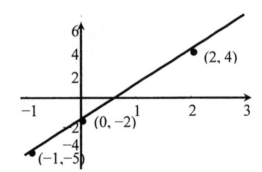

EXAMPLE 3

Graph $y = \dfrac{1}{2}x - 1$

x	y
0	−1
2	0
−2	−2

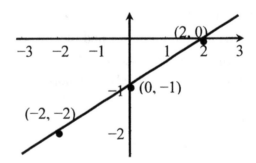

The graphed line crosses the x-axis at (2, 0). This is called the **x-intercept**.

STANDARD FORM

Recall from Section 7.1 that a linear equation in two variables is an equation that can be written in the form $Ax + By = C$ where A and B are not both 0. This form is called **standard form**.

EXAMPLE 4

Graph $3x + 4y = 12$

SOLUTION

$3x + 4y = 12$ Solve for y.

$4y = -3x + 12$ Subtract $3x$ from each side of the equation.

$y = -\dfrac{3}{4}x + 3$ Divide each side of the equation by 4.

Find three ordered-pair solutions of the equation.

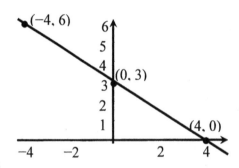

x	y
0	3
4	0
-4	6

Graph the ordered pairs and then draw a line through the points.

Another method of graphing an equation of the form $Ax + By = C$ is to find the x- and y-intercepts. In general, to find the x-intercept of a line, let $y = 0$ or $f(x) = 0$ since any point on the x-axis has a y-coordinate of 0. To find the y-intercept of the graph of an equation not in the form $y = mx + b$, let $x = 0$ since any point on the y-axis has an x coordinate of 0.

EXAMPLE 5

Graph $x - 3y = 6$ by plotting intercepts.

SOLUTION

Let $y = 0$ to find the x-intercept and $x = 0$ to find the y-intercept.

If $y = 0$	then	If $x = 0$	then
	$x - 3(0) = 6$		$0 - 3y = 6$
	$x - 0 = 6$		$-3y = 6$
	$x = 6$		$y = -2$

The *x*-intercept is (6, 0) and the *y*-intercept is (0, −2). We find a third ordered pair solution to check our work. If we let $y = -1$, then $x = 3$. Plot the points (6, 0), (0, −2), and (3, −1). The graph of $x - 3y = 6$ is the line drawn through these points, as shown.

x	y
6	0
0	−2
3	−1

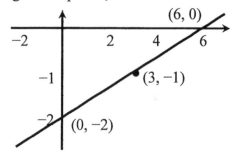

EXAMPLE 6

Find the intercepts for the graph of $2x + y = 4$. Draw the graph.

SOLUTION

Find the *y*-intercept by letting $x = 0$; find the *x*-intercept by letting $y = 0$.

x	y
0	4
2	0
1	2

$$2x + y = 4$$
$$2(0) + y = 4$$
$$0 + y = 4$$
$$y = 4$$

$$2x + y = 4$$
$$2x + 0 = 4$$
$$2x = 4$$
$$x = 2$$

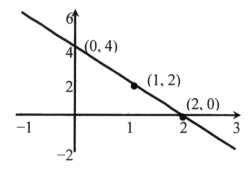

The *y*-intercept is (0, 4). The *x*-intercept is (2, 0). The graph with the two intercepts shown in color is given below. We get a third point as a check. For example, choosing $x = 1$ gives $y = 2$. These three ordered pairs are shown in the table below. Plot (0, 4), (2, 0), and (1, 2), then draw a line through them.

The graph of a linear equation with one variable missing is a vertical or horizontal line. The graph of $x = c$ is a vertical line passing through (c, 0). The graph of $y = c$ is a horizontal line passing through (0, c).

EXAMPLE 7

Graph $y = -3$

SOLUTION

The equation $y = -3$ can be written as $0x + y = -3$. For any *x*−value chosen, *y* is −3. If we choose 4, 0, and −2

312

as x-values, the ordered pair solutions are $(4, -3)$, $(0, -3)$, and $(-2, -3)$. We will use these ordered pairs to graph $y = -3$.

The graph is a horizontal line with y-intercept $(0, -3)$ and no x-intercept.

EXERCISE 7.2

Complete the ordered pairs for each equation. Then graph the equation by plotting the points and drawing a line through them.

1) $x + y = 5$ $(0, \ )$ $(\ , 0)$ $(2, \ )$

2) $y = x - 3$ $(0, \ )$ $(\ , 0)$ $(5, \ )$

3) $y = x + 4$ $(0, \ )$ $(\ , 0)$ $(-2, \ )$

4) $y + 5 = x$ $(0, \ )$ $(\ , 0)$ $(6, \ )$

5) $y = 3x - 6$ $(0, \ )$ $(\ , 0)$ $(3, \ )$

6) $x = 2y + 1$ $(0, \ )$ $(\ , 0)$ $(3, \ )$

7) $2x + 5y = 20$ $(0, \ )$ $(\ , 0)$ $(5, \ )$

8) $3x - 4y = 12$ $(0, \ )$ $(\ , 0)$ $(8, \ )$

9) $x + 5 = 0$ $(\ , 2)$ $(0, \ )$ $(\ , -3)$

10) $y - 4 = 0$ $(3, \ )$ $(0, \ )$ $(-2, \ )$

Find the intercepts for each equation.

11) $2x + 3y = 6$ 12) $7x + 2y = 14$ 13) $3x - 5y = 9$ 14) $6x - 5y = 12$

15) $2y = 5x$ 16) $x = -3y$ 17) $-2x = 8$ 18) $6y = 12$

Graph each linear equation.

19) $x - y = 2$ 20) $x + y = 6$ 21) $y = x + 4$ 22) $y = x - 5$

23) $x + 2y = 6$ 24) $3x - y = 6$ 25) $4x = 3y - 12$ 26) $5x = 2y - 10$

27) $3x = 6 - 2y$ 28) $2x + 3y = 12$ 29) $2x - 7y = 14$ 30) $3x + 5y = 15$

31) $3x + 7y = 14$ 32) $6x - 5y = 30$ 33) $y = 2x$ 34) $y = -3x$

35) $y + 6x = 0$ 36) $x - 4y = 0$ 37) $x + 2 = 0$ 38) $y - 3 = 0$

39) $y = -1$ 40) $x = 2$ 41) $x = 0$ 42) $y = 0$

7.3 LINES AND THEIR SLOPES

The slope of a line is a measure of the slant of a line. Given a line passing through points (x_1, y_1) and (x_2, y_2), the slope m of the line through the points (x_1, y_1) and (x_2, y_2), is given by

$$m = \frac{y_2 - y_1}{x_2 - x_1} \text{ or } m = \frac{y_1 - y_2}{x_1 - x_2}$$

Provided $x_1 \neq x_2$. (If $x_1 = x_2$, either the two points coincide or the slope is undefined.)

EXAMPLE 1

Find the slope of the line through the points (1, 3) and (5, 4).

SOLUTION

We use the formula for slope with $(x_1, y_1) = (1, 3)$ and $(x_2, y_2) = (5, 4)$. Therefore,

$$m = \frac{4 - 3}{5 - 1} = \frac{1}{4}$$

EXAMPLE 2

Find the slope of the line through the points (1, −5) and (−2, 3).

SOLUTION
According to the formula,

$$m = \frac{3 - (-5)}{-2 - (-1)} = -8$$

slope of a line from its equation,

Step 1 Solve the equation for y
Step 2 The slope is given by the coefficient of x.

Since the slope of a line is always given by the coefficient of x (when the equation is solved for y), any two ordered pairs that satisfy the equation can be used to find the slope of the line. The result is always the same slope.

EXAMPLE 3

Find the slope of the line $2x - 5y = 4$.

SOLUTION

Solve the equation for y.

$2x - 5y = 4$

$-5y = -2x + 4$

$y = \dfrac{2}{5}x - \dfrac{4}{5}$

The slope is given by the coefficient of x, so the slope $m = \dfrac{2}{5}$.

EXAMPLE 4

Find the slope of the line $8x + 4y = 1$

SOLUTION

Solve the equation for y.

$8x + 4y = 1$

$4y = -8x + 1$

$y = -2x + \dfrac{1}{4}$

The slope of this line is given by the coefficient of $m = -2$.

EXAMPLE 5

Find the slope of the line containing the points whose coordinates are $(-2, 2)$ and $(4, 2)$.

SOLUTION

Let $P_1 = (-2, 2)$ and $P_2 = (4, 2)$.

$$m = \frac{y_2 - y_1}{x_2 - x_1} = \frac{2 - 2}{4 - (-2)} = \frac{0}{6} = 0$$

A horizontal line has a zero slope.

316

EXAMPLE 6

Find the slope of the line through (6, 2) and (6, −9).

SOLUTION

$$\text{slope} = \frac{-9-2}{6-6} = \frac{-11}{0} \qquad \text{Undefined}$$

Division by 0 is not possible, so the slope undefined. Since all points on a vertical line have the same *x*-value, the slope of any vertical line is undefined.

Horizontal lines, with equations of the form $y = k$, have a slope of 0.
Vertical lines, with equations of the form $x = k$, have an undefined slope.

EXAMPLE 7

Find the slope of the line through (−8, 4) and (2, 4).

SOLUTION

Use the definition of slope.

$$= \frac{4-(4)}{2-(-8)} = \frac{0}{10} = 0$$

Since all points on the line have the same *y*-value, the slope of any horizontal line is 0.

Parallel and Perpendicular Lines	Two lines with the same slope are parallel; two lines that have slopes with $(x_1, y_1) = (1,3)$ *and* $(x_2, y_2) = (5,4)$ a product of −1 are perpendicular.

EXAMPLE 8

Decide whether each pair of lines is parallel, perpendicular, or neither.

a) $x + 2y = 7$ b) $3x - y = 4$ c) $4x + 3y = 6$

    $-2x + y = 3$ $6x - 2y = 9$ $2x - y = 5$

SOLUTION

a) Find the slope of each line by first solving each equation for y:

$$x + 2y = 7 \qquad\qquad -2x + y = 3$$
$$2y = -x + 7 \qquad\qquad y = 2x + 3$$
$$y = -\frac{1}{2}x + \frac{7}{2}$$

Slope is $-\dfrac{1}{2}$ Slope is 2.

Since the slopes are not equal, the lines are not parallel. Check the product of the slopes: $(-\dfrac{1}{2})2 = -1$. The two lines are perpendicular because the product of their slopes is -1.

b) Find the slopes. Both lines have a slope of 3, so the lines are parallel.

c) Here the slopes are $-\dfrac{4}{3}$ and 2, these lines are neither parallel nor perpendicular.

318

EXERCISE 7.3

Find the slope of the line going through each pair of points. Round to the nearest thousandth in exercises 1 through 11.

1) $(-4, 1), (2, 8)$ 2) $(3, 7), (5, 2)$ 3) $(-1, 2), (-3, -7)$ 4) $(5, -4), (-5, -9)$

5) $(8, 0), (0, 5)$ 6) $(0, -3), (2, 0)$ 7) $(-1, 6), (4, 6)$ 8) $(5, 3), (5, -2)$

9) $(-9, 1), (-9, 0)$ 10) $(1.23, 4.80), (2.56, -3.75)$

11) $(0.03, 1.57), (3.54, -2.01)$

Find the slope of each of the following lines.

12) $y = 5x + 2$ 13) $y = -x + 4$ 14) $y = x + 1$ 15) $y = 6 - 5x$

16) $y = 3 + 9x$ 17) $2x + y = 5$ 18) $4x - y = 8$ 19) $-6x + 4y = 1$

20) $3x - 2y = 5$ 21) $2x + 5y = 4$ 22) $9x + 7y = 5$ 23 $y + 4 = 0$

Decide whether each pair of lines is parallel, perpendicular, or neither.

24) $\begin{array}{l} x + y = 5 \\ x - y = 1 \end{array}$ 25) $\begin{array}{l} y - x = 3 \\ y - x = 5 \end{array}$ 26) $\begin{array}{l} y - x = 4 \\ y + x = 3 \end{array}$ 27) $\begin{array}{l} 2x - 5y = 4 \\ 4x - 10y = 1 \end{array}$

28) $\begin{array}{l} 3x - 2y = 4 \\ 2x + 3y = 1 \end{array}$ 29) $\begin{array}{l} 3x - 5y = 2 \\ 5x + 3y = -1 \end{array}$ 30) $\begin{array}{l} 4x - 3y = 4 \\ 8x - 6y = 0 \end{array}$ 31) $\begin{array}{l} 3x - 2y = 4 \\ 2x + 3y = 1 \end{array}$

32) $\begin{array}{l} 8x - 9y = 2 \\ 3x + 6y = 1 \end{array}$ 33) $\begin{array}{l} 5x - 3y = 8 \\ 3x + 5y = 10 \end{array}$ 34) $\begin{array}{l} 6x + y = 12 \\ x - 6y = 12 \end{array}$ 35) $\begin{array}{l} 2x - 5y = 11 \\ 4x + 5y = 2 \end{array}$

Find the slope of the following lines:

36) $\frac{2}{3}y = \frac{5}{4}x - 3$ 37) $\frac{3}{4}y - \frac{2}{5}x = 6$ 38) $\frac{y}{2} + \frac{x}{4} = 12$ 39) $\frac{5}{4}x + \frac{1}{4}y = -3$

7.4 GENERAL FORM OF THE EQUATION OF A LINE

Any linear equation that is written in the form $Ax + By = C$, where A, B, and C are constants, is said to be written in **general form**.

In the previous section, it was assumed that the graphs of linear equations are straight lines. This section develops various forms for the equation of a line and shows that all these forms, including $f(x) = ax + b$. have a straight-line graph.

Point-Slope Form

The equation of the line passing through $P(x_1, y_1)$ and with slope m is
$$y - y_1 = m(x - x_1).$$

EXAMPLE 1

Find the equation of the line with slope $\dfrac{1}{3}$, going through the point $(-2, 5)$.

SOLUTION

Use the point-slope form of the equation of a line, with $(x_1, y_1) = (-2, 5)$ $m = \dfrac{1}{3}$.

$$y - y_1 = m(x - x_1)$$

$$y - 5 = \frac{1}{3}\left[x - (-2)\right] \qquad \text{Let } y_1 = 5, \ m = \frac{1}{3}, \ x_1 = -2.$$

$$y - 5 = \frac{1}{3}(x + 2)$$

$$3y - 15 = x + 2 \qquad \text{Multiply by 3.}$$

or $\qquad\qquad x - 3y = -17 \qquad \text{Combine terms.}$

EXAMPLE 2

Find an equation of the line through the points $(-4, 3)$ and $(5, -7)$.

SOLUTION

First find the slope, using the definition.

$$m = \frac{-7 - 3}{5 - (-4)} = -\frac{10}{9}$$

321

Use either (−4, 3) or (5, −7) as (x_1, y_1) in the point-slope form of the equation of a line. If (−4, 3) is used, then $-4 = x_1$ and $3 = y_1$.

$$y - y_1 = m(x - x_1) \qquad \text{Point-slope form.}$$

$$y - 3 = -\frac{10}{9}[x - (-4)] \qquad \text{Let } y_1 = 3, m = -\frac{10}{9}, x_1 = -4.$$

$$y - 3 = -\frac{10}{9}(x + 4)$$

$$9(y - 3) = -10(x + 4) \qquad \text{Multiply by 9.}$$

$$9y - 27 = -10x - 40 \qquad \text{Distributive property.}$$

$$10x + 9y = -13 \qquad \text{Standard form.}$$

Slope-Intercept Form

The equation of the line with slope m and y-intercept (0, b) is called the **slope-intercept form**,

$$y = mx + b$$

EXAMPLE 3

Use the slope-intercept form to write the equation of the line with slope 4 that passes through the point $P(5, 9)$.

SOLUTION

Since we are given that $m = 4$ and that the ordered pair (5, 9) satisfies the equation, we can substitute 5 for x, 9 for y, and 4 for m in the equation $y = mx = b$ and solve for b.

$$y = mx + b$$

$$9 = 4(5) + b \qquad \text{Substitute 9 for } y, 4 \text{ for } m, \text{ and } 5 \text{ for } x.$$

$$9 = 20 + b \qquad \text{Simplify.}$$

$$-11 = b \qquad \text{Subtract 20 from both sides.}$$

Because $m = 4$ and $b = -11$, the equation is $y = 4x - 11$.

EXAMPLE 4

Find an equation of the line with slope $-\dfrac{4}{5}$ and y-intercept (0, −2) in standard form.

SOLUTION

Here $m = -\dfrac{4}{5}$ and $b = -2$. Substitute these values into the slope-intercept form.

$$y = mx + b$$
$$y = -\frac{4}{5}x - 2$$

$5y = -4x - 10 \qquad$ Multiply by 5.

$4x + 5y = -10 \qquad$ Standard form.

EXERCISE 7.4

Write an equation for each line given its slope and *y*-intercept.

1) $m = 3$, *y*-intercept 5   2) $m = -2$, *y*-intercept 4   3) $m = -1$, *y*-intercept -6

4) $m = \dfrac{5}{3}$, *y*-intercept $\dfrac{1}{2}$ 5) $m = \dfrac{2}{5}$, *y*-intercept $-\dfrac{1}{4}$ 6) $m = 8$, *y*-intercept 0

7) $m = 0$, *y*-intercept -5 8) $m = -2.15$, *y*-intercept .832

9) $m = 4.61$, *y*-intercept -2.38

Write an equation for the line passing through the given point and having the given slope. Write the equation in the form $Ax + By = C$.

10) (1, 4), $m = -4$ 11) (2, −8), $m - 2$ 12) (−1, 7), $m = -4$

13) (3, 5), $m = \dfrac{2}{3}$ 14) (2, −4), $m = \dfrac{4}{5}$ 15) (−3, −2), $m = -\dfrac{3}{4}$

16) *x*-intercept −8, $m = -\dfrac{5}{9}$ 17) *x*-intercept 6, $m = -\dfrac{8}{11}$

Write equations of the lines passing through each pair of points. Write the equation in the form $Ax + By = C$.

18) (7, 4), (8, 5)   19) (−2, 1), (3, 4)   20) (−8, −2), (−1, −7)

21) (3, −4), (−2, −1)   22) (−7, −5), (−9, −2)   23) (0, 2), (3, 0)

24) (4, 0), (0, −2)   25) (2, −5), (−4, 7)

Summary

The slope of the line through the points (x_1, y_1) and (x_2, y_2) is

$$m = \frac{change\ in\ y}{change\ in\ x} = \frac{y_2 - y_1}{x_2 - x_1}$$

Linear Equation		
	$Ax + By = C$	General form (Neither A nor B is 0) Slope is $-\dfrac{A}{B}$ x-intercept is $\dfrac{C}{A}$ y-intercept is $\dfrac{C}{B}$
	$x = k$	Vertical line Slope is undefined x-intercept is k
	$y = k$	Horizontal line Slope is 0 y-intercept is k
	$y = mx + b$	Slope-intercept form Slope is m y-intercept is b
	$y - y_1 = m(x - x_1)$	Point-slope form Slope is m Line goes through (x_1, y_1)

CHAPTER 7 REVIEW EXERCISES

Decide whether the given ordered pair is a solution of the given equation.

1) $x + y = 7$; $(3, 4)$ 　　　2) $2x + y = 5$; $(1, 4)$ 　　　3) $x + 3y = 9$; $(1, 3)$

4) $2x + 5y = 7$; $(1, 1)$ 　　　5) $3x - y = 4$; $(1, -1)$ 　　　6) $5x - 3y = 16$; $(1, -2)$

Complete the given ordered pairs for each equation.

Equation	Ordered Pairs
7) $y = 3x - 2$	$(-1, \ )$ $(\ , -2)$ $(\ , 5)$
8) $2y = 4x + 1$	$(0, \ )$ $(\ , 0)$ $(\ , 2)$
9) $x + 4 = 0$	$(\ , -3)$ $(\ , 0)$ $(\ , 5)$
10) $y - 5 = 0$	$(-2, \ )$ $(0, \ )$ $(8, \ )$

Graph each linear equation.

11) $y = 2x + 3$ 　　　12) $x + y = 5$ 　　　13) $2x - y = 5$ 　　　14) $x + 2y = 0$

15) $y + 3 = 0$ 　　　16) $x - y = 0$

Find the intercepts for each equation.

17) $y = 2x - 5$ 　　　18) $2x + y = 7$ 　　　19) $5x - 2y = 0$

Find the slope of each line.

20) Through $(2, 3)$ and $(-1, 1)$ 　　　21) Through $(0, 0)$ and $(-1, -2)$.

22) Through $(2, 5)$ and $(2, -2)$ 　　　23) $y = 3x - 1$

24) $y = 8$ 　　　25) $x = 2$ 　　　26) $5x - 2y = 3$

Decide whether each pair of lines is parallel, perpendicular, or neither.

27) $\begin{array}{l} 3x + 2y = 5 \\ 6x + 4y = 12 \end{array}$ 　　28) $\begin{array}{l} x - 3y = 8 \\ 3x + y = 6 \end{array}$ 　　29) $\begin{array}{l} 4x + 3y = 10 \\ 3x - 4y = 12 \end{array}$ 　　30) $\begin{array}{l} x - 2y = 3 \\ x + 2y = 3 \end{array}$

Write an equation for each line in the form of $Ax + By = C$.

31) $m = 3$, y-intercept -2 　　　32) $m = -1$, y-intercept $\dfrac{3}{4}$

33) $m = \dfrac{2}{3}$, y-intercept 5 　　　34) Through $(5, -2)$, $m = 1$

35) Through $(-1, 4)$, $m = \dfrac{2}{3}$

36) Through $(1, -1)$, $m = \dfrac{-3}{4}$

37) Through $(5, -2)$ and $(-2, 2)$

38) Through $(-2, 6)$ and $(3, 6)$

CHAPTER 7 TEST

Choose the best answer.

Find the two ordered pairs that are both solutions of the given equation.
1) $5x = 2y + 6$

    a) $(0, 3)$ and $(4, 7)$ b) $(\frac{6}{5}, 0)$ and $(6, -12)$

    c) $(0, -3)$ and $(2, 2)$ d) None of these

2) $x - 4 = 0$

    a) $(4, 0)$ and $(4, 4)$ b) $(-4, 0)$ and $(-4, 1)$

    c) $(4, 2)$ and $(0, 0)$ d) None of these

Graph each linear equation.
3) $2x - 3y = 6$

a) 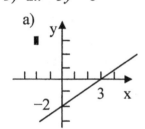   b) 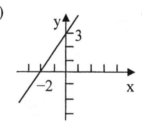   c)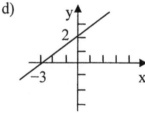

d)

4) $x - 2y = -4$

a) 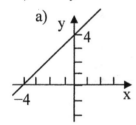   b) 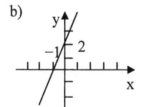   c) 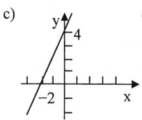   d)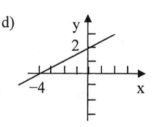

Choose the best answer.
Find the two ordered pairs that are both solutions of the given equation.
5) $3x - 2y = 7$

    a) $(\frac{7}{3}, 0)$ and $(-1, 2)$ b) $(0, \frac{7}{2})$ and $(1, -2)$

    (c) $(0, -\frac{7}{2})$ and $(1, -2)$ d) None of these

6) $x = -4y$

    a) $(4, 1)$ and $(0, 0)$ b) $(1, -4)$ and $(0, 0)$

329

c) (2; −1) and (−4, 1)   d) None of these

7) $x + 5 = 0$
   a) (−5, 0) and (−5, −5)   b) (0, −5) and (−3, −5)
   c) (5, 0) and (5, −5)   d) None of these

Graph each linear equation.

8) $x - y = 4$
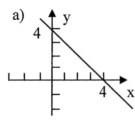
   a)   b)   c)   d)
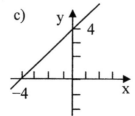
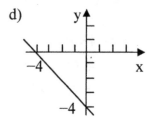

9) $3x + 4y = 12$
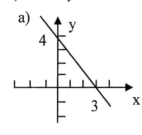
   a)   b)   c)   d)
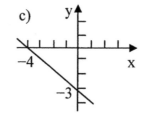
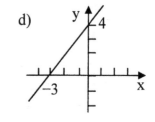

10) $4y - x = 0$
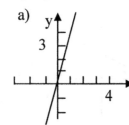
   a)   b)   c)   d)
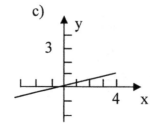
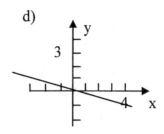

11) $x + 3 = 0$
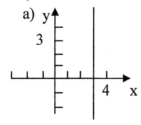
   a)   b)   c)   d)
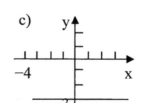
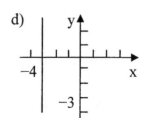

Find the slope of each line (12 and 13).

12) Through $(6, -3)$ and $(-4, -1)$

    a) $-\dfrac{2}{5}$ b) $-\dfrac{1}{5}$ c) $\dfrac{1}{5}$ d) None of these

13) $y - 3 = 0$

    a) 0   b) Undefined   c) 3   d) -3

Graph each linear equation (14 and 15).

14) $3x + y = 0$

a)   b)   c)   d)

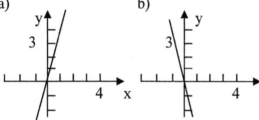 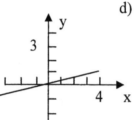 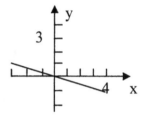

15) $y + 2 = 0$

a)   b)   c)   d)

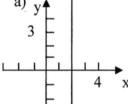 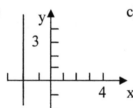 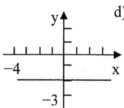 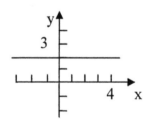

Find the slope of each line.

16) Through $(-4, 6)$ and $(2, -3)$

    a) $\dfrac{3}{2}$ b) $-\dfrac{2}{3}$ c) $-\dfrac{3}{2}$ d) None of these

17) $3x = -2y$

    a) 0   b) Undefined   c) $-\dfrac{2}{3}$ d) $-\dfrac{3}{2}$

18) $4x + 7y = 28$

    a) $-\dfrac{4}{7}$ b) $\dfrac{4}{7}$ c) $\dfrac{7}{4}$ d) $-\dfrac{7}{4}$

For each line, write an equation in the form $Ax + By = C$.

19) Through $(-3, -5)$; $m = \dfrac{3}{4}$

    a) $3x - 4y = -3$ b) $3x - 4y = -11$ c) $3x - 4y = 11$ d) $4x - 3y = 3$

20) $m = \dfrac{1}{7}$, y-intercept $(0, 3)$

    a) $x - 7y = 3$ b) $7x - y = 21$ c) $x - 7y = -3$ d) None of these

21) Through $(-3, -4)$ and $(1, 0)$

    a) $x + y = -7$ b) $x - y = 1$ c) $x - y = -1$ d) $x + y = 1$

22) Through $(-6, 4)$; $m = \dfrac{2}{3}$

    a) $2x - 3y = 0$ b) $2x - 3y = -24$ c) $3x - 2y = -24$ d) None of these

23) Slope undefined; x-intercept $(-4, 0)$

    a) $x = -4$ b) $y = -4$ c) $x + y = -4$ d) $x = 4$

24) Through $(-7, 2)$ and $(-4, 5)$,

    a) $x + y = -5$ b) $3x - 11y = -43$ c) $3x - 11y = 1$ d) $x - y = -9$

8 FUNDAMENTALS AND APPLICATIONS OF GEOMETRY

CHAPTER

In this chapter we will study the arithmetic applications of common geometric figures as they relate to applications in these fields.

The word geometry means "measurement of the earth." The original purpose of geometry was to measure land. Today, geometry is used in various fields such as chemistry, geology, and physics.

8.1 FUNDAMENTAL CONCEPTS

A **line** is made up of points. A line extends indefinitely in both directions.

Shown above is line $\overleftrightarrow{EF}$.

A **ray** (see $\overrightarrow{KL}$ above) is a part of a line that extends indefinitely in one direction.

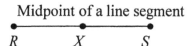

A **line segment** is part of a line that consists of two points and all the points in between those two points.

The **midpoint of a line segment** is the point that divides a line segment into two equal parts. In the figure above, X is the midpoint of the line segment $\overline{RS}$ since RX = XS.

A **plane** is a set of points forming a flat surface with no end in any direction. This is read plane BCD.

Parallel lines are two lines in a plane that do not intersect.

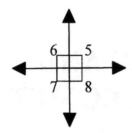

Perpendicular lines are lines that intersect to form right angles. The figure above is an example of perpendicular lines. ∠5, ∠6, ∠7, and ∠8 are right angles.

Angles

An **angle** is a set of points consisting of two rays with the same endpoint. The common endpoint is called the **vertex** of the angle and the two rays are called **sides** of the angle.

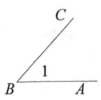

You can see, in the figure above, that B is the vertex. The symbol for the word angle is ∠. An angle can be named in one of three ways:

1. By the point at the vertex: ∠B.
2. By three letters, with the vertex in the middle: ∠ABC or ∠CBA.
3. By the number inside the angle: ∠1.

Classification of Angles

Angles are measured in degrees (°). Angles may be classified based upon the number of degrees they contain.

> **Right angles:** Angles containing 90°.
>
> **Straight angles:** Angles containing 180°.
>
> **Acute angles:** An angle whose degree measure is greater than 0 but less than 90°.
>
> **Obtuse:** An angle whose degree measure is greater than 90 but less than 180°.

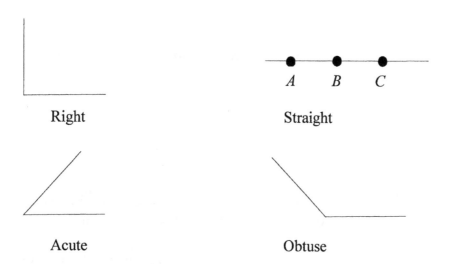

Right

Straight

Acute

Obtuse

Polygons

A **polygon** is a simple closed curve formed by line segments such that no two endpoints are part of the same line.

The line segments are called sides of the polygon and the points of the intersection of the sides are the vertices (plural of vertex) of the polygon.

Some common polygons are:

Number of Sides	Name of Polygon
3	triangle
4	quadrilateral
5	pentagon
6	hexagon
7	heptagon
8	octagon
9	nonagon
10	decagon

Triangles

A **triangle** is a figure that has three sides. One way to classify triangles is by the lengths of the sides.

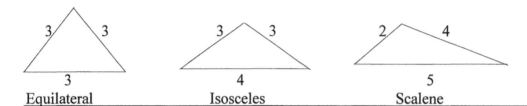

Equilateral   Isosceles   Scalene

Equilateral: All three sides are equal and all angles are equal.

Isosceles: Two sides are equal and the angle opposite the equal sides is equal.

Scalene: No two sides have the same length and no two angles are equal.

Equiangular: All angles are equal.

Obtuse: One angle of the triangle is between 90 and 180 degrees.

Acute: One angle of the triangle is between 0 and 90 degrees.

EXERCISE 8.1

Complete the following sentences/statements.

1) Lines in a plane that are always the same distance apart are called _____.

2) A right triangle has a 60° angle. Find the measure of the other angles.

3) Name the quadrilateral in which opposite sides are parallel and equal.

4) Name the solid in which all the faces are rectangles.

5) Name the parallelogram with four right angles.

6) Name the rectangles with four equal sides.

7) Name the solid in which all points are the same distance from the center.

8) A _____ is formed by two rays with a common endpoint.

9) Part of a line that extends indefinitely in one direction. _____

10) The measure of a right angle is _____.

11) The measure of a straight angle is _____.

12) Two lines that intersect to form right angles are _____.

13) An isosceles triangle has _____ equal sides.

14) A triangle with all equal angles is said to be _____.

15) A polygon with six sides is called a _____.

16) Angles having measures greater than 0 and less than 90 degrees are called _____.

Classify the following angles as right, straight, acute, or obtuse.

17)

18)

19)

20) $\angle ABC = 66°$

21) $\angle ABC = 117°$

22) $\angle ABC = 180°$ 23) $\angle ABC = 90°$

Classify the triangles as equilateral, isosceles, or scalene.

24)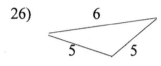
9
4
7

25)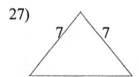
5
2
5

26)
6
5   5

27)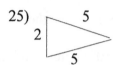
7   7

8.2 ANGLE MEASUREMENTS

Pairs of angles whose sum equals 90 degrees or 180 degrees have special names.

Complementary angles are two angles whose sum equals 90°.

In the figure above, $\angle 1$ and $\angle 2$ are complementary since their sum equals 90°. Also, two angles with measures of 47° and 43° respectively are complementary.

Supplementary angles are two angles whose sum equals 180°.

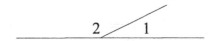

$\angle 1$ and $\angle 2$ are supplementary. Also, two angles of 20° and 160° are supplementary since their sum is 180°.

EXAMPLE 1

Find the complement of a 42° angle.

SOLUTION

The sum of two complementary angles is 90°. Since one angle is 42°, the other angle is 90° − 42° = 48°.

EXAMPLE 2

Two angles are supplementary and one is 20° more than three times the other. Find the two angles.

SOLUTION

a) Represent the two angles.

  Let x = the smaller angle.
  $3x + 20$ = the larger angle since it is 20° more than three times the smaller.

b) Write an equation for the problem.

The sum of two supplementary angles is 180°, thus the sum of the smaller and larger must equal 180°

$$x + (3x + 20) = 180^\circ$$

c) Solve the equation.

$$\begin{aligned} x + (3x + 20) &= 180^\circ \\ 4x + 20^\circ &= 180^\circ \\ 4x &= 160^\circ \\ x &= 40^\circ \text{ (smaller angle)} \end{aligned}$$

The larger angle equals $3x + 20$, substituting gives
$$3(40^\circ) + 20^\circ = 140^\circ \text{ (larger angle)}$$

Angle Measurements of Interior and Exterior Angles

Angles and their measurements are important in the study of triangles. Over the years, many properties have been discovered. One of the most basic of these properties is the following:

> The sum of the measures of the interior angles in any triangle is 180°.

EXAMPLE 3

Find the missing angle measure in the triangle below.

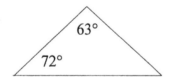

SOLUTION

The sum of the interior angles of a triangle is 180°. Since the sum of the two angles, 72° + 63° is 135°, the third angle must be 45° ($180 - 135 = 45$).

Exterior Angles

Another property of triangles states the following:

> The measure of an exterior angle of a triangle equals the sum of the measures of the two opposite interior angles.

In the figure below, angle *ABD* is not an angle within triangle ABC.

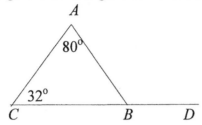

Since one of the sides of the angle is a side of the triangle and the other side is an extension of one side of the triangle, angle ABD is an exterior angle of the triangle.

In the previous triangle, angle A has a measure of $80°$ and angle C has a measure of $32°$. Thus, the exterior angle ABD has a measure of $32° + 80° = 112°$.

EXAMPLE 4

Find the value of *x* in the following figure.

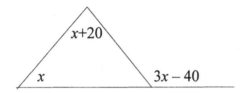

SOLUTION

By definition, the measure of the exterior angle $(3x - 40)$ equals the sum of the measure of the opposite interior angles (x and $x + 20$). Combining the interior angles and equating the sum to the exterior angle gives,

$x + (x + 20) = 3x - 40$	Remove parentheses and combine like terms.
$2x + 20 = 3x - 40$	Subtract $2x$ from both sides of the equation.
$20 = x - 40$	Add 40 to both sides of the equation.
$60 = x$	One angle.

The other angles are: $x + 20 = 60 + 20 = 80$

$$3x - 40 = 3(60) - 40 = 180 - 40 = 140.$$

EXERCISE 8.2

Perform the indicated operations.

1) How many degrees are in one complete revolution?

2) Two lines that intersect and form right angles are called what kind of lines?

3) Find the supplement of a $162°$ angle.

4) Find the complement of a $68°$ angle.

5) Find the supplement of a $128°$ angle.

6) Find the complement of a $72°$ angle.

7) A triangle has $60°$ and $104°$ angles. Find the measure of the other angle.

Find the number of degrees in both the complement and supplement of $\angle DEF$, if:

8) $\angle DEF = 60°$ 9) $\angle DEF = 22\frac{1}{2}°$ 10) $\angle DEF = 15.5°$

Find the measure of the missing angle in each of the following.

11) One angle of a right triangle is $37°$. What is the measure of the other two angles?

12) A triangle has angles of $36°$ and $42°$. What is the measure of the third angle?

13) The complement of a $57°$ angle is ___?

14) The supplement of a $123°$ angle is ___?

In each of the following figures, find the measure of the third angle.

15)
45°   58°

16)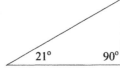
21°   90°

Find all indicated angles in each of the following figures.

17)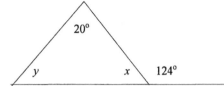

18)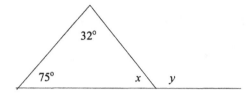

19)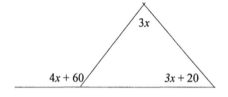

20)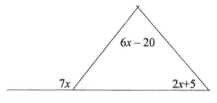

PERIMETER OF GEOMETRIC FIGURES

The sum of the distance around a plane geometric figure is called its perimeter. The formulas for the perimeter of several common geometric figures are given below.

NAME	FORMULA	FIGURE
RECTANGLE	$P = 2L + 2W$	
SQUARE	$P = 4S$	
TRIANGLE	$P = a + b + c$	

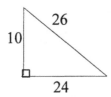

The following problems will illustrate how we can apply the formulas for the perimeters of certain geometric figures.

Try to sketch the geometric figure before attempting to work the examples.

EXAMPLE 1

Find the perimeter of the right triangle below.

SOLUTION

The formula for the perimeter of a triangle is $P = a + b + c$
$$= 10 + 24 + 26$$
$$= 60$$

EXAMPLE 2

Calculate the perimeter of a rectangle 17 feet long and 15 feet wide.

17 ft

15 ft

SOLUTION

The formula for the perimeter of a rectangle is $P = 2L + 2W$

Substituting, $\qquad P = 2(17 \text{ feet}) + 2(15 \text{ feet})$
$$= 34 \text{ feet} + 30 \text{ feet}$$
$$= 64 \text{ feet}.$$

EXERCISE 8.3

Solve each of the following problems.

1) Find the perimeter of the triangle with sides measuring 23 centimeters, 16 centimeters, and 23 centimeters.

2) Calculate the perimeter of a rectangle whose length measures 5 feet and width 3 feet.

3) Find the perimeter of a square whose sides measure 5.8 inches.

4) If the distance between the bases of a little-league baseball field diamond is 60 feet, how many yards does a batter run when he or she hits a home run?

5) Find the perimeter of a triangle whose sides measure 4 feet, 12 feet, and 13 feet.

6) Find the length of one side of a square whose perimeter is 25 feet.

7) One side of a square room is 9 yards long. Find the number of yards of molding needed to go around the room.

8) The length of a rectangular table is 10 feet and the width is 3 feet. Find the perimeter of the table.

9) What is the perimeter in inches of a square which measures 4 feet on each side?

10) A man wants to fence in a square flower bed measuring 4 feet on each side and a rectangular flower bed 24 inches wide and 96 inches long. How many feet of the fence will he need?

11) Calculate the perimeter of a rectangle whose length measures 5 feet and width 3 feet.

12) Pat Brown wants to put weather-stripping around the two windows in her kitchen. Each window is 36 inches wide and 42 inches tall. How many feet of weather-stripping does she need?

13) Calculate the perimeter of a square with one side 6 cm long.

14) Calculate the perimeter of an equilateral triangle with one side 12" long.

8.4 AREAS OF GEOMETRIC FIGURES

The **area** of any geometric figure is the number of square units contained in the surface. Observe the figure below. There are a total of 15 square units in the figure; therefore, its area is 15. Square feet, square inches, square yards, and square miles are the common measures of area in our English measurement system. The common measures in our metric system are square centimeters and square meters. Always express all linear units in the same denominations when computing the area of geometric figures.

The formulas for the areas of several common geometric figures are given below.

NAME	FORMULA	FIGURE
Rectangle	$A = LW$	W ☐ L
Square	$A = S^2$	☐ S
Parallelogram	$A = bh$	h
Triangle	$A = \dfrac{1}{2}bh$	h b
Trapezoid	$A = \dfrac{(B+b)h}{2}$	b h B

EXAMPLE 1

Find the area of a square whose side is 40 centimeters.

SOLUTION

The formula for the area of a square is $A = s^2$,

Substituting $s = 40$ in the formula gives $A = (40)^2 = 1600$ square centimeters.

349

EXAMPLE 2

Find the cost for a rug of 9 feet by 12 feet for a room if the cost per square yard is $8.75.

SOLUTION

First calculate the area in square feet, and then convert to square yards.

The room is rectangular, so $A = LW = 9 \times 12 = 108$ square feet.

To convert to square yards, divide by 9, thus 108 sq. feet $\div 9 = 12$ sq. yards.

Cost of the rug = number of square yards × cost per square yard

$$= \quad 12 \quad \times \quad \$8.75 \quad = \quad \$105.00$$

EXAMPLE 3

Find the area of the trapezoid with base 20 feet and 16 feet having a height of 15 feet.

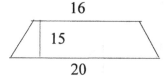

SOLUTION

The formula for the area of a trapezoid is $A = \dfrac{(B+b)h}{2}$. A trapezoid has two bases, the longer side is big "B" and the shorter side is little "b".

Substituting $B = 20$, $b = 16$, and $h = 15$, gives $A = \dfrac{(20+16)15}{2} = \dfrac{(36)(15)}{2}$

$$= 18 \times 15 = 270 \text{ square feet.}$$

EXAMPLE 4

A concrete walk is built around a pool as shown in the figure. The dimensions of the pool are 14 ft. × 34 feet. What is the area of the walk?

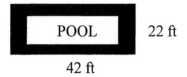

POOL   22 ft

42 ft

SOLUTION

The sketch shows that the figure is made up of two rectangles. To find the area of the walk, we find the area of the larger rectangle and subtract the area of the pool. The remainder is the area of the walk.

Area of larger rectangle = 42 ft × 22 ft = 924 square feet

Area of pool = 14 ft × 34 ft = 476 square feet

Area of walk = Area of larger rectangle minus area of pool
          =   924 square feet   −   476 square feet = 448 square feet

EXERCISE 8.4

Find the area of the following problems (1 – 7).

1) A rectangle 15 inches long and 11 inches wide.

2) A rectangle 9 ft. wide and 17 ft. long.

3) A square 5 feet on each side.

4) A square 9 inches on each side.

5) A triangle with base 5 and height 10.

6) A triangle with base 14 and height 8.

7) A trapezoid having 8 inches as the height, 9 inches as the upper base and 18 inches as the lower base.

8) What is the area of a field shaped like a parallelogram having 140 inches as its base and 155 inches as its height?

9) Find the cost of cementing a runway 10 feet by 15 feet at $3.85 per square feet.

10) Find the cost of a carpet covering a floor 25 feet by 25 feet at $12.25 per square yard.

11) A rectangle has an area of 87 square feet and a width of 5 feet. Calculate the length.

12) Find the area of a rectangle with a length 12 feet and width 13 feet.

13) Find the cost of plastering the walls of a house 22 feet long, 25 feet 6 inches wide and 8 feet high. Subtract 120 square feet for windows and doors. The cost is $2.75 per square foot.

14) Mrs. Zimmerman is putting new wood flooring in the spare room of her house. The room is 9 feet wide and 12 feet long. How much flooring will she need for the room?

15) Bennie wants to put 2-foot by 2-foot carpet tiles on the floor of his living room. The room is 18 feet long and 10 feet wide. Find the smallest number of carpet tiles needed to completely cover the room.

16) A gallon of floor paint will cover about 200 square feet of concrete. The floor of Charles Ervin's basement is 25 feet wide and 40 feet long. How many gallons of paint does he need to put one coat of paint on the basement floor?

17) Calculate the area of the circle with a diameter of 12 inches.

18) Calculate the area of a triangle with base 4 centimeters and height 10 centimeters.

8.5 CIRCLES AND COMPOSITE FIGURES

A circle is the set of all points in a plane that are an equal distance from a fixed point. The **diameter** (*d*) (Figure 1) of a circle is a line that crosses a circle through its center. The **radius** (*r*) (Figure 2) of a circle is a line from the center of the circle to the curve of the circle. **The radius is half the diameter**.

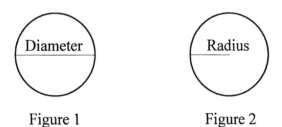

Figure 1   Figure 2

The **circumference** (*C*) is the distance around a circle. The formulas for finding the circumference and the area of circles contain a special number that is represented by the Greek letter π (pi). In calculations we let π equal to 3.14. Pi is obtained by dividing the circumference of the circle by its diameter. We may use the value $\frac{22}{7}$ or π.

Formulas for area and circumference of a circle:

Circumference	Area
$C = \pi d$ or $C = 2\pi r$	$A = \pi r^2$

The examples below illustrate how to find the area and circumference of circles.

EXAMPLE 1

Find the circumference of a circle with a diameter of 21 inches.

Use $\pi = \frac{22}{7}$.

SOLUTION

In the formula for the circumference of a circle, $C = \pi d$, replace π with $\frac{22}{7}$ and *d* with 21 inches.

$$C = \pi d = \frac{22}{7}(21) = 3(22) = 66 \text{ inches.}$$

EXAMPLE 2

Find the area of a circle with a radius of 3 inches. Use $\pi = 3.14$.

SOLUTION

Replace π with 3.14 and r with 3 inches in the formula $A = \pi r^2$.

Thus, $A = 3.14 \times (3)^2 = 3.14 \times 9 = 28.26$ square inches.

EXAMPLE 3

Find the circumference and area of the circle below.

SOLUTION

Circumference   Area

$C = 2\pi r$ $A = \pi r^2$

$\quad = 2(3.14)(5)$ $\quad = (3.14)(5)^2$

$\quad = 31.4$ inches $\quad = 78.5$ square inches

Composite Geometric Figures

Geometric figures made up of two or more geometric figures are called **composite figures**. To find the area or perimeter of these figures, we find the area or perimeter of the different figures that make up the composite and then combine the results for the different geometric figures.

The example below shows how to find the area of a composite figure.

EXAMPLE 4

If the largest possible circular piece is cut from a 2' square piece of wood, how much is waste? Use $\pi = 3.14$.

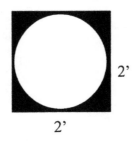

2'

2'

SOLUTION

The shaded area represents the amount of waste.

Step 1. Find the area of the square: $A = s^2 = 2^2 = 4$ square feet.

Step 2. Find the area of the circle: $A = \pi r^2 = 3.14(1)^2 = 3.14$ square feet.
       (NOTE: The radius must be half the diameter or 1')

Step 3. Subtract the area of the circle from the area of the square.
               $4 - 3.14 = .86$ square feet.

EXERCISE 8.5

Solve each problem.

1) Find the circumference and area of a circle whose radius is 6 feet.

2) Find the circumference and area of a circle whose diameter is 14 cm.

3) Find the area and circumference of the circle below.

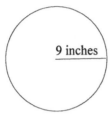

4) Find the area and circumference of the circle below.

5) The unshaded part of the diagram at the right shows the walk around the pools at a local community center. Find the area of the walk.

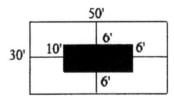

6) The diagram below shows the measurements of the floor of the Hugine's living room and dining room. How many square yards of carpet do they need to cover the floors of both rooms?

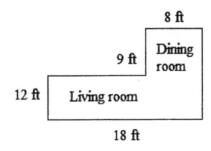

359

7) Find the area of the shaded part of the figure below. Use $\pi = \dfrac{22}{7}$.

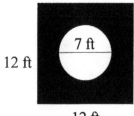

12 ft

7 ft

12 ft

8) Find the area of the figure shown below.

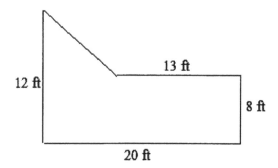

12 ft

13 ft

8 ft

20 ft

9) Find the area of the figure below.

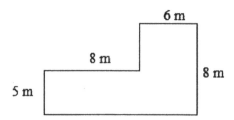

6 m

8 m

8 m

5 m

10) Find the area of the roller skate rink with the following dimensions.

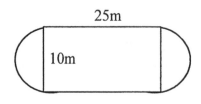

25m

10m

11) Find the area of the region that is not shaded in the figure below.

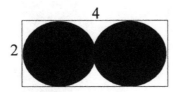

12) Find the area of the shaded region in the figure below.

8.6 SOLID GEOMETRY

The **volume** is the measure of the amount of space inside a three-dimensional figure, called a solid. Illustrations and formulas for finding the volume of common geometric solids are given below.

Rectangular Solid

A **rectangular solid** is a box. The volume of a rectangular box is equal to the length times the width times the height. Formula: $V = LWH$

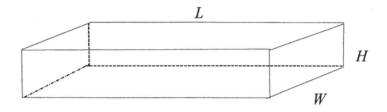

Cube

A **cube** is a rectangular box for which the length, width, and height are the same. Formula: $V = s^3$ (s = any side of the cube)

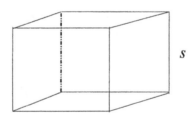

Cylinder

A circular **cylinder** is a geometric figure in which the bottom and top form square corners with the side. The volume of a cylinder is equal to π times the radius squared times the height. You can think of a cylinder as two circles with height in between. Formula: $V = \pi r^2 h$

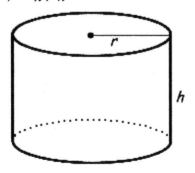

Sphere

A **sphere** is a round object similar to a ball. The volume of a sphere is equal to $\frac{4}{3}$ times π times the cube of the radius. Formula: $V = \frac{4}{3}\pi r^3$.

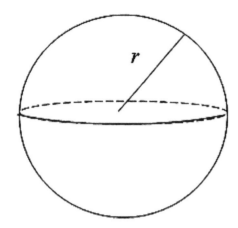

The examples below illustrate how the volume formulas can be applied.

EXAMPLE 1

Find the volume of the rectangular box 8 feet long, 3 feet wide, and 9 feet high.

SOLUTION

A box is a rectangular solid. Using the formula $V = LWH$ and substituting, we get,

$$V = 8 \times 3 \times 9 = 216 \text{ cubic feet}$$

EXAMPLE 2

Calculate the volume of a cube whose side is 15 meters.

SOLUTION

A cube has all equal sides, therefore the formula for the volume of a cube is $V = s^3$. Substituting gives,

$$V = (15)^3 = 3375 \text{ cubic meters.}$$

EXAMPLE 3

Find the volume of a cylinder 40 feet high with its base having a radius of 15 feet.

SOLUTION

We basically calculate the area of a square and multiply this by the height. Substituting in the formula $V = \pi r^2 h$ gives,

$$V = (3.14)(15)^2(40) = 28,260 \text{ cubic feet.}$$

EXAMPLE 4

The base of a rectangular box measures 2 feet by 3 feet. How many feet deep should the box be so that the volume is 24 cubic feet?

SOLUTION

In the formula we know that $V = 24$, $L = 2$, and $W = 3$. To find H, we substitute the known values and solve the equation.

$$V = LWH$$
$$24 = 2 \times 3 \times H$$
$$24 = 6H \qquad \text{Divide both sides by 6.}$$
$$H = 4$$

EXAMPLE 5

How many gallons of water are needed to fill an aquarium that is 16" high, 32" long, and 12" wide? (NOTE: 1 gallon = 231 cubic inches.)

SOLUTION

An aquarium is a rectangular solid. Using the formula for finding the volume,
$V = LWH$, we get

$$V = 16 \times 32 \times 12 = 6144 \text{ cubic inches.}$$

Since each gallon equals 231 cubic inches, we divide 6144 by 231 $\left(6144 \div 231\right)$ to get 26 gallons.

EXERCISE 8.6

Solve the following problems:

1) A cube has one side that measures 1.2 centimeters. Find the volume of the cube.

2) Find the volume of a rectangular box that is $5\frac{1}{4}$ inches long, $3\frac{1}{2}$ inches wide, and $2\frac{1}{4}$ inches high.

3) Find the volume of a cylinder with a radius of 1 meter and a height of 7 meters. Let $\pi = \frac{22}{7}$.

4) Find the volume of a cube that measures 8 inches on one side.

5) Find the volume of a rectangular solid with length 15 inches, width 12 inches, and height 9 inches.

6) How much concrete must be poured to make a slab that is 30 feet long, 5 feet wide, and $\frac{1}{2}$ foot high?

7) Find the volume of a sphere with a radius of 3 inches.

8) A cylindrical water tank has a radius of 9 feet and a height of 35 feet. Find its volume. Let $\pi = \frac{22}{7}$.

9) A rectangular box has dimensions of 6 inches by 6 inches by 10 inches. What would be the diameter of the largest ball that could fit in the box?

10) For the construction of a new building, a hole was dug measuring 30 yards long, 20 yards wide, and 4 yards deep. The trucks used to carry away the dirt can each hold 30 cubic yards. How many truckloads were needed to carry away all the dirt from the hole?

11) Find the volume of a rectangular solid with a length of 17 inches, width of 12 inches, and height of 15 inches.

12) Find the surface area of the rectangular solid in problem 11.
Surface Area Formula $= 2LW + 2LH + 2WH$.

13) Find the volume of a cylinder with a radius of 8 inches and a height of 12 inches.

14) Find the volume of a beach ball with radius of 4 cm.

15) Find the volume of a soup can with radius 3.2 cm and height 9.5 cm.

16) Find the volume of a pork and beans can with a diameter of 1.2 cm and height of 10.5 cm.

17) How many cubic feet of sand are needed to fill a box 15' by 9' by 12' to within 3' of the top?

8.7 RIGHT TRIANGLES

A triangle that has one right angle is called a **right triangle**. Many important relationships and applications are derived from right triangles. One such relationship is expressed in the Pythagorean Theorem.

Hypotenuse

The rule of Pythagorous, (an ancient Greek mathematician), explains the relationship of the sides in a right triangle. The side opposite the right angle is called the **hypotenuse**. The other two sides are called **legs**. It is common convention to let the letter c represent the hypotenuse and the length of the other two legs are identified as a and b, as seen in the figure below.

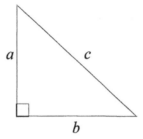

Pythagorean Theorem

The **Pythagorean Theorem** states that, in every right triangle, the square of the hypotenuse is equal to the sum of the squares of the other two legs. The formula derived from this theorem is the following: $c^2 = a^2 + b^2$.

EXAMPLE 1

Find the hypotenuse of a right triangle if one leg is 4 inches and the other is 3 inches.

SOLUTION

From the Pythagorean Theorem $c^2 = a^2 + b^2$, we have $c = ?$, $a = 4$ and $b = 3$.

Substituting, $c^2 = a^2 + b^2$

$$c^2 = 4^2 + 3^2$$

$$c^2 = 16 + 9$$

$$c^2 = 25$$

$$\sqrt{c^2} = \sqrt{25}$$
$$c = 5 \text{ inches}$$

EXAMPLE 2

Calculate the length of the leg represented by *a* in the triangle.

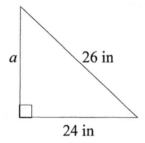

SOLUTION

From the Pythagorean Theorem, $c^2 = a^2 + b^2$, we have in the given triangle, $b = 24$ inches, $c = 26$ inches and a is the unknown. Substituting in the Pythagorean Theorem we get,

$$c^2 = a^2 + b^2$$

$$(26)^2 = a^2 + (24)^2$$

$$676 = a^2 + 576$$

$$100 = a^2 \qquad \text{Subtract 576 from both sides}$$

$$\sqrt{100} = \sqrt{a^2}$$

$$10 \text{ inches} = a$$

EXERCISE 8.7

1) Find the length of the third side of the right triangle where $a = 3$, and $b = 2$.

2) The hypotenuse of a triangle is 5 inches and one side is 4 inches. Find the other side.

3) Find the length of the hypotenuse (correct to the nearest tenth) with $a = 5$ feet and $b = 4$ feet.

4) Decide if the three numbers in each example could be the hypotenuse and length of sides of a right triangle. The greatest number is the length of the hypotenuse.

    a. 24, 10, 26   b. 10, 8, 6

5) Could a triangle with sides 36, 39, and 15 be a right triangle? Please explain.

6) Find the hypotenuse of a right triangle whose sides are 7 centimeters and 9 centimeters, respectively.

7) The foot of a ladder is 5 feet from the side of a building. If the ladder is 12 feet long, how high on the side of the building does the ladder stretch?

8) A truck is driven 9 miles east and then 4 miles south. How far is the truck from the starting point?

9) A ladder 8 meters long is leaning against a house. How high on the house will the ladder reach when the bottom of the ladder is 3 meters from the house?

8.8 SIMILAR TRIANGLES

Similar triangles are triangles that have the same shape but not necessarily the same size. In order for two triangles to be similar, they must have corresponding angles equal in measure and corresponding sides in the same ratio.

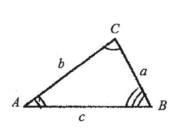

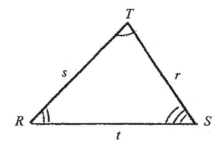

In the figures above, in order for $\triangle ABC$ to be similar to $\triangle RST$, it is necessary that:
$\angle A = \angle R$, $\quad \angle B = \angle S$, $\angle C = \angle T$, $\quad$ and
$\dfrac{b}{s} = \dfrac{a}{r} = \dfrac{c}{t}$.

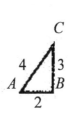

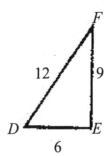

The two triangles ABC and DEF are similar. The ratios of corresponding sides are equal.

$$\frac{AB}{DE} = \frac{2}{6} = \frac{1}{3}, \ \frac{BC}{EF} = \frac{3}{9} = \frac{1}{3}, \ \frac{AC}{DF} = \frac{4}{12} = \frac{1}{3}$$

The ratio of corresponding sides $= \dfrac{1}{3}$,

Since the ratios of corresponding sides are equal, three proportions can be formed,

$$\frac{AB}{DE} = \frac{BC}{EF}, \ \frac{BC}{EF} = \frac{AC}{DF}, \text{ and } \frac{AC}{DF} = \frac{AC}{DF}.$$

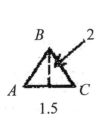

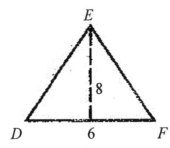

The ratio of corresponding heights equals the ratio of corresponding sides. Ratio of corresponding sides $= \dfrac{1.5}{6} = 4$.

Ratio of heights $= \dfrac{2}{8} = \dfrac{1}{4}$.

373

Congruent objects have the same shape *and* the same size.

The two triangles at the right are congruent. They have exactly the same size.

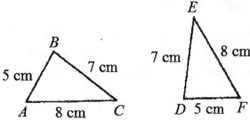

For triangles, congruent means that the corresponding sides *and* angles of the triangles must be equal, unlike similar triangles which just have corresponding angles equal but not necessarily the corresponding sides.

Here are two major rules that can be used to determine if two triangles are congruent.

Side-Side-Side Rule (SSS)

The corresponding sides of congruent triangles are equal.

$AB = DE$, $AC = DF$, $BC = EF$

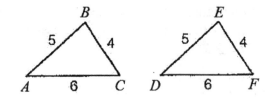

Side-Angle-Side Rule (SAS)

Two triangles are congruent when two sides and the included angle of one triangle equal the corresponding sides and angle of the second triangle.

$AB = EF$, $AC = DE$, Angle BAC = Angle DEF

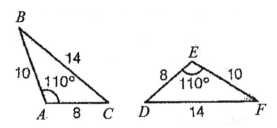

EXAMPLE 1

A vertical yardstick casts a shadow 2' long. At the same time a building casts a shadow 42' long. Find the height of the building.

SOLUTION

Sketch a figure of the problem situation and label the figure.

BC represents the yardstick.
AB represents its shadow.
ST represents the building.
RS represents its shadow.

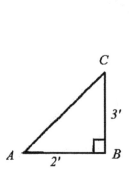

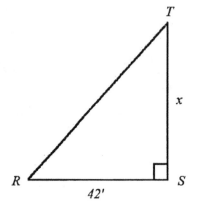

These two triangles are similar. We may use the idea of proportionality to find the height of the building.

$$\frac{AB}{RS} = \frac{CB}{x}$$

$$\frac{2}{42} = \frac{3}{x}$$

$2x = 42(3)$ Cross multiply.

$2x = 126$ Divide both sides by 2.

$x = 63'$

EXAMPLE 2

Triangles *ABC* and *DEF* are similar. Find side *EF*.

SOLUTION

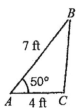

$$\frac{AB}{DE} = \frac{BC}{EF}$$

$$\frac{8m}{12m} = \frac{6m}{EF}$$

$8 \times EF = 12 \times 6m$
$8 \times EF = 72m$
$EF = 72m \div 8$
$EF = 9m$

EXAMPLE 3

Determine if triangle *ABC* is congruent to triangle *DEF*.

SOLUTION

Since $AB = DF$, $AC = EF$, and
Angle *BAC* = Angle *DFE,* the triangles are
congruent.

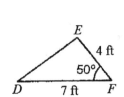

EXAMPLE 4

Triangles *ABC* and *DEF* are similar. Find
the height of *FG*.

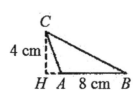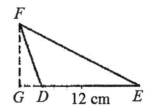

SOLUTION

To find the height of *FG:*
 Solve a proportion to find the height.

$$\frac{AB}{DE} = \frac{\text{height } CH}{\text{height } FG}$$

$$\frac{8\text{cm}}{12\text{cm}} = \frac{4\text{cm}}{\text{height}}$$

$$8 \times \text{height} = 12 \times 4\text{cm}$$

$$8 \times \text{height} = 48\text{cm}$$
$$\text{height} = \frac{48\text{cm}}{8}$$

height = 6cm
The height *FG* is 6 cm.

EXERCISE 8.8

Find the ratio of corresponding sides for the similar triangles.

1)

2)

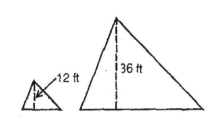

3)

4)

Determine if the two triangles are congruent

5)
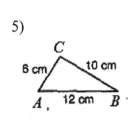

6)
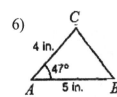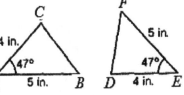

7)
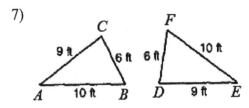

8)
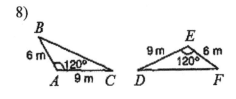

Triangles *ABC* and *DEF* are similar. Find the indicated distance. Round to the nearest tenth.

9)
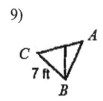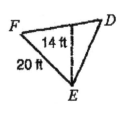

10)
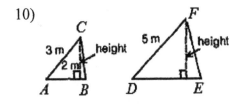

11)
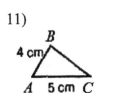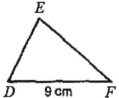

12)
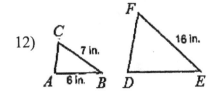

Triangles *ABC* and *DEF* are similar.

13) Find the area of triangle *DEF*.

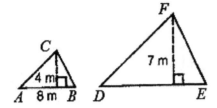

16) Find the area of triangle *DEF*.

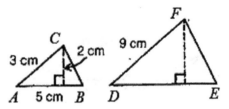

14) Find the area of triangle *ABC*.

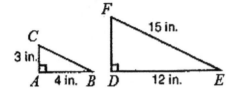

17) Find the perimeter of triangle *ABC*.

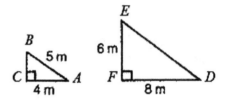

18) Find the area of triangle *DEF*.

15) Find the perimeter of triangle *ABC*.

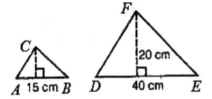

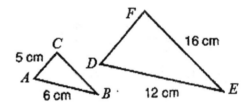

CHAPTER 8 REVIEW EXERCISES

Complete the following sentences or statements.

1) A triangle with no equal sides is called a _____?

2) An angle containing between 90° and 180°. _____

3) A four-sided figure in which two of the sides are parallel and two of the sides are not parallel. _____

4) A five-sided figure. _____

5) Two lines which intersect at right angles. _____

6) Two triangles which have corresponding sides and corresponding angles that are equal. _____

7) The total number of degrees in any triangle. _____

8) The supplement of a 108° angle. _____

9) Which pair of lines is parallel?

   a) ⟋   b) ⊥   c) ⟍   d) None

10) Which figure is a trapezoid?

   a) ▱   b) ⌐   c) ⏢

Work the following problems.

11) Each side of a square measures 3.2 meters. Find the perimeter of the square.

12) Find the area, perimeter, and hypotenuse of the triangle below.

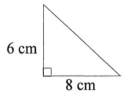

6 cm

8 cm

13) The hall in Ron's house is 3 feet wide and 18 feet long. He wants to put carpet on the hall floor. At $16 a square yard, how much will the carpet for the hall cost?

14) Charles Ervin wants to put fencing around the vegetable garden in his yard. The garden is 20 feet long and 8 feet wide. He will leave a four-foot opening for a walkway into the garden. How many feet of fencing does he need?

15) Find the circumference of a circle with a diameter of 30 feet. Use $\pi = 3.14$.

16) What is the area of a circle with a radius of 35 inches? Use $\pi = \dfrac{22}{7}$.

17) How many degrees are in the complement of a 70° angle?

18) What kind of angles are $\angle$ AOB and $\angle$ BOC in the figure below?

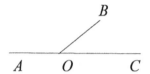

19) Find the area of the kite in the figure at right.

20) Sonnie is building a garage. The garage floor will be made of poured concrete. It will be 22 feet long, 18 feet wide, and 6" deep. How many cubic feet of concrete does Sonnie need to purchase for the floor?

21) In triangle XYZ, $\angle$ X = 33° and $\angle$ Y = 57°. What kind of triangle is triangle XYZ?

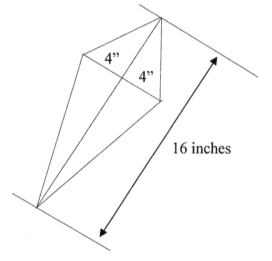

22) The vertex angle of an isosceles triangle measures 102°. How many degrees are there in each base angle?

23) An 8-foot ladder is leaning against the ledge of a window. The ladder makes an angle of 35° with the ground. How high is the ledge?

CHAPTER 8 TEST

Circle the letter which represents the correct answer.

1) Two angles of a triangle measure 40° and 101°. Find the measure of the other angle.
   a) 39°
   b) 31°
   c) 147°
   d) 90°

2) Find the area of a triangle with a base of 1.2 feet and a height of 0.8 feet.
   a) 4
   b) 2.44
   c) 0.48
   d) 0.96

3) Find the volume of a sphere with a radius of 7 inches.
   a) $359\frac{1}{3}$ cubic inches
   b) $1437\frac{1}{3}$ cubic inches
   c) 1078 cubic inches
   d) none of these

4) Find the area of a rose garden 13 feet 6 inches wide and 22 feet long.
   a) 297 square feet
   b) 286 feet
   c) 286 square feet
   d) 29.7 square feet
   e) none of these

5) What would it cost to carpet Leon's living room that is 5 feet 4 inches wide and 24 feet 6 inches long, if the carpet costs $6.50 per square yard?
   a) $94.37
   b) $130.66
   c) $849.33
   d) $264

   e) none of these

6) What is the perimeter of a lot that is a rectangle 140 feet long and 55 feet wide?
   a) 195 feet
   b) 185 feet
   c) 370 feet
   d) 390 feet

7) The opening of a pipe is a circle with a radius of 15 meters. What is the area of the opening?
   a) 225 square meters
   b) 707 square meters
   c) 730 square meters
   d) 47 square meters

8) What is the volume of a cylinder 16 centimeters high and 7 centimeters in radius?
   a) 2160 cubic meters
   b) 2460 cubic meters
   c) 730 cubic meters
   d) none of these

9) What is the perimeter of a rectangular lot 84 inches wide and 210 inches long?
   a) 294 inches
   b) 598 inches
   c) 588 inches
   d) 568 inches

10) What is the area of a circular pipe with a radius of 25 meters?
    a) 625 square meters
    b) 1963 square meters
    c) 50 square meters
    d) 1960 square meters

11) If a wall is 18 feet long and 9 feet high and contains 3 windows, each

of which is 3 feet high and 2 feet wide, how many square feet of wallpaper are needed to paper the wall, leaving the windows uncovered?
a) 124
b) 148
c) 180
d) 162
e) 144

12) How long is side AB of right triangle ABC?

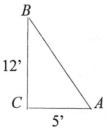

a) 13 feet
b) 17 feet
c) 15 feet
d) 20 feet
e) 14 feet

13) How many cubic feet of sand are needed to fill a box 15 feet by 3 inches by 12 feet?
a) 540 cubic feet
b) 45 cubic feet
c) 60 cubic feet
d) 16 cubic feet
e) 12 cubic feet

14) Each of the angles of a triangle measures 60 degrees. One side of this triangle is 6 inches long. The perimeter therefore is
a) 6 inches
b) 9 inches
c) 12 inches
d) 15 inches
e) 18 inches

15)

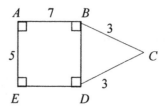

In the diagram above, which of the following is true?
a) ∠1 equals ∠4
b) ∠1 plus ∠2 equals ∠3
c) ∠1 plus ∠2 plus ∠3 equals 180 degrees
d) ∠1 plus ∠2 equals ∠4

16) In one hour, the minute hand rotates through 360°. In two hours it rotates through .
a) 720°
b) −720°
c) 360°
d) 120°

17) The perimeter of figure ABCDE is

a) 18
b) 25
c) 38
d) 44
e) 45

18) Find the perimeter of a rectangle with a width of 8 feet and length of 3 feet.
a) 11 ft.
b) 22 ft.
c) 24 ft.
d) none of these

19) What is the diameter of a circle that has a radius of 4?
a) 4
b) 8
c) 16
d) none of these

20) Two angles of a triangle are 30° and 45°. How large is the third angle?
a) 105°
b) 15°
c) 75°
d) 95°

21) An isosceles triangle is a triangle with
a) two equal sides
b) no equal sides
c) three equal sides
d) three equal angles

22) What is the supplement of an angle of 37°?
a) 37°
b) 53°
c) 143°
d) 153°

23) What is the complement of an angle of 41°?
a) 41°
b) 49°
c) 139°
d) 90°

24) A right triangle has sides 5, 12, and 13. Find the area.
a) 60
b) 30
c) 78
d) 65

25) A circle has a radius of 21". Its circumference is

a) 1,386"
b) 66"
c) 132"
d) 441"

26) A polygon with 6 sides is called a(n)
a) quadrilateral
b) pentagon
c) hexagon
d) octagon

27) Three sides of a triangle are 3, 5, and 6. Find the longest side of a similar triangle whose shortest side is 2.
a) 8
b) 3
c) 5
d) 4

28) Find to the nearest tenth of an inch the hypotenuse of a right triangle whose legs are 5" and 8".
a) 9.2"
b) 8.9"
c) 9.4"
d) 9.5"

29) In triangle ABC, $\angle C = 90°$. If AC = 6" and BC = 8", then AB =
a) 2"
b) 9.2"
c) 8"
d) 10"

30) The vertex angle of an isosceles triangle is 80°. Each of the other angles of the triangle is
a) 50°
b) 100°
c) 20°
d) 10°

31) How many square tiles of 9" on a side will be needed to tile a floor 9' by 12'?
a) 108
b) 192
c) 144
d) 156

32) A patio 9' by 12' is to be 6" thick. How many cubic yards of cement are needed to construct it?
a) 2
b) 648
c) 54
d) 72

33) There are 231 cubic inches in a gallon. A cylindrical can has a radius of 7" and a height of 15". How many gallons does it contain?
a) 15
b) 12
c) 13
d) 10

34) A box measures 24" by 12" by 9". How many cubic inches does it contain?
a) 45
b) 324
c) 2,592
d) 2,692

35) Two circles have the same center. If their radii are 7" and 10", find the area included between them.
a) 3 sq. in.
b) 9 sq. in.
c) 9π sq. in.
d) 51π sq. in.

36) A baseball diamond is 90' (90' on each side, not 90 sq. ft.). How many feet, to the nearest foot, is it from home plate to the second base?

a) 180
b) 126
c) 115
d) 127

37) Which of the following is a parallelogram?
a)
b)
c)
d)

38) The measure of angle ACD (in degrees) of the figure below is?

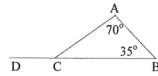

a) 75
b) 105
c) 15
d) 90

39) The perimeter of an equilateral triangle with one side of 15 is
a) 15
b) 30
c) 45
d) none of these

40) The area of the figure below is?

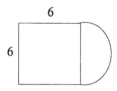

a) 95.52
b) 27.42
c) 64.26
d) 50.13

384

41) A car travels 20 miles east then 15 miles south. What is the shortest distance between the starting point of the car and its present position?
a) 20 miles
b) 35 miles
c) 25 miles
d) 5.9 miles

42) The length of side x in the figure below is?

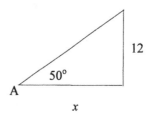

a) 15.66
b) 10.07
c) 18.67
d) Cannot be determined

CHAPTER 9 MEASUREMENT

The notion of measurement probably dates back to prehistoric times. It developed in response to the need to convey or describe the size of different objects. In the absence of a standard unit of measure, various parts of the body were used to describe the size of the objects. The disadvantage of these units was that the measurement depended upon the size of the person making the measurement. To avoid confusion in measurement, specific standards were eventually developed to provide for uniformity. The first such system of measurement was called the customary system, whose development is credited to the English. Some of the units in this system include the inch, foot, mile, ounce, pound, quart, and gallon. Another system of measurement, the metric system, was developed in France around 1795. In the metric system measurements are made in meters, liters, or grams.

THE METRIC SYSTEM

Compared to the customary system of measurement, the metric system is more consistent, uniform, and logical. The metric system of measurement uses three base units. The **meter** is the base unit for measuring length, the **liter** is the base unit for measuring liquid volume or capacity, and the **gram** is the base unit of measure for mass or weight.

The size of the base unit is determined by standard decimal prefixes which give the size of the unit in relation to the base. For example, hecto means one hundred; therefore, a hectometer would be 100 times the base unit meter, a hectoliter would be 100 times the base unit liter, and a hectogram would be 100 times the base unit grams. This information allows us to determine the size of the measure in the metric system by looking at the prefix and the type of measure by looking at the base.

Common Metric System Prefixes

Prefix	Symbol	Numerical Meaning	Meaning in Exponential Notation
Kilo	K	1000	10^3
Hecto	H	100	10^2
Deka	Da or DK	10	10^1
Base Unit			
Meter	M	1	10^0
Liter	L	1	10^0
Gram	G	1	10^0
Deci	D	1/10 or .1	10^{-1}
Centi	C	1/100 or .01	10^{-2}
Milli	M	1/1000 or .001	10^{-3}

The common prefixes used and their numerical meanings and symbols are given above.

In the table, the prefixes are arranged in order, by powers of ten, from largest to smallest. For any two consecutive units, either (1) the unit of a quantity is 10 times the next consecutive smallest unit of the same quantity, or (2) each unit is .1 of the next consecutive largest unit. For example, a kilometer and a hecto meter are consecutive units with the kilometer larger than the hectometer, therefore 1 kilometer = 10 hectometers. We can use these observations to place the prefix values on a metric number line. We will then be able to convert from one unit of measure to the another by simply moving the decimal point in the value of the measurement the number of places, and in the direction, that corresponds to its position on the metric number line.

The metric number line is given below.

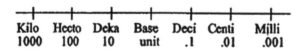

```
  +----+----+----+----+----+----+----+
 Kilo Hecto Deka Base Deci Centi Milli
 1000  100   10  unit  .1   .01   .001
```

EXAMPLE 1

Convert 7.34 kilometers to meters.

SOLUTION

Kilometer means 1000 times a meter; therefore, to convert 7.34 kilometers to meters, we multiply 7.34 × 1000. The answer is 7,340 meters.

An alternative approach is to use the metric number line. Starting at the left, and moving from kilo to the base unit on the line, requires moving over three places to the right. So we move the decimal in 7.34 kilometers three places to the right as well. Answer =7340 meters.

EXAMPLE 2

Convert 7.245 decigrams to kilograms.

SOLUTION

To get from deci to kilo, we must go four units to the left. Similarly, we move the decimal in 7.245 four places to the left. The answer is .0007245 kilograms.

To add, subtract, multiply, or divide using metric measurements, first convert all units to the same base unit and then perform the indicated operation.

EXAMPLE 3

Add 59.8 ml and 3.46 liters.

SOLUTION

First, change each unit to the same base unit. In this instance, let us use liters.

$$
\begin{array}{r}
59.8 \text{ ml } = .0598 \text{ L} \\
+ 3.46 \text{ L} = \underline{3.46 \text{ L}} \\
\hline
3.5198 \text{ L}
\end{array}
$$

EXERCISE 9.1

Convert each of the following.

1) 3 L to kl
2) 11 cm to mm
3) 5261 mm to km
4) 2000 mg to kg
5) 2.5 kl to L
6) 5000 cm to m
7) 25.91 g to kg
8) 3 m to dam
9) 978 km to mm
10) 428 kg to cg
11) .068 m to dm
12) 68 m to dam
13) 2.6×10^3 dg to hg
14) 3.4×10^{-2} dam to cm

Insert <, >, or = to make a true statement.

15) 1 dm   1 meter
16) 1 km   1 mm
17) 2.1 kl   2100 L
18) 1 cg   1 mg
19) .14 dm   5 dam
20) 4.75 L   475 dl

Perform the indicated operation(s).

21) 345 m + .89 km
22) 2.5 kl + .89 dl
23) 45.6 dm − 123.56 mm
24) 3.25 dm × 2.6 mm
25) 30.9 kg ÷ 3 mg
26) $(3.6 \times 10^6$ kg) + 10 dg

DENOMINATE NUMBERS (MEASUREMENT)

If you wish to describe how tall you are, how old you are, how much you weigh, or how much money you have in your pocket you use certain standard units of measure. For example, you may be 68 inches tall, weigh 170 pounds, be 20 years old, and have 25 dollars in your pocket. Here the standard units of measure are inches, pounds, years, and dollars.

Numbers expressed in standard units of measure, such as inches, pounds, years, dollars, etc., are called denominate numbers. A **denominate number** is a number that specifies a given measurement.

Tables 9.2 and 9.3, on the next page, may be used for conversions for the English and metric systems.

Table of Measures (English Measures)

Length
1 foot (ft or ') = 12 inches (in or ")
1 yard (yd) = 36 inches
1 yard = 3 feet

1 rod (rd) = $16\frac{1}{2}$ feet

1 mile (mi) = 5280 feet
1 mile = 1760 yards
1 mile = 320 rods

Liquid Measure
1 cup (c) = 8 fluid ounces (fl oz)
1 pint = 4 gills (gi)
1 quart (qt) = 2 pints
1 gallon (gal) = 4 quarts

Weight
1 pound (lb) = 16 ounces (oz)
1 ton (T) = 2000 pounds

Dry Measure
1 quart = 2 pints (pt)
1 bushel (bu) = 4 pecks

Area
1 square foot (ft^2) = 144 square inches (in^2)
1 square yard (yd^2) = 9 square feet

Volume
1 cubic foot (ft^3 or cu ft) = 1728 cubic inches
1 cubic yard (yd^3 or cu yd) = 27 cubic feet
1 gallon = 231 cubic inches

English-Metric Conversions (Approximate)

English to Metric	**Metric to English**
1 quart = .9467 liters	1 centimeter = .39 inches
1 inch = 2.54 centimeters	1 meter = 3.28 feet
1 fluid quart = .945 liters	1 liter = 1.06 liquid quart
1 yard = .914 meters	1 meter = 1.1 yards
1 mile = 1.61 kilometers	1 kilometer = .621 miles
1 ounce = 28 grams	1 kilogram = 2.2 pounds
1 pound = 454 grams	
1 fluid ounce = 30 milliliters	

Conversions in the English System of Measurement

To convert to smaller units of measure

1) Find the number of the smaller denomination that is equivalent to one unit of the larger denomination.

2) Multiply the given number of units of the larger denomination by this number which is called the conversion factor.

EXAMPLE 1

Convert 7 yards to inches.

SOLUTION

1 yard = 36 inches (from Table 9.2)
36 is our conversion factor.
$7 \times 36 = 252$ inches

Therefore, 7 yards = 252 inches.

EXAMPLE 2

Convert 5 pounds to ounces.

SOLUTION

16 ounces = 1 pound (from Table 9.2)
16 is our conversion factor.
$16 \times 5 = 80$ ounces

Therefore, 5 pounds = 80 ounces.

EXAMPLE 3

Convert 8 square yards to square feet.

SOLUTION

9 square feet = 1 square yard (from Table 9.2)
9 is the conversion factor.
$9 \times 8 = 72$ square feet

Therefore, 8 square yards = 72 square feet.

To convert to larger units of measure

   a) Find the number of units of the smaller denomination that is equivalent to one unit of the larger denomination.

   b) Divide the given number of units of the smaller denomination by this conversion factor.

EXAMPLE 4

Convert 48 inches to feet.

SOLUTION

1 foot = 12 inches (conversion factor)

48 inches ÷ 12 = 4 feet

Therefore, 48 inches = 4 feet.

If there is a remainder it is expressed in terms of the smaller unit of measure.

EXAMPLE 5

Convert 40 months to years and months.

SOLUTION

12 months = 1 year
12 is our conversion factor.

$$
\begin{array}{r}
3 \text{ years} \\
12\overline{)40 \text{ months}} \\
36 \\
\hline
4 \text{ months}
\end{array}
$$

Therefore, 40 months = 3 years, 4 months.

EXAMPLE 6

Convert 35 ounces to pounds and ounces.

SOLUTION

1 pound = 16 ounces (conversion factor)

$$\begin{array}{r} 2 \text{ pounds} \\ 16\overline{)\ 35 \text{ ounces}} \\ \underline{-32} \\ 3 \text{ ounces} \end{array}$$

Therefore, 35 ounces = 2 pounds 3 ounces.

Conversions between the English and Metric Systems

The examples which follow should help you learn to think metrically a little more comfortably. They should also help you to make better approximations of the equivalent measurements in the two systems.

In general, to convert from a metric measure to an English measure or from an English measure to the metric system, find how many units of the desired measure are equal to one unit of the given measure.

EXAMPLE 7

Convert 4 kilometers to miles.

SOLUTION

1 kilometer = .62 miles. Therefore, 4 km = 4 × .62 or 2.48 miles.

EXAMPLE 8

Convert 8 kilograms to pounds.

SOLUTION

1 kilogram = 2.2 pounds. Therefore, 8 kg = 8 × 2.2 or 17.6 pounds.

EXAMPLE 9

Which of the following is the best estimate of the weight of an average size automobile?
  a) 1450 kg   b) 145 kg   c) 14.5 kg

SOLUTION

A kilogram is a little more than 2 pounds. If we multiply each estimate in kilograms by 2, it will give us the approximate weight of the car in the English system. For 1450 kg we get 2900 lb; for 145 kg we get 290 lb; and for 14.5 kg we get 29 lb. Obviously, an average size car weighs more than 29 lb or 290 lb, so the best estimate is 2,900 lb. The answer is a.

EXAMPLE 10

If you were traveling 120 km per hour, how many mph would you be traveling?

SOLUTION

A kilometer is a little more than a half-mile. Multiplying $(1/2) \times 120$, we get roughly 60 mph.

EXAMPLE 11

Gasoline sells for approximately .26 cents per liter in Canada. What is the cost of this gasoline per gallon in the United States?

SOLUTION

A liter is a little more than a quart and there are 4 quarts in a gallon. Thus, the gasoline would cost .26 cents per quart or $4 \times .26 = \$1.04$ per gallon in the United States.

EXERCISE 9.2 A

Convert each of the following to the unit indicated. Answers should be rounded to the nearest tenth, if possible.

1) 15 yards = ___ feet

2) $12\frac{3}{4}$ pounds = _____ ounces

3) 29 feet = _____ yards

4) 6 gallons = _____ quarts

5) 22 pints = ____ quarts

6) 208 cubic feet = _____ cubic yards

7) 40 yards = _____ feet

8) 11 quarts = ____ gallons

9) 40 inches = _____ yards

10) 3 gallons = ____ pints

11) 8 feet = _____ inches

12) 3 pounds = _____ ounces

13) 16 quarts= ____ gallons

14) 30 pints = _____ quarts

15) 7 yards= _____ feet

16) 64 fluid ounces = _____ pints

17) 9 gallons = _____ quarts

18) 167 inches = _____ yards

19) 34 quarts = _____ gallons

20) 7,854 feet = ____ miles

Insert < or > in each of the following to make a true statement.

21) 1 in   1 cm

22) 2 mi   1 km

23) 1 yd   1 m

24) 1 ft   .5m

25) 20 km   14 miles

26) 1 lb   1 kg

27) 1 quart   1 L

28) 5 lb   3 kg

29) 5 kg   10 lb

30) 3 ft   1 meter

Pick the best estimate for each of the following.

31) The distance from the gym floor to the rim of a basketball hoop.
    a) 3 dm   b) 3m
    c) 3 km

32) A carpenter repaired a bookcase 1,800 mm tall. What is its height in feet?
    a) 4 feet   b) 5 feet
    c) 6 feet

33) A package contains 2.268 kg of sugar. How many pounds of sugar is this?
    a) 2 lb   b) 5 lb
    c) 10 lb

34) A glass pitcher contains 750 ml of orange juice. How many liters of orange juice is this?
    a) 75   b) 7.5
    c) .75

35) A package of drink mix can make 2 quarts of lemonade. How many liters is this?
    a) 1   b) 2
    c) 3

36) How many milliliters are contained in a 12-ounce can of soda?
    a) 3.6   b) 36
    c) 360

37) What would be the size of a standard sheet of $8\frac{1}{2} \times 11$ inch notebook paper?
    a) 1.7 × 2.2 cm   b) 17 × 22 cm
    c) 170 × 220 cm

38) The width of a 35mm film in inches is approximately
    a) 1.5 inches   b) 3.5 inches
    c) 4 inches

Give the name of the metric unit which is most commonly used to measure each object.

39) The length of a basketball court

40) The distance from Charleston to Columbia

41) The dimensions of a room

42) The width of your foot

43) Your weight

44) The weight of a penny

45) The amount of gas in a car's tank

Complete the following sentences with the appropriate metric measurement.

46) Each day, T. J. drinks one _____ of milk.

47) The speed limit on the interstate highway is approximately 110 _____ per hour.

48) The distance between Baltimore and New York City is approximately 325 _____.

49) David is on the basketball team. His height is approximately two _____.

50) A microwave package of popcorn (prior to popping) has a weight of about 99

   _____.

51) A container of aspirin reads "30 tablets, 80 _____ each."

52) A can of soda contains about 355 _____.

53) A bedroom is approximately 30 square _____.

EXERCISE 9.2 B

Solve the following problems.

1) 66 in = _____ yd

2) 120 in = _____ yd

3) 5 yd = _____ in

4) $2\frac{1}{3}$ yd = _____ in

5) $4\frac{1}{2}$ ft = _____ yd

6) 16 ft = _____ yd

7) $4\frac{1}{2}$ yd = _____ ft

8) 13 yd = _____ ft

9) 64 in = _____ ft

10) 30 in = _____ ft

11) 9 ft = _____ in

12) 6 ft = _____ in

13) 7 gal = _____ qt

14) $2\frac{1}{4}$ gal = _____ qt

15) 10 qt = _____ gal

16) 22 qt = _____ gal

17) 12 pt = _____ qt

18) $2\frac{1}{2}$ pt = _____ c

19) 5 c = _____ pt

20) 8 c = _____ pt

21) $2\frac{1}{2}$ c = _____ fl oz

22) 3 c = _____ fl oz

23) 48 fl oz = _____

24) 60 fl oz = _____ c

25) $2\frac{5}{8}$ lb = _____ oz

26) $1\frac{1}{2}$ lb = _____ oz

27) 90 oz = _____ lb

28) 66 oz = ____ lb   29) $1\frac{1}{4}$ tons = ____ lb.   30) 6 tons = ____ lb

31) 7000 lb = ____ tons   32) 3200 lb = ____ tons   33) 7 lb = ____ oz.

34) 4 lb = ____ oz   35) 36 oz = ____ lb   36) 64 oz = ____ lb

EXERCISE 9.2 C

In problems 1–28, fill in the blank with the appropriate number. (Assume the given numbers are all exact.)

1)   78 cm = _____ yd

2)   100 cm = _____ yd

3)   2 yd = _____ cm

4)   3 yd = _____ cm

5)   1 in³ = _____ cm³
      (to the nearest hundredth)

6)   1 cm³ = _____ in³
      (to the nearest hundredth)

7)   75 cm = _____ ft

8)   800 m = _____ ft

9)   8.1 liters = _____ qt

10)   11 liters = _____ qt

11)   5 qt = _____ liters

12)   6.1 qt = _____ liters

13)   5 kg = _____ lb

14)   1.2 kg = _____ lb

15)   6 lb = _____ kg

16)   8 lb = _____ kg

17)   3.7 km = _____ mi

18)   14 kg = _____ lb

19)   4 mi = _____ km

20)   6.1 mi = _____ km

21)   3.7 m = _____ yd

22)   4.5 m = _____ yd

23)   51 yd = _____ m

24)   1.2 yd = _____ m

25)   12 cm = _____ in

26)   25 cm = _____ in

27)   8 in. = _____ cm

28)   5.2 in. _____ cm

29)   The cost of gasoline is $1.47 per gallon. Find the cost per liter.

30)   Paint costs $9.80 per gallon. Find the cost per liter.

31)   Peaches cost $.69 per pound. Find the cost per kilogram.

32) Bacon costs $1.69 per pound. Find the cost per kilogram.

33) Express 30 mph in kilometers per hour.

34) Express 65 mph in kilometers per hour.

35) The winning long jump at a track meet was 29 ft 2 in. Convert this distance to meters.

36) Find the number of milliliters in 1 c.

37) Find the number of liters in 14.3 gallons of gasoline.

38) How many kilograms does a 12-pound ham weigh?

39) Find the number of liters in 1 gal of punch.

40) Find the height in meters of a person 6 ft 4 in.

41) Find the weight in kilograms of a 135-pound person.

42) Convert the 100-yard dash to meters.

OPERATIONS WITH DENOMINATE NUMBERS

Addition of Denominate Numbers

To add denominate numbers, arrange them in columns by column unit, and then add each column. If necessary, simplify the answer, starting with the smallest unit.

EXAMPLE 1

Add: 2 gal 3 qt 1 pt, 1 gal 3 qt 1 pt, and 2 gal 2 qt 1 pt.

SOLUTION

2 gal	3 qt	1 pt
1 gal	3 qt	1 pt
2 gal	2 qt	1 pt
5 gal	8 qt	3 pt

	1 qt	
5 gal	9 qt	1 pt
2 gal		
7 gal	1 qt	1 pt

EXAMPLE 2

Add: 1 yd 2 ft 8 in, 2 yd 2 ft 10 in, and 3 yd 1 ft 9 in.

SOLUTION

1 yd	2 ft	8 in
2 yd	2 ft	10 in
3 yd	1 ft	9 in
6 yd	5 ft	27 in

	2 ft	
6 yd	7 ft	3 in
2 yd		
8 yd	1 ft	3 in

Subtraction of Denominate Numbers

To subtract denominate numbers: arrange them in columns by common unit, and then subtract each column starting with the smallest unit. If necessary, borrow to increase the number of a particular unit.

EXAMPLE 3

Subtract 4 hr. 50 min. from 9 hr. 10 min.

SOLUTION

$$9 \text{ hr. } 10 \text{ min.} = 8 \text{ hr. } 70 \text{ min.}$$
$$- \ 4 \text{ hr. } 50 \text{ min.} = 4 \text{ hr. } 50 \text{ min.}$$
$$4 \text{ hr. } 20 \text{ min.}$$

Note that 1 hr. was borrowed from 9 hr. since 1 hr. = 60 min.,
9 hr. 10 min. = 8 hr. 70 min.

EXAMPLE 4

Subtract 2 da. 9 hr. 40 min. from 4 da. 7 hr. 20 min.

SOLUTION

$$4 \text{ da. } 7 \text{ hr.} = 3 \text{ da. } 31 \text{ hr.}$$
$$- \ 2 \text{ da. } 9 \text{ hr.} = 2 \text{ da. } 9 \text{ hr.}$$
$$1 \text{ da. } 22 \text{ hr.}$$

Multiplication of Denominate Numbers

To multiply a denominate number by a given number:

a) If the denominate number contains only one unit, multiply the numbers and write the unit.

b) If the denominate number contains more than one unit of measurement, multiply the number of each unit by the given number and simplify the answer if necessary.

EXAMPLE 5

Multiply 4 yd 2 ft by 5

SOLUTION

$$
\begin{array}{r}
4 \text{ yd } 2 \text{ ft} \\
\times \ 5 \\
\hline
20 \text{ yd } 10 \text{ ft} \\
\underline{3 \text{ yd}} \\
23 \text{ yd } 1 \text{ ft}
\end{array}
$$

EXAMPLE 6

Multiply 4 gal 2 qt 1 pt by 7.

SOLUTIION

	4 gal	2 qt	1 pt
×			7
	28 gal	14 qt	7 pt

	3 qt	
28 gal	17qt	1 pt
4 gal		
32 gal	1 qt	1 pt

Division of Denominate Numbers

To divide a denominate number by a given number: convert all units to the smallest unit, then divide. Simplify the answer if necessary.

EXAMPLE 7

Divide 17 lb 3 oz by 5.

SOLUTION

```
      3 lb
  5⟌ 17 lb 3 oz
     15 lb
      2 lb 3 oz
```

Since 2 1b 3 oz is not divisible by 5, change 2 lb 3 oz to 35 oz, and then divide.

```
     3 lb 7oz
  5⟌17 lb 3 oz
     2 lb  3 oz = 35 oz
                  35 oz
                   0 oz
```

EXAMPLE 8

Divide 9 yd 2 ft 5 in by 4

SOLUTION

```
         2 yd 1 ft  4 in
     4 / 9 yd 2 ft  5 in
         8 yd
         1 yd 2ft  = 5 ft
                      4 ft
                      1 ft 5 in = 17 in
                             16 in
                              1 in        remainder
```

EXERCISE 9.3

Add the following denominate numbers.

1) 3 lb   9 oz
    5 lb   10 oz

2) 9 ft   7 in
    3 ft   10 in

3) 7 gal   3 qt
   1 gal   3 qt

4) 6 gal   3 qt
    5 gal   1 qt
    4 gal   1 qt

5) 7 lb   8 oz
    9 lb   10 oz
    4 lb   11 oz

6) 3 ft   4 in
    7 ft   9 in
    6 ft   5 in

Subtract the following.

7) 7 ft   11 in
    2 ft   4 in

8) 9 lb   8 oz
    5 lb   15 oz

9) 3 hr.   30 min.
    1 hr.   50 min.

10) 4 yd   1 ft   6 in
     2 yd   2 ft   8 in

11) 9 lb   4 oz
    4 lb   10 oz

12) 17 ft   7 in
      6 ft   9 in

Multiply the following.

13) 3 × 20 oz

14) 5 × 10 in

15) 3 × (5 ft 5 in)

16) 3 × (12 yd 2 ft)

17)   4 gal 2 qt 1 pt
     ×   5

18)   5 mi 1,789 ft
    ×   7

Divide the following.

19) (4 ft 6 in) ÷ 2

20) (12 lb 15 oz) ÷ 3

21) 25 hr. 40 min.
     ─────────────────
            5

22) 22 lb   14 oz
     ─────────────────
            6

23) 3) 29 yrs. 9 mos.

24) 5) 12 gal 3 qt

25) The first floor of a 10-story building is 21 ft 5 in high. The other 8 floors are each 12 ft 4 in high and the last floor is 14 ft 7 in high. How high is the building?

26) Thirty years ago the height of a tree was 4 ft 8 in. Today its height is 63 ft 4 in. How much did it grow?

9.4 TEMPERATURE UNITS: FAHRENHEIT, CELSIUS, AND KELVIN

Three temperature scales are commonly used today. The **Fahrenheit scale** is used in the United States. It is defined so that water freezes at 32°F and boils at 212°F.

Internationally, temperature is usually measured on the **Celsius scale**, which places the freezing point of water at 0°C and the boiling point at 100°C. In science, temperature is usually measured on the **Kelvin scale**, which is the same as the Celsius scale except for its zero point. A temperature of 0°K is the coldest possible temperature, known as **absolute zero**. It is approximately –273.15°C or –459.67°F. The following box summarizes conversions among the three scales.

TEMPERATURE CONVERSIONS

The conversions are given both in words and with formulas in which C, F, and K are Celsius, Fahrenheit, and Kelvin temperatures, respectively.

To Convert from	Conversion in Words	Conversion Formula
Celsius to Fahrenheit	Multiply by 1.8 (or $\frac{9}{5}$). Then add 32.	$F = 1.8C + 32$
Fahrenheit to Celsius	Subtract 32. Then divide by 1.8 (or multiply by $\frac{5}{9}$).	$C = \dfrac{F - 32}{1.8}$
Celsius to Kelvin	Add 273.15.	$K = C + 273.15$
Kelvin to Celsius	Subtract 273.15.	$C = K - 273.15$

Where do these conversion formulas come from? For conversions between Kelvin and Celsius, note that the scales are the same except for the 273.15° difference in their zero points. Thus, we simply add 273.15 to a Celsius temperature to get a Kelvin temperature. For conversions between Celsius and Fahrenheit, look at the differences in temperature between the freezing and boiling points of water. The Celsius scale has 100°C between these points, while the Fahrenheit scale has 212°F – 32°F = 180°F between them. Thus, each Celsius degree represents $\frac{180}{100} = 1.8$ Fahrenheit degrees, which explains the factor of 1.8 (or $\frac{9}{5}$) in the conversions. The 32 appears in the Celsius-Fahrenheit conversions to account for the difference in their zero points.

EXAMPLE 1

Bennie had the flu. His temperature was 104°F. What is that on the Celsius scale?

SOLUTION

We replace F by 104 in the formula

$$C = \frac{5(F-32)}{9} \text{ to get } C = \frac{5(104-32)}{9} = \frac{5(72)}{9} = 40$$

Thus, his temperature was 40°C.

EXAMPLE 2

The melting point of gold is 1000°C. What is that on the Fahrenheit scale?

SOLUTION

We substitute 1000 for C in the formula $F = \frac{9C}{5} + 32$ to find

$$F = \frac{9(1000)}{5} + 32$$
$$= 1800 + 32 = 1832$$

Hence, the melting point of gold is 1832°F.

EXAMPLE 3

The average human body temperature is 98.6°F. What is it in Celsius and Kelvin?

SOLUTION

We convert from Fahrenheit to Celsius by subtracting 32 and then dividing by 1.8:

$$C = \frac{F-32}{1.8} = \frac{98.6-32}{1.8} = \frac{66.6}{1.8} = 37.0$$

We find the Kelvin equivalent by adding 273.15 to the Celsius temperature:

$$K = C + 273.15 = 37 + 273.15 = 310.15°K$$

Normal human body temperature is 37°C or 310.15°K.

EXERCISE 9.4

Perform the indicated operations.

Celsius-Fahrenheit conversions. Convert Celsius temperatures into Fahrenheit or Fahrenheit temperature into Celsius.

1) a) 45°F   b) 20°C   c) –15°C   d) −30°C   e) 70°F

2) a) –8°C   b) 15°F   c) 15°C   d) 75°F   e) 20°F

Celsius-Kelvin conversions. Convert Celsius temperatures into Kelvin or Kelvin temperature into Celsius.

3) a) 50°K   b) 240°K   c) 10°C

4) a) –40°C   b) 400°K   c) 125°C

5) Dry ice changes from a solid to a vapor at –78°C. Express this temperature in degrees Fahrenheit.

6) During the second quarter of the play-offs in Cincinnati, the wind-chill factor reached –58°F. (It got worse later.) What is the equivalent Celsius wind-chill factor?

7) What Celsius temperature should we use for the following?
   a) Cool to 41°F
   b) Boil at 212°F

8) Tungsten, which is used for the filament in electric light bulbs, has a melting point of 4310°C. What is that on the Fahrenheit scale?

9) The average normal human body temperature is 98.6°F. What is that on the Celsius scale?

10) For a very short time in September, the temperature in Orangeburg rose to 70°C. What is that on the Fahrenheit scale?'

9.5 CALIBRATED SCALES AND SCALE DRAWINGS

The process of measuring an object results in assigning a number to the object. This is accomplished by choosing a unit and comparing it with the object measured. The unit used for a measurement must be specified. Different types of units are used for measurements of length, volume, area, weight, and so forth. For the measurement of length, which is of concern in this section, a line segment is used. To make a linear measurement, we mark-off the segment to be measured into unit lengths and then count the number of units marked off, usually with the aid of a ruler or tape measure.

EXAMPLE 1

Give the number of unit segments needed to measure the following lengths.

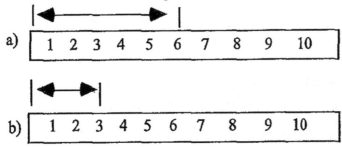

SOLUTION

Count the number of units marked off. In a), 6 units are marked off and in b), 3 units are marked off.

In the examples above, the number of units required to measure the length was exact. This very seldom occurs. Most times, the units must be estimated using fractions. As a result, measuring devices usually have units subdivided into halves, fourths, eighths, and so forth.

The line segment below is measured with three different rulers. As the marked off units get smaller, the measurements become more exact.

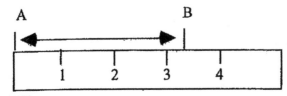

$\overline{AB}$ = 3 to the nearest inch.

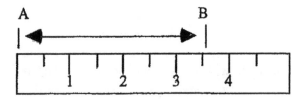

$\overline{AB} = 3\dfrac{1}{2}$ to the nearest

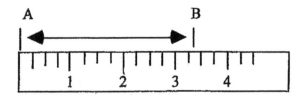

$\overline{AB} = 3\dfrac{1}{4}$ to the nearest

A line segment may also be measured using a metric ruler. Since the metric system is based on ten, the metric ruler is usually subdivided into tenths.

Below is a line segment marked off in millimeters and centimeters on a metric ruler. The smaller units are millimeters and the larger units are centimeters.

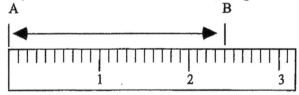

Length $\overline{AB}$ is: 2 centimeters to the nearest cm.
24 millimeters to the nearest mm.

EXAMPLE 2

Approximate the length of the line segment AB to the nearest inch, 1/2 inch, 1/4 inch and 1/8 inch.

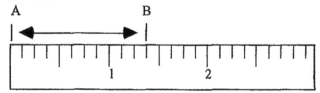

SOLUTION

The ruler is marked off in major units of inches. Counting the number of sub-units between each inch, we get 8. Thus, half of an inch would be (1/2) × 8 or 4 units; a fourth of an inch would be (1/4) × 8 or 2 units; and an eighth of an inch would be (1/8) × 8 or 1. Counting these sub-units, line segment AB is:

1 inch to the nearest inch.

$1\frac{1}{2}$ inch to the nearest 1/2 inch.

$1\frac{1}{4}$ inch to the nearest 1/4 inch.

$1\frac{3}{8}$ inch to the nearest 1/8 inch.

Scale Drawings

One application where estimates of measurements are useful is in scale drawing. A scale drawing is a drawing that has been reduced proportionally in size from a larger drawing to a smaller drawing by a definite fractional part. A line in a scale drawing may be one-half of the line it represents, one-fourth of the line it represents, one-hundredth, one thousandth, or any definite part. For example, in a scale drawing the scale may be written:

$\frac{1''}{4} = 1'$. This means that every 1/4 inch length on the scale drawing represents 1 foot on the original object.

Examples of scale drawings are maps printed in textbooks and blueprints of houses.

The examples below illustrate how to interpret scale drawings.

EXAMPLE 3

Find the distance between Columbia, South Carolina, and Atlanta, Georgia, using the map given.

SOLUTION

417

With a ruler, measure the line segment on the map between Columbia and Atlanta. To the nearest inch, the length of the line segment is 1 inch. Now compare this length to the scale given. On the scale each 1/4 inch represents 55 miles. Since there are 4 one-fourth inch line segments in 1 inch, $4 \times 55 = 220$ miles.

EXAMPLE 4

Find the length of the silhouette shown below.

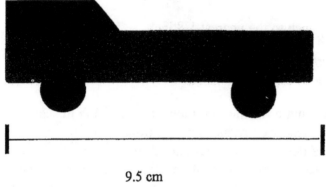

9.5 cm

Scale: 1 cm = .5 meter

SOLUTION

Each centimeter represents .5 meter. There are 9.5 cm, thus $9.5 \times .5 = 4.75$ meters.

EXAMPLE 5

Two cities are $2\frac{1}{4}$ inches apart on a map drawn to a scale of 3/8 inch = 40 miles. What is the distance between the two cities?

SOLUTION

The scale is given in 3/8 of an inch; therefore, we must find the number of 3/8 inches in $2\frac{1}{4}$ inches. Dividing, $2\frac{1}{4} \div 3/8 = 6$. Since each 3/8 inch equals 40 miles we multiply 6×40 to get 240 miles.

EXAMPLE 6

An architect wants to represent 3 meters on a blueprint. If 1 cm = .6 meter, how many centimeters must be used?

SOLUTION

Each .6 of a meter is represented by 1 cm, therefore, we must find the number of .6 meters in 3 meters. Dividing, $3 \div .6 = 5$. The answer is 5 cm.

An alternative strategy is to express the relationship as a proportion. If 1 cm = .6 meters, then how many centimeters represent 3 meters? The proportion would be:

$$\frac{\text{Centimeters}}{\text{Meters}} \qquad \frac{1}{x} = \frac{.6}{3}$$

$.6x = 3$ Cross multiply

$x = 5$ cm   Divide both sides by .6

EXERCISE 9.5

Using the inch ruler below, find the length of each segment to the accuracy indicated.

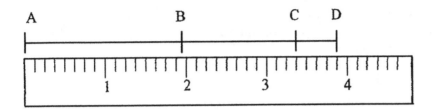

1) $\overline{AB}$ to the nearest inch

2) $\overline{AD}$ to the nearest 1/4 inch

3) $\overline{AC}$ to the nearest 1/8 inch

4) $\overline{AC}$ to the nearest 1/2 inch

Using the metric ruler below, find the length of each segment to the accuracy indicated.

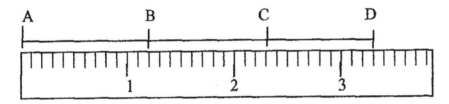

5) $\overline{AB}$ to the nearest cm

6) $\overline{AC}$ to the nearest mm

7) $\overline{AD}$ to the nearest cm

8) $\overline{AD}$ to the nearest mm

If a scale of 1/4 inch = 1 ft is used in a blueprint of a house, give the dimensions in feet of the rooms below.

9) Living room   3" by 5"

10) Master bedroom $3\frac{1}{2}$ inches by $4\frac{1}{2}$ inches

If a scale of 1 cm = .4 meter, give the number of centimeters needed to represent the following.

11) 8 meters   12) 3.6 meters   13) .066 km

14) Two cities are $3\frac{1}{2}$ inches apart on a map drawn to a scale of 3/8 inch = 30 miles. What is the distance between the two cities?

421

15) Two cities are 120 km apart. If 1 cm = 20 km, how many centimeters would it take to represent the distance on a map?

16) In statistics, the value of the correlation coefficient, r, is between -1 and 1 inclusive. The scale below shows the degree of correlation, that is, low, moderate, or strong. Give the estimated value of the coefficient for the following degree of correlation:

a) strong

b) low

Low Moderate Strong Low Moderate Strong

-1 0   1

CHAPTER 9 REVIEW EXERCISES

Fill in the blanks (1 – 3).

1) In the metric system the basic unit for weight is _____, for length or distance is _____, and for volume is _____.

2) In the metric system, for small measures, we use the prefixes _____ and _____ for larger measures we use the prefixes _____ .

3) Give the numerical value for each of the following prefixes:
   milli _____ , deci _____ , hecto _____ .

4) Convert the following to the unit of measurement indicated.
   a) 100 yd to m   b) 100 ft to m   c) 25 gal to L   d) 3.9 m to in
   e) 135 mi to km   f) 2 qt to cl   g) 18 lb to kg   h) 34 km to mi
   i) 59.4 L to qt

5) If hamburger meat costs 99 cents per pound, how much does it cost per kilogram?

6) A roast that weighs 5.2 kilograms is to be cut into 4 equal pieces. How much will each piece weigh?

7) The scale of a map is 1 inch = 80 miles. Find the distance between two cities $8\frac{1}{2}$ inches apart.

8) Find the number of half-pints in 19 gallons of milk.

9) Convert the following to the unit of measure indicated.

   a) 15 yards to feet   b) 108 inches to yards

   c) 5 quarts to gallons   d) 47 ounces to pounds

10) Convert the following to the unit of measure indicated.

   a) 15.8 mm to cm   b) 169 L to kl

11) Perform the indicated operation.
   a) 5 ft 9 in + 4 ft 7 in   b) 3 × (4 ft 7 in)
   c) 8 lb 4 oz − (5 lb 13 oz)   d) (11 lb 4 oz) ÷ 5

12) Perform the indicated operation.
   a) 964 dl − 354 dl   b) 5 m + 600 mm

13) The line below is a straight angle with the segments as shown. The indicator points to how many degrees?

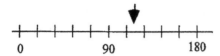

0   90   180

14) Use a ruler to measure each line segment in
   a) inches, b) centimeters, and c) millimeters.

A   B   C   D

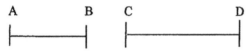

15) A market sells oranges by weight. In the metric system, the oranges should be weighed in which of the following units?
   a) liters   b) kilograms   c) meters

16) A person is buying some paint to paint a house. In the metric system, the paint would be measured in which of the following?
   a) grams   b) centimeters   c) liters

17) Write the power of 10 that corresponds to the following prefixes:
   a) centi   b) milli   c) kilo

18) A steel rod is $\frac{3}{4}$ m long. How many centimeters is that?

19) Which of the following is most nearly correct? Ten inches is about:
   a) 30 cm   b) 25 cm   c) 20 cm

20) The sides of a triangle are measured to be 16.3, 18.9, and 15.46 cm long, respectively. How many centimeters long is the perimeter of the triangle?

21) The side of a square is 3.5 in long. How many centimeters long is the perimeter of the square?

22) The sides of a triangle are found to be 152, 178, and 135 m long. How many kilometers long is the perimeter of the triangle?

23) A wine bottle contains 750 ml of wine.
   a) How many liters is this?
   b) How many quarts?

24) Pat weighs 52 kg. How many pounds is this?

25) In Orangeburg, the posted speed limit is 40 km/hr. How many miles per hour is this?

26) The temperature in Georgetown was $20°C$. What is that in degrees Fahrenheit?

27) The temperature in Lamar, SC was $77°F$. What is that in degrees Celsius?

28) A baby weighs 4.1 kg at birth. How many grams is that?

29) Reggie bought a 250-g can of tuna fish. How many kilograms is that?

30) Convert 275 mg to kilograms.

31) A market prices onions at $1 per kilogram. How many cents per gram is that?

32) A 2-carat diamond weighs 0.2 g. How many milligrams does a 2-carat diamond weigh?

33) A liter of wine will fill about ten large wine glasses. How many milliliters does one of these glasses hold?

34) A container holds 19 ml of liquid. How many centiliters is that?

35) A bottle with a capacity of 92 cl holds how many liters?

36) A water tank holds 8.3 kl of water. How many liters is that?

37) A piece of a straight line is 0.082 m long. How many millimeters is that?

38) The sides of a triangle are 3 cm, 4 cm, and 6 cm long, respectively. What are these lengths in millimeters?

39) A track is 0.9 km long. How many meters is that?

CHAPTER 9 TEST

Circle the letter which corresponds to the correct answer.

1) John is 5 ft 6 in tall and Peter is 4 ft 11 in tall. How much taller is John than Peter?
   a) 1 ft 5 in   b) 7 in
   c) 1 ft
   d) none of these

2) James places 9 bricks one on top of each other. If each brick is 3 ½ inches high, how high is the pile in feet and inches?
   a) 2 ft 7 in   b) 5 ft
   c) 63 in   d) 2 ft

3) How many quart bottles of milk can be filled from three 10-gallon cans of milk?
   a) 40   b) 30
   c) 120
   d) none of these

4) If Catherine was born in 1896 and died in 1942, how old was she when she died?
   a) 4   b) 10
   c) 40   d) 46

5) A man needs five lengths of board, each 2 ft 3 in long. How many feet of board will he need?
   a) 11' 3"   b) 10'
   c) 12'   d) 14'

6) A carpenter has an eight-foot long piece of board. He cuts pieces from it 1' 5", 2' 3", and 3' 2". How much is left over?
   a) 6'10"   b) 1' 2"
   c) 2'
   d) none of these

7) $\frac{1}{4}$ of a yard is ___ meter(s).
   a) 2.1   b) .225
   c) 6.75   d) 6

8) The number of yards equal to 367 inches is
   a) 10.18   b) 122.3
   c) 3
   d) none of these

9) The sum of 5 ft $\frac{3}{4}$ in, 4.5 ft, and 8 ft is
   a) 17.84 ft   b) 17 ft
   c) 17.25 ft
   d) 18 ft 3 in

10) The difference of 6.25 lb and 1 lb 12 oz is
   a) 5 lb   b) 5 lb 8 oz
   c) 4 lb 8 oz
   d) none of these

11) Two cities are $3\frac{1}{2}$ inches apart on a map whose scale is $\frac{2}{3}$ inch = 30 miles. How far apart are the two cities?
   a) 105 miles   b) 5.25 miles
   c) 157.5 miles   d) 525 miles

12) A figure has sides of 1 inch, 3 inches, 2 ½ inches, and 2 inches. Find the actual perimeter of the figure is the scale is 1" = 20 ft.
   a) 20 ft   b) 340 ft
   c) 170 ft
   d) none of these

427

13) The distance between Columbia, South Carolina, and Savannah, Georgia would be measured in
a) centimeters b) millimeters
c) meters d) kilometers

14) Assuming that 1 liter = 1.06 quarts, three gallons would most nearly equal which of the following?
a) 3.18 liters b) 12.72 liters
c) 11.32 liters d) 2.8 liters

15) Assuming that 1 kilometer = .62 mile, a car traveling 70 mph would most nearly equal which of the following?
a) 43 km/h b) 113 km/h
c) 35 km/h d) 70 km/h

16) If 1 meter = 3.28 feet, then 3 meters would equal approximately how many inches?
a) 118 b) 10
c) 1 d) 6

17) The sum of 18.9 dg, 6 kg, 4 g, and 86 cg is
a) 109.5 grams b) 606.75 grams
c) 60 grams
d) none of these

18) How many grams are equal to 3600 mg?
a) 360 b) 36
c) 3.6 d) .36

19) Which of the following pairs are equivalent?
a) 1 cm and $\frac{2}{5}$ in
b) 1 km and 6 mi
c) 1 kg and 22 lb
d) 1 yard and 4 feet

20) Two kg of fertilizer are used for every eight trees in a grove. Use a proportion to find the amount of fertilizer needed for 1000 trees.
a) 4000 kg b) 400 kg
c) 24 kg d) 250 kg

21) Divide: (12 kg 450 g) ÷ 15
a) 0.83 kg b) 8.3 kg
c) 0.083 kg d) 0.803 kg

22) Convert: 5 kg 50 g = ___ kg
a) 5.50 kg b) 55 kg
c) 5.005 kg d) 5.05 kg

23) A log 4 m 70 cm long is cut into five equal pieces. Find the length in meters of each piece.
a) 0.814 m b) 8.14 m
c) 0.85 m d) 0.94 m

24) Subtract:
4 km 420 m − 1 km 892 m
a) 2.528 km b) 7.312 km
c) 2.538 km d) 3.528 km

25) Convert: 18 m 75 cm = ___ m
a) 18.075 m b) 18.75 m
c) 1875 m d) 187.5 m

26) A block weighs 2 kg 350 g. Find the weight in kilograms of a load of 500 blocks.
a) 1017.5 kg b) 1675 kg
c) 1155 kg d) 1175 kg

27) Add: 3 g + 672 mg
a) 3.0672 g b) 3.672 g
c) 36.72 g d) 367.2 g

28) Convert 1270 mg to grams.
a) 1.27 g b) 12.70 g
c) 127 g d) 0.127 g

29) A man needs 18 rafters each 3 m 72 cm long. Find the total length in meters of the rafters needed.
a) 55.35 m b) 675 m

428

c) 67.5 m   d) 553.5 m

30) Subtract: 32 m − 42 cm
   a) 342 cm   b) 31.58 m
   c) 34.2 m   d) 3042 m

31) Convert 32.5 km to meters.
   a) 216 m   b) $1\frac{1}{2}$ m
   c) $4\frac{1}{2}$ m   d) 32,500 m

32) Convert 18 pt to gallons.
   a) $2\frac{1}{4}$ gal   b) $4\frac{1}{2}$ gal
   c) $\frac{4}{9}$ gal   d) 3 gal

33) Subtract: $4\frac{1}{3}$ qt $- 1\frac{5}{6}$ qt
   a) $6\frac{1}{6}$ qt   b) $2\frac{1}{2}$ qt
   c) $3\frac{1}{2}$ qt   d) $2\frac{1}{6}$ qt

34. Multiply: $1\frac{1}{2}$ c $\times 12$
   a) 8 c   b) 36 c
   c) 24 c   d) 18 c

35) If one serving contains 1 ½ c, how many servings can be made from 3 qt?
   a) 12 servings   b) 8 servings
   c) 18 servings   d) 6 servings

36) Convert: 7 qt = ___ gal ___ qt
   a) 2 gal 1 qt   b) 1 gal 1 qt
   c) 1 gal 2 qt   d) 1 gal 3 qt

37) Sixty-four bricks, each 9 in long, are laid end-to-end to make the base for a wall. Find the length of the wall in feet.
   a) 48 ft   b) $84\frac{2}{3}$ ft
   c) 576 ft   d) $5\frac{1}{3}$ ft

38) A board $7\frac{1}{2}$ ft long is cut into four equal pieces. How long is each piece?
   a) $1\frac{7}{8}$ ft   b) 30 ft
   c) $1\frac{3}{4}$ ft   d) $\frac{8}{15}$ ft

39) Add: $3\frac{2}{5}$ in $+ 5\frac{1}{3}$ in
   a) $2\frac{11}{15}$ in   b) $8\frac{3}{8}$ in
   c) $8\frac{11}{15}$ in   d) $8\frac{1}{15}$ in

40) Convert 18 in to feet.
   a) 216 ft   b) 1 ½ ft
   c) $4\frac{1}{2}$ ft   d) 6

CHAPTER 10 STATISTICS AND PROBABILITY

The word "statistics" brings to the minds of many people an image of a mass of numerical data. Statistics determine what television programs we watch, what products are marketed, how much we pay in taxes, and decisions by business and government that affect the way we live.

The word statistics refers to recorded facts or numbers. Examples of a statistic include the number of births, the number of highway fatalities, or the number of students enrolled in a college.

This collection of information without being placed in some order would be meaningless. Highs and lows would not be evident and trends would be difficult to identify. The science of statistics gives us a way to present information in a clear and meaningful fashion.

By definition, **statistics** is the science of collecting, classifying, presenting, and interpreting numerical data.

Statistics may be roughly divided into two major branches: descriptive and inferential.

Descriptive statistics involves collecting information or data and presenting it in some meaningful way.

In **inferential statistics**, the purpose is not to describe the data that has been collected, but to generalize or make inferences about it.

In presenting and interpreting statistical data or information, we make use of statistical or summary measures. In statistics, the two most commonly used summary measures are measures of central tendency and measures of variability.

MEASURES OF CENTRAL TENDENCY

Measures of central tendency are numerical values that attempt to locate the "middle" of a set of data. These measures summarize data in terms of one number. The commonly used measures of central tendency are the mode, median, and the mean. Each is defined and the steps in calculating the measures are given below.

> The **mean** of a set of numbers is the sum of the numbers divided by the number of elements in the set. The mean is usually denoted by the symbol $\bar{x}$ (read, x bar).

EXAMPLE 1

Ten basketball teams scored the following number of points: 71, 69, 68, 69, 73, 68, 69, 72, 70, and 71. Find the mean (average) of the points scored.

SOLUTION

$$\bar{x} = \frac{71 + 69 + 68 + 69 + 73 + 68 + 69 + 72 + 70 + 71}{10}$$

$$= \frac{700}{10} = 70$$

EXAMPLE 2

Find the mean of 10, 9, 8, 7, 6, 2.

SOLUTION

$$\bar{x} = \frac{10 + 9 + 8 + 7 + 6 + 2}{6} = \frac{42}{6} = 7$$

> The **mode** of a set of numbers is that number of the set that occurs most often.

EXAMPLE 3

Find the mode of Example 1.

SOLUTION

The score that occurred most often is 69 (three times). Thus, the mode is 69.

EXAMPLE 4

Find the mode for the scores 1, 3, 6, 9, 9, 10, 23.

SOLUTION

Since 9 is the most frequently occurring score, the mode is 9.

> The **median** is the middle score when the scores or data are arranged in order from smallest to largest.

EXAMPLE 5

Find the median for 3, 6, 9, 8, 23, 10, and 1.

SOLUTION

First arrange the scores in order from smallest to largest.

$$1, 3, 6, 8, 9, 10, 23$$

Note that there are an odd number of scores, therefore, the middle score, 8, is the median.

EXAMPLE 6

Find the median for 1, 3, 6, 9, 10, and 12.

SOLUTION

As before, arrange the scores in order from smallest to largest. Since two scores are in the middle, take the average of the two.

$$1, 3, 6, 9, 10, 12$$

The two middle scores are 6 and 9. Add them together, $6 + 9 = 15$, and divide by 2.

$$15 \div 2 = 7.5$$

The median is 7.5.

EXAMPLE 7

Marshall has been exercising for the past 10 days. He must exercise if he wants to keep his weight down. Listed below are 10 different activities with the corresponding hourly energy expenditure (in calories) for a 230-pound person.

433

Activity	Calories		Activity	Calories
Wood chopping	400		Standing	150
Volleyball	350		Sitting	100
Swimming	300		Running	900
Squash	600		Golf	250
Square dancing	350		Fencing	300

a) Find the mean of these numbers.
b) Find the median number of calories spent in these activities.
c) Find the mode of these numbers.

SOLUTION

a) The mean is obtained by adding all the numbers and dividing the sum by 10. The sum of the numbers in the first column and of the numbers in the second column is 3700.

$$\bar{x} = \frac{3700}{10} = 370 \text{ calories per hour}$$

b) To find the median, we must first arrange the numbers in order of smallest to highest.

Sitting	100
Standing	150
Golf	250
Fencing	300
Swimming	300
Square dancing	350
Volleyball	350
Wood chopping	400
Squash	600
Running	900

$$\leftarrow \text{Median} = \frac{300+350}{2} = 325$$

c) The mode is the number with the greatest frequency. In this case, the numbers 300 and 350 both occur twice, while all other numbers occur just once. Thus, there are two modes, 300 and 350.

SELECTING A MEASURE OF CENTRAL TENDENCY

In many cases we must decide between the mode, median, and the mean as a measure of central tendency. In making this decision, there are a number of simple principles we can use to guide our decision. Let us take an example to illustrate these principles.

Suppose we are given the following distribution.

$$1, 3, 4, 7, 9, 11, 14$$

Both the mean and the median of the distribution are 7. Now let us change the two end scores and note what happens to the mean and median.

$$-1, 3, 4, 7, 9, 11, 500$$

The median of the set of scores is still 7. However, the mean is 76.14. This score is meaningless since there are no scores in the distribution even close to 76.14. Note also that the median was not affected by the change of extreme scores.

In summary, the following guidelines are used in determining the appropriateness of the mean versus the median as a measure of central tendency.

1. The mean utilizes all of the data in its calculation and therefore is a much more powerful measure of central tendency than the median.

2. Since the mean utilizes all of the data in its calculation, it is very responsive to extreme scores and will move in that direction.

3. The median as a measure of central tendency denotes position (the middle score) and, therefore, is not affected by extreme scores.

EXAMPLE 8

In reporting the "average" cost of homes in the United States, which measure of central tendency should be used?

SOLUTION

Values of homes in the United States vary significantly. There are homes costing $50,000 to $100,000, and there are mansions valued in the millions of dollars. Because of these extremes, the median would be the better measure of central tendency.

EXERCISE 10.1

Find the three measures of central tendency (the mean, median, and mode)

1) 1, 2, 3, 45

2) 17, 18, 19, 20

3) 103, 104, 105, 106, 107

4) 3, 5, 8, 13, 21

5) 79, 90, 96, 95, 95

6) 5, 4, 1, 2, 3, 3, 3

7) 21, 16, 3, 4, 2, 1, 1, 0

8) 81, 70, 95, 79, 85

9) 4, 9, 1, 16, 25

10) 765, 767, 766, 768, 769

11) 18, 17, 19, 20,21

12) Find the mean, the median, and the mode of the following salaries of employees of the Moe D. Lawn Landscaping Company:

Salary	Frequency
$25,000	4
28,000	3
30,000	2
45,000	1

13) Find the mean, the median, and the mode of the following: G. Thumb, the leading salesperson for the Moe D. Lawn Landscaping Company, turned in the following summary of sales contacts for the week of October 23–28:

Date	Number of Clients contacted by G. Thumb	Date	Number of clients contacted by G. Thumb
Oct 23	12	Oct 26	16
Oct 24	9	Oct 27	10
Oct 25	10	Oct 28	21

14) Find the mean, median, and mode of the following scores:

Test score	Frequency
90	1
80	3
70	10
60	5
50	2

15) Out of 10 possible points, a class of 20 students made the following text scores:

0, 0, 1, 2, 4, 5, 5, 6, 6, 6, 7, 8, 8, 8, 8, 9, 9, 9, 10, 10

Find the mean, the median, and the mode. Which of these three measures do you think is the least representative of the set of scores?

16) Find the mean and the median of the set of numbers:

0, 3, 26, 43, 45, 60, 72, 75, 79, 82, 83

17) Find the mean and the median for each set of numbers:
   a)   1, 5, 9, 13, 17
   b)   1, 3, 9, 27, 81
   c)   1, 4, 9, 16, 25
   d)   For which of these sets are the mean and the median the same? Which measure is the same for all three sets? Which (if any) of the sets has a mode?

18) Show that the median of the set of numbers 1, 2, 4, 8, 16, 32 is 6. How does this compare with the mean?

19) A math instructor gave a short test to a class of 25 students and found the scores on the basis of 10 to be as follows:

SCORE	3	4	5	6	7	8	9	10
NUMBER OF STUDENTS	2	1	3	2	6	4	4	3

The instructor asked two students, Derrick and Travil, to calculate the average score. Derrick made the calculation

$$\frac{3+4+5+6+7+8+9+10}{8} = \frac{52}{8} = 6.5$$

and said the average score is 6.5. Travil calculated a weighted average as follows:

$$\frac{2\cdot3+1\cdot4+3\cdot5+2\cdot6+6\cdot7+4\cdot8+4\cdot9+3\cdot10}{25} = \frac{177}{25} = 7.08$$

He then said the average was 7.08. Who is correct, Derrick or Travil? Why?

20) Here are the temperatures at 1-hour intervals in Georgetown, from 1 P.M. on a certain day to 9 A.M. the next day:

1 P.M.   90   8 P.M.   81   3 A.M.   66

2 P.M.	91	9 P.M.	79	4 A.M.	65
3 P.M.	92	10 P.M.	76	5 A.M.	66
4 P.M.	92	11 P.M.	74	6 A.M.	64
5 P.M.	91	12 A.M	71	7 A.M.	64
6 P.M.	89	1 A.M.	71	8 A.M.	71
7 P.M.	86	2 A.M.	69	9 A.M.	75

a. What was the mean temperature? The median temperature?

b. What was the mean temperature from 1 P.M. to 9 P.M.? The median temperature?

c. What was the mean temperature from midnight to 6 A.M.? The median temperature?

21) A mathematics professor lost a test paper belonging to one of his students. He remembered that the mean score for the class of 20 was 81, and that the sum of the 19 other scores was 1560. What was the grade of the paper he lost?

22) The mean score of a test taken by 20 students is 75; what is the sum of the 20 test scores?

23) The mean salary for the 20 women in company A is $90 per week, while in company B the mean salary for its 30 men is $80 per week. If the two companies merge, what is the mean salary for the 50 employees of the new company?

24) If in Problem 7 the mean was 82 and the sum of the 19 other scores was still 1560, what was the grade on the lost paper?

25) A student has a mean score of 88 on five tests taken. What score must he obtain in his next test to have a mean (average) score of 80 on all six tests?

26) A test to determine the braking distance of a truck yielded the following results: 220 feet, 208 feet, 216 feet, 219 feet, and 227 feet. Find the average braking distance of the truck.

27) A bus driver's records show the number of gallons of gasoline purchased each day on the job last week. Find the average number of gallons of gasoline purchased.

Wed	Thu	Fri	Sat	Sun
9.4	9.3	11.8	10.3	9.7

28) Mary's paychecks for four months are given in the table below. Find the average monthly check. Round to the nearest cent.

Jan	Feb	Mar	Apr
$1200	$1350	$1190	$1220

439

29) The number of hotdogs sold during five lunch hours at a fast-food restaurant was 252, 286, 245, 292, and 285. Find the average number of hotdogs sold at the restaurant per lunch hour.

30) Kenneth received grades of 86, 92, 88, 94, and 95 on five history exams. Find the average grade of Kenneth's history exams.

31) During the past year, six homes in a small town sold for the following prices: $75,400; $89,450; $116,295; $247,600; $82,500; and $176,300. Find the average price of a home in this town.

32) A car representative recorded the number of miles traveled for each of five days of a business trip. Find the average number of miles traveled each day.

Mon	Tue	Wed	Thu	Fri
125	212	188	231	189

33) A car repair shop's records show the number of requests for service each day last week. Find the average number of requests for service.

Mon	Tue	Wed	Thu	Fri	Sat
24	30	28	26	25	35

34) The prices of 1 pound of steak at six different grocery stores were $2.58, $2.62, $2.49, $2,75, $2.66, and $2.68. Find the average price of steak at the six grocery stores.

35) The closing prices of a stock for five days were $36.25, $35.75, $36.50, $36.00, and $35.40. Find the average closing price of the stock.

36) The total college enrollment in the United States from 1920 to 1970 is shown below. Find the median enrollment for these years.

1920	238,000
1930	355,000
1940	598,000
1950	1,101,000
1960	1,494,000
1970	2,659,000

10.2   MEASURES OF VARIABILITY

Most of the time people want to know more about a set of numbers than we can learn from a measure of central tendency.

Measures of central tendency alone are inadequate to describe a set of data or information. The need for an additional measure may be seen from the comparison of the following distribution of test scores:

Class 1:   10, 50, 50, 50, 50, 50, 50, 50, 50, 90

Class 2:   10, 10, 20, 40, 50, 60, 70, 70, 80, 90

The median for both classes is 50 and the mean for both classes is also 50. Yet, the two class distributions are quite different. In Class 1, 8 of the 10 scores are bunched together at the mean value of 50; while in Class 2, the scores are reasonably well spread out. We need some way of describing this "spread" among the scores.

Statistical measures that describe the amount of spread or deviation of one score or data point from another are called **measures of variability**. The common measures of variability are the range and the standard deviation.

> The **range** is the difference between the highest and lowest score when scores are arranged in order from smallest to largest.

EXAMPLE 1

Find the range of the scores, 1, 3, 6, 9, 10, and 23.

SOLUTION

The scores are already arranged in order from smallest to largest. The largest score is 23 and the smallest is 1. The difference between 23 and 1 is 22; therefore, the range is 22.

The two sets of numbers {7, 5, 3} and {10, 5, 0} have ranges, $7 - 3 = 4$ and $10 - 0 = 10$. Because the range is determined by only two numbers of the set, one can see that it gives little information about the other numbers of the set.

Another measure of dispersion is called the standard deviation (SD). The easiest way to define the standard deviation is by the means of the formula.

Let n number be denoted $x_1, x_2, x_3, ... x_n$, and the mean of these numbers be denoted by $\bar{x}$. Then, the **standard deviation** a is given by

441

$$a = \sqrt{\frac{(x_1 - \bar{x})^2 + (x_2 - \bar{x})^2 + (x_3 - \bar{x})^2 + \ldots + (x_n - \bar{x})^2}{n}}$$

In order to find the standard deviation, we have to find:

1) The mean, $\bar{x}$, of the set of numbers
2) The difference (deviation) between each number of the set and the mean
3) The squares of these deviations
4) The mean of these squares
5) The square root of this last mean, which is the number a.

As we shall learn, the number a gives a good indication of how the data are spread about the mean.

EXAMPLE 2

The ages of 5 children were found to be 7, 9, 10, 11, and 13. Find the standard deviation a for this set of ages.

SOLUTION

We follow the five steps given above, making a table as shown below.

Calculation of the Standard Deviation

Age x	DIFFERENCE FROM MEAN $x - \bar{x}$	SQUARE OF DIFFERENCE $(x - \bar{x})^2$
7	-3	9
9	-1	1
10	0	0
11	1	1
13	3	9
50 Sum of ages		20 Sum of squares
$\bar{x} = \dfrac{50}{5} = 10$ Mean of ages		$\dfrac{20}{5} = 4$ Mean of squares
		$\sqrt{4} = 2 = a$

1) The mean of the five ages is

$$\bar{x} = \frac{7 + 9 + 10 + 11 + 13}{5} = \frac{50}{5} = 10 \quad \text{Column 1}$$

2) We now find the difference (deviation) between each number and the mean (column 2).

3) We square the numbers in column 2 to get column 3.

4) We find the mean of the squares in column 3:

$$\frac{9+1+0+1+9}{5} = \frac{20}{5} = 4$$

5) The standard deviation is the square root of the number found in step 4.
   Thus, $a = \sqrt{4} = 2$.

EXAMPLE 3

Patricia checks the price of 1 dozen large eggs at 10 chain stores, with the following results:

STORE NUMBER	1	2	3	4	5	6	7	8	9	10
PRICE (Cents)	70	68	72	60	63	75	66	65	72	69

Find the mean, median, mode, and standard deviation.

SOLUTION

In Example 3, the data are arranged in order of magnitude. The mean is found to be 68¢. The median is the average of the two middle prices, 68¢ and 69¢, which is 68.5¢. The mode is 72¢. The calculation of the standard deviation is shown in the table below. The result is $a = 4.34$.

x	$x - \bar{x}$	$(x - \bar{x})^2$
60	-8	64
63	-5	25
65	-3	9
66	-2	4
68	0	0
69	1	1
70	2	4
72	4	16
72	4	16
75	7	49
680		188

$$\bar{x} = \frac{680}{10} = 68 \qquad \frac{188}{10} = 18.8$$

$$a = \sqrt{18.8} \approx 4.34$$

We can also use a graphing calculator to find the standard deviation.

EXAMPLE 4

Find the standard deviation for 18, 25, 31, 19, 21, 22, and 27 using a calculator.

SOLUTION

On the TI-35 enter the data using the following keystrokes:

$$[3rd] \ [STAT \ 1] \ 18 \ [\Sigma +] \ 25 \ [\Sigma +] \ 31 \ [\Sigma +] \ \text{and so forth.}$$

When all the data has been entered, press [2nd] [STAT] move cursor to DATA [2] [ENTER] (this clears the calculator) [2nd] [STAT] move cursor to DATA [1]. In XI enter 18 X2 enter 25, X3 enter 31, X4 enter 19, X5 enter 21, X6 enter 22, and X7 enter 27. [2nd] [STAT] [Enter] Display reads 4.29997627=4.30. The standard deviation is 4.30.

Percentile

When you took the SAT test your score was described as a measure of position. **Measures of position** are often used to make comparisons of scores of individuals from different populations. For example, if a student's score is at the 88th percentile, this means that 88% of the students taking the exam scored below that student and 12% scored above. A **percentile** divides the set of data into 100 equal parts and a **quartile** divides it into four parts. To designate a percentile we use a "P" with the percentage expressed as subscript. The 67th percentile would be written P_{67}. For quartile we use a "Q" with the same designation. The 3rd quartile would be written Q_3.

To find a percentile for ungrouped data, we order the data from smallest to largest, just as we did for the median. Next we calculate a **Score Position Index (SPI)** by multiplying the percentile, expressed as a percent, by the number of scores. If the SPI is a whole number, find the average of that score position and the next. If the SPI is not a whole number, use the next whole number position to identify the percentile.

EXAMPLE 5

For the distribution, 19, 31, 41, 14, 13, 28, 13, 22, 7, 7, 13, 6 find

a) 67th percentile P_{67}
b) Q_3

SOLUTION

Arrange the data in order from smallest to largest.

6, 7, 7, 13, 13, 13, 14, 19, 22, 28, 31, 41

a) Calculate the SPI

SPI = percentage × number of scores = .67 × 12 = 8.04
Since it is not a whole number we go to score position 9 in the distribution.

6, 7, 7, 13, 13, 13, 14, 19, 22, 28, 31, 41

The 67th percentile is 22, the ninth score.

b) Q_3 is the same as the 75th percentile, therefore,

SPI = percentage × number of scores = .75 × 12 = 9.

Since 9 is a whole number, we use the average of the 9th and 10th score positions as the percentile.

6, 7, 7, 13, 13, 13, 14, 19, 22, 28, 31, 41

These scores are 22 and 28. Averaged, 22 and 28 is 25. Thus $Q_3 = 25$.

EXERCISE 10.2

Find the range and the standard deviation (count to two decimal places)

1) 4, 5, 1, 3, 2

2) 104, 103, 106, 105, 107

3) 8, 5, 3, 13, 21

4) 95, 96, 95, 75, 90

5) 3, 3, 1, 1, 4, 5

6) 17, 19, 18, 20, 21

7) 769, 768, 766, 767, 765

8) 1, 4, 9, 16, 25

9) 70, 81, 79, 95, 85

10) Use a calculator to find the mean and standard deviation of the following numbers.
   a) 25, 34, 22, 41, 39, 30, 27, 31
   b) 318, 326, 331, 308, 316, 322, 310, 319, 342, 330

11) For the Distribution 11,14,18,14,21,17,13,21,25,19,17,13,28,13,17,18 find the following:
   a) Q_1 b) Q_2 c) Q_3 d) P_{45} e) P_{89}

THE NORMAL DISTRIBUTION—INTERPRETING AND UNDERSTANDING THE STANDARD DEVIATION

The **standard deviation** is a measure of dispersion or spread in the data. It has been defined as a value calculated with the use of a number of steps. But one wonders what it really is. The standard deviation is a kind of "yardstick" by which we can compare one set of data with another.

NORMAL CURVE

To make comparisons, we use a bell-shaped curve, called a **normal distribution**. The normal distribution is symmetric about the mean. If you were to fold the curve down the middle, the left side would fit exactly on the right side. In a normal distribution the mean, median, and mode all have about the same value. A sketch of the normal curve is given below.

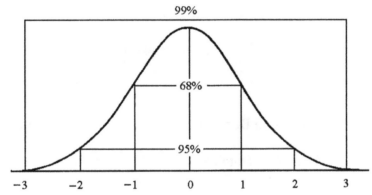

Since the curve is symmetric, 50% of the data always falls above (to the right of) the mean and 50% of the data falls below (to the left of) the mean. Additionally, every normal distribution has approximately 68% of the data between the value that is one standard deviation below the mean and the value that is one standard deviation above the mean (see figure above); 95% between the value that is two standard deviations below to two standard deviations above the mean; and approximately 99% falls within three standard deviations below to three standard deviations above.

Properties of the normal distribution summarized

---

1) The total area under the curve is 1 or 100%.

2) The mean divides the curve in half so that 50% of the scores are above the mean and 50% of the scores are below the mean.

3) The percentages in the normal curve are cumulative, therefore:

a) Roughly 2/3 or 68% of the scores are within one standard deviation below (-1) to one standard deviation above $(+1)$ the mean.

b) Roughly 95% of the scores are within two standard deviations below (-2) to two standard deviations above $(+2)$ the mean.

c) Roughly 99% of the scores are within three standard deviations below (-3) to three standard deviations above $(+3)$ the mean.

Thus if a normal distribution has mean of 100 and a standard deviation of 10, then the following would hold true.

Percentage of Scores	Would Fall Between	Score Interval
68%	Mean -1 S.D. to Mean $+1$ S.D. $100 - 10$ to $1000 + 10$	90 to 110
95%	Mean -2 S.D. to Mean $+2$ S.D. $100 - 20$ to $100 + 20$	80 to 120
99%	Mean -3 S.D. to Mean $+3$ S.D. $100 - 30$ to $100 + 30$	70 to 130

z-Scores

There are many normal distributions, each determined by its mean and standard deviation. While this makes the normal distribution very useful in describing many real world problems, it would be very difficult to provide tables of areas for each normal distribution. To avoid this problem, we use a **z-score or standard score** to determine how far, in terms of standard deviations, a given score is from the mean. For example, a score that has a z-value of 1.5 means that the score is 1.5 standard deviations above the mean. The formula for the z-score follows.

$$z = \frac{x - \bar{x}}{\text{s.d.}} = \frac{\text{score} - \text{mean}}{\text{standard deviation}} \text{ where}$$

z is the standard deviation value or z-score

x is the score of interest

$\bar{x}$ is the mean of the normal distribution

s.d. is the standard deviation of the normal distribution

Let us use the formula to find the z-values of several scores.

EXAMPLE 1

A normal distribution has a mean of 90 and a standard deviation of 9. Find the z-scores for the following values.

a) 99 b) 108 c) 90 d) 78

SOLUTION

The mean is 90 and standard deviation is 9. Substitute these numbers and the given score in the formula for each value. The formula for the z-score is

$$z = \frac{x - \bar{x}}{\text{s.d.}} = \frac{\text{score} - \text{mean}}{\text{standard deviation}}.$$

a) $z_{99} = \dfrac{x - \bar{x}}{\text{s.d.}} = \dfrac{99 - 90}{9} = \dfrac{9}{9} = 1$. A score of 99 is one standard deviation above the mean.

b) $z_{108} = \dfrac{x - \bar{x}}{\text{s.d.}} = \dfrac{108 - 90}{9} = \dfrac{18}{9} = 2$

c) $z_{90} = \dfrac{x - \bar{x}}{\text{s.d.}} = \dfrac{90 - 90}{9} = \dfrac{0}{9} = 0$. The mean always has a z-score of 0.

d) $z_{78} = \dfrac{x - \bar{x}}{\text{s.d.}} = \dfrac{78 - 90}{9} = \dfrac{-12}{9} = -1.33$. A score of 78 is 1.33 standard deviations below the mean.

Using the Standard Normal Distribution Table

To find the percentage of scores between any two intervals in a normal distribution we use the **standard normal distribution table** (Table 10.1). The table gives the percentage of scores falling below a given z-score for the normal distribution. The percentage corresponding to any observation from a normal distribution can be found by converting the observation to a corresponding z-score and then looking in the table. For example, suppose a z-score on a test is 1.5. What percent of the students scored below this score? Since the table gives values falling below a given z-score, we look up 1.5 in the table and the area corresponding to it is .9332. To convert this to a percent we simply multiply by 100 $(.9332 \times 100)$ to get 93.32%. More detailed tables in statistics books show z-scores carried out to more decimal places. Therefore, we will round our z-scores to the nearest tenth.

Table 10.1 Table of Areas Under the Normal Curve (Standard Normal Distribution Table)

z-score	Area below	z-score	Area below	z-score	Area below
−3.0	.0013	−1.0	.1587	1.1	.8643
−2.9	.0019	−0.9	.1841	1.2	.8849
−2.8	.0026	−0.8	.2119	1.3	9032
−2.7	.0035	−0.7	.2420	1.4	.9192
−2.6	.0047	−0.6	.2743	1.5	.9332
−2.5	.0062	−0.5	.3085	1.6	.9452
−2.4	.0082	−0.4	.3446	1.7	.9554
−2.3	.0107	−0.3	.3821	1.8	.9641
−2.2	.0139	−0.2	.4207	1.9	.9713
−2.1	.0179	−0.1	.4602	2.0	.9773
−2.0	.0227	0.0	.5000	2.1	.9821
−1.9	.0287	0.1	.5398	2.2	.9861
−1.8	.0359	0.2	.5793	2.3	.9893
−1.7	.0446	0.3	.6179	2.4	.9918
−1.6	.0548	0.4	.6554	2.5	.9938
−1.5	.0668	0.5	.6915	2.6	.9953
−1.4	.0808	0.6	.7258	2.7	.9965
−1.3	.0968	0.7	.7580	2.8	.9974
−1.2	.1151	0.8	.7781	2.9	.9981
−1.1	.1357	0.9	.8159	3.0	.9987
		1.0	.8413		

EXAMPLE 2

Two hundred students took a mathematics final examination. The mean score was 75 and the standard deviation was 10.

a) Use the properties of the normal curve to determine the intervals in which 68%, 95%, and 99% of the scores fall.

b) What percent of the scores are between 75 and 85?

c) How many students made scores between 70 and 90?

d) How many of the scores are more than 1.5 standard deviations above the mean?

e) How many scores are greater than 95?

SOLUTION

a) If a normal distribution has a mean of 75 and a standard deviation of 10, then the following would hold true.

Percentage of Scores	Would Fall Between	Score Interval
68%	Mean -1 S.D. to Mean $+1$ S.D.	65 to 85
	$75-10$ to $75+10$	
95%	Mean -2 S.D. to Mean $+2$ S.D.	55 to 95
	$75-10$ to $75+20$	
99%	Mean -3 S.D. to Mean $+3$ S.D.	45 to 105
	$75-30$ to $75+30$	

The results are displayed in the normal curve graph sketched below.

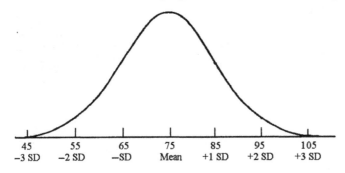

45	55	65	75	85	95	105
-3 SD	-2 SD	$-$SD	Mean	$+1$ SD	$+2$ SD	$+3$ SD

b) First, find the z-score for each value.

The z-score for 85 is $z_{85} = \dfrac{x-\overline{x}}{\text{s.d.}} = \dfrac{85-75}{10} = \dfrac{10}{10} = 1.$

The z-score for 75 is $z_{75} = \dfrac{x-\overline{x}}{\text{s.d.}} = \dfrac{75-75}{10} = \dfrac{0}{10} = 0.$

Second, look up the area in the table. The graph below shows that we want the area between 0 and 1, which we find by subtracting.

453

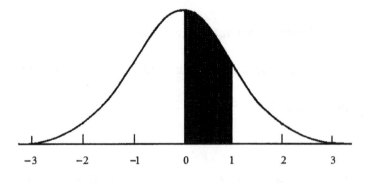

For $z = 1$, the area is .8413
For $z = 0$, the area is .5000
Subtracting gives .3413 or 34.13%

c) First, find the z-score for 70 and 90.

$$z_{90} = \frac{x - \bar{x}}{\text{s.d.}} = \frac{90 - 75}{10} = \frac{15}{10} = 1.5 \text{ and } z_{70} = \frac{x - \bar{x}}{\text{s.d.}} = \frac{70 - 75}{10} = \frac{-5}{10} = -.5$$

Second, find the area for each z-score and subtract to find the area between.

$z = 1.5$, the area is .9332
$z = -.5$ the area is .3085
            .6247

Third, to find the number of students making scores in this range, multiply .6247 by 200 to get approximately 125 students.

d) We want the area to the right of $z = 1.5$. See the graph below. Since the total area under the curve is 1, we subtract this area from 1 to find the area to the right of $z = 1.5$.

Total area under the curve   1.0000
Area for $z = 1.5$ .9332
Area greater than $z = 1$ .0668

To find the proportion of scores greater than a z-score of 1.5, multiply 6.68% by 200 to get 13.36 or approximately 13 students.

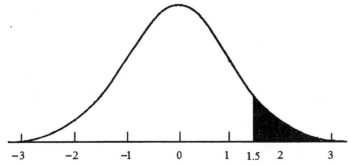

e) To find the number of scores greater than 95 we convert the value to a z-score and look up the area.

$$z_{95} = \frac{x - \bar{x}}{\text{s.d.}} = \frac{95 - 75}{10} = \frac{20}{10} = 2.$$

A z-score of 2 has an area of .9773. To find the scores greater than this, subtract the area from 1 $(1 - .9773)$ to get .0227 or 2.27%. Multiplying 200 by .0227 we get 4.54 or approximately 5.

Percentiles

The normal curve may be used to find percentiles and percentile ranks. Recall that a percentile is a score that divides a distribution into two parts. For example, the 87th percentile is the score at which 13% of the scores are above and 87% are below.

EXAMPLE 3

Ron made a score of 87 on an exam where the mean was 75 and the standard deviation was 8. In which percentile is his score of 87?

SOLUTION

Since the Normal Distribution Table gives the cumulative area up to the z-score, we find the z-score of the score and then note the area. When this area is converted to a percent, we have found the percentile.

1) Find the z-score for 87. $\quad Z_{87} = \dfrac{x - \bar{x}}{\text{s.d.}} = \dfrac{87 - 75}{8} = \dfrac{12}{8} = 1.5$

2) Look up the z-score. The area for a z-score of 1.5 is .9332.

Therefore, a score of 87 is at the 93.32 percentile.

EXAMPLE 4

Find the score associated with the 67th percentile for IQ scores with a mean of 100 and a standard deviation of 10.

SOLUTION

Find the z-score associated with an area of .67 in the Normal Distribution Table. That z-score is .44. Now substitute in the z-score formula and solve for x.

$$z = \frac{x - \bar{x}}{s.d.} \qquad .44 = \frac{x - 100}{10} \qquad \text{Cross multiply}$$

$$x - 100 = (.44)(10)$$

$$x - 100 = 4.4 \qquad \text{Add 100 to both sides}$$

$$x = 104.4$$

EXERCISE 10.3

1) The table below shows the percentile rank of scores achieved on a national test.

Score	Percentile
800	99
725	84
650	50
575	50
500	42
425	29
350	18
275	10
200	1

a) What is the mean?
b) What is the mode?
c) What percent of scores are between 425 and 275?

2) The GPA for students at a certain college and the percentile rank are given in the following table.

GPA	Percentile rank
4.00	99
3.50	88
3.00	68
2.50	50
2.00	26
1.50	16
1.00	12
0.50	5

a) What is the mean?
b) What is the mode?
c) What percent of scores are less than 2.00?

What percent of the total population is found between the mean and the z-score given in Problems 3 – 14?

3) $z = 1.4$

4) $z = 1.86$

5) $z = -2.33$

6) $z = -1.19$

7) $z = 0.3$

8) $z = 3.25$

9) $z = -0.50$

10) $z = -2.22$

11) $z = 2.43$

12) $z = -0.6$

13) $z = -0.46$

14) $z = -3.41$

In Problems 15 – 18, suppose that people's heights (in centimeters) are normally distributed, with a mean of 170 and a standard deviation of 5. We find the heights of 50 people.

15) a) How many would you expect to be between 165 and 175 cm tall?

b) How many would you expect to be taller than 168 cm?

16) a) How many would you expect to be between 170 and 180 cm?
   b) How many would you expect to be taller than 176 cm?

17) What is the probability that a person selected at random is taller than 163 cm?

18) What is the variance in heights for this experiment?

In Problems 19 – 23, suppose that, for a certain exam, a teacher grades on a curve. It is known that the mean is 50 and the standard deviation is 5. There are 45 students in the class.

19) How many students should receive a C?

20) How many students should receive an A?

21) What score would be necessary to obtain an A?

22) If an exam paper is selected at random, what is the probability that it will be a failing paper?

23) What is the variance in scores for the exam?

PROBABILITY

"There is a 60% chance it will rain today." "If you toss a coin in the air, there is a 50% chance it will land on tails." These statements describe the probability of an event. **Probability** is the science that estimates the likelihood of the occurrence of an event. For example, suppose an urn contains 4 red balls, 6 blue balls, and 5 yellow balls. Each of the 15 balls that can be selected from the urn represents an outcome. Let A be the event of selecting a yellow ball from the urn without looking. The probability that event A will occur, denoted $P(A)$, is equal to the number of favorable outcomes (5 yellow balls) divided by the total number of possible outcomes (15 balls). Therefore,

$$P(A) = \frac{\text{number of yellow balls}}{\text{total number of balls in urn}} = \frac{5}{15} = \frac{1}{3}.$$

In general, if all outcomes are equally likely to occur, then the probability of an event, for example, A, will be defined as follows:

$$P(A) = \frac{\text{number of favorable outcomes for event } A}{\text{total number of possible outcomes}}.$$

We call the total number of possible outcomes the **sample space**, denoted **S**.

Probabilities may be expressed as fractions, ratios, decimals, or percents.

Properties of Probability

1) For any event A, the probability of event A occurring, denoted $P(A)$ is some number between 0 and 1.

2) The closer to 0 the probability is, the less likely the event will occur. The closer to 1, the more likely it is that the event will occur.

3) The probability of the sample space is 1, $P(S) = 1$. That is, if an event is sure to happen, then its probability is 1.

4) The probability that an event cannot occur is 0.

Finding the Probability of an Event

EXAMPLE 1

A six-sided die is rolled once. What is the probability of getting an even number?

SOLUTION

First, determine the possible number of outcomes, the sample space.
$S = \{1, 2, 3, 4, 5, 6\}$.

Second, find the number of elements in the event. Let A be the event, the die lands on an even number, then A = {2, 4, 6}.

The probability of getting an even number then is,

$$P(A) = \frac{\text{number of successful outcomes, even numbers}}{\text{total number of possible outcomes}} = \frac{3}{6} = \frac{1}{2}$$

EXAMPLE 2

Suppose that there is an equally likely chance that the spinner will stop on any one of the eight numbers. What is the probability that it will stop on a number less than 6?

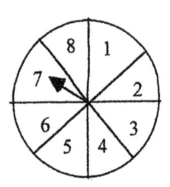

SOLUTION

Determine the sample space and number of elements in the event.

Sample space, $S = \{1, 2, 3, 4, 5, 6, 7, 8\}$ the total possible outcomes.

Event, numbers less than 6, $A = \{1, 2, 3, 4, 5\}$

$$P(A) = \frac{\text{number of successful outcomes, numbers less than 6}}{\text{total number of possible outcomes}} = \frac{5}{8}$$

EXAMPLE 3

The enrollment in a statistics class consisted of the following majors:

Major	Number of students
Mathematics	14
Computer Science	10
Engineering	3
Science	7
Social work	8

What is the probability of selecting?
a) a science major
b) a non-social work major?

SOLUTION

Determine the sample space and the number of elements in the event.

a) Sample space equals total enrollment in the class, 42 students.
Event A, the number of science majors, is 7.

$$P(A) = \frac{\text{number of successful outcomes, science majors}}{\text{total number of possible outcomes, class enrollment}} = \frac{7}{42} = \frac{1}{6} = .17$$

b) Sample space = {total class enrollment of 42}

Event A is non-social work majors, so that would be students majoring in mathematics, computer science, engineering, or science $A = \{34\}$

$$P(A) = \frac{\text{number of successful outcomes, non-social work majors}}{\text{total number of possible outcomes, class enrollment}} = \frac{34}{42} = .81$$

Multiple Trials—Product Tables

When the experiment consists of multiple trials, for example, tossing a coin three times or a pair of dice, we can find the sample space by using a product table or a tree diagram.

EXAMPLE 4

What is the probability of forming an even two-digit number using the digits?
$\{1, 2, 3\}$

SOLUTION

This task consists of multiple trials. We choose the first digit, which can occur in any of three ways (1, 2, 3), and then we choose the second digit, which can occur in any of three ways (1, 2, 3). We can use a product table to make the work easier.

		First Digit		
		1	2	3
	1	1 1	1 2	1 3
Second Digit				
	2	2 1	2 2	2 3
	3	3 1	3 2	3 3

There are a total of 9 possible outcomes. Of the nine, 3 numbers are even, so,

$$P(A) = \frac{\text{number of successful outcomes, even numbers}}{\text{total number of possible outcomes}} = \frac{3}{9} = \frac{1}{3}.$$

EXAMPLE 5

What is the probability of getting two heads and a tail in the toss of three coins?

SOLUTION

For each toss, the possible outcome is heads (H) or tails (T). Form the cross product for coin 1 and coin 2, and then use this result with the cross product of coin 3.

Coin 1 Result of Coins 1 and 2

		H	T
	H	HH	HT
Coin 2			
	T	TH	TT

		HH	HT	TH	TT
	H	HHH	HHT	HTH	HTT
Coin 3					
	T	THH	THT	TTH	TTT

S = {HHH HHT HTH HTT THH THT TTH TTT}

Event A = {getting two heads and a tail} = {HHT< HTH< THH}

$$P(A) = \frac{\text{number of successful outcomes, two heads \& a tail}}{\text{total number of possible outcomes}} = \frac{3}{8} = .375 = .38$$

EXERCISE 10.4

Find the probability for the following.

1) A jar contains balls with the numbers 1, 2, 3, 4, 5, 6, 7, and 8. If a ball is selected at random, what is the probability of?
   a) selecting a ball with an odd number
   b) selecting a ball with a number greater than 2
   c) selecting a ball with a number less than 9

2) Two four-sided dice with the numbers 1, 2, 3, and 4 on each are rolled. What is the probability of?
   a) getting doubles (same number on both)
   b) getting an even number on the first and an odd number on the second
   c) getting an odd number on both dice
   d) getting an even sum
   e) getting a sum greater than 5

3) A couple wants to have three children. Assuming that there are no multiple births, what is the probability of the couple having?
   a) all three girls
   b) two girls and a boy
   c) two boys first and then a girl
   d) having a boy, a girl, and a boy

4) The U.S. Centers for Disease Control and Prevention reported the following statistics for deaths per thousand by cause for 1990.

Cause of Death	Number of Thousands
Smoking	434
Alcohol (including drunk driving)	105
Car Accidents (including drunk driving)	49
Fires	4
AIDS	31
Suicide	31
Homicide	22
Illegal Drugs	6

What is the probability that a person died from
a) smoking
b) illegal drugs
c) AIDS

5) On a multiple choice test, each question has five possible answers. If you randomly guessed on the last question, what is the probability that you answered the question?
   a) correctly
   b) incorrectly

6) In a class, the distribution of students by classification is as follows:

Freshmen	8
Sophomores	12
Juniors	18
Seniors	20

What is the probability of randomly selecting a junior or senior?

7) On a class exam, there were 4 A's, 6 B's, 12 C's, 5 D's, and 3 F's. What is the probability that a randomly selected student
   a) made an A
   b) made a C
   c) passed the exam, if a D or better is considered passing

8) Two balanced dice are rolled. What is the probability of obtaining the following?
   a) sum of 4
   b) sum of 9
   c) even sum
   d) sum of 10 or more
   e) sum of 2, 3, 11, or 12
   f) sum of 20

9) An ordinary deck of playing cards consists of 52 cards. These 52 cards are divided into four suits of 13 cards each (spades, hearts, diamonds, clubs). If a single card is drawn at random from the deck, find the probability of getting
   a) an ace
   b) a heart
   c) a spade
   d) a red card
   e) a red card or a black card

10) A coin is tossed two times, what is the probability of getting
   a) two heads
   b) one head
   c) no heads

COMBINING EVENTS AND ODDS

In many applications of probability, there is often a need to combine the probabilities of related events. Generally, events formed using the word "or" can be combined by using the addition rule and those formed using the word "and" can be combined using the multiplication rule.

Addition Rule Is Used For "Or"

> The probability of event A occurring or event B occurring is given by the formula:
> $$P(A \text{ or } B) = P(A) + P(B) - P(A \text{ and } B)$$
> $$P(A \cup B) = P(A) + P(B) - P(A \cap B)$$

EXAMPLE 1

In the toss of a six-sided die, what is the probability of getting an even number or a 5?

SOLUTION

The possible outcomes (sample space) is 1, 2, 3, 4, 5, and 6.

Let event A be getting an even number, $A = \{2,4,6\}$ and $P(A) = \dfrac{3}{6}$

Let event B be getting a 5, $B = \{5\}$ and $P(B) = \dfrac{1}{6}$

Since no elements are in common for sets A and B, $A \cap B = \{\ \}$ and $P(A \cup B) = 0$

We want the probability of A or B so we use the formula:
$$P(A \cup B) = P(A) + P(B) - P(A \cap B)$$
$$= \frac{3}{6} + \frac{1}{6} - \frac{0}{6} = \frac{4}{6} = \frac{2}{3}$$

EXAMPLE 2

Suppose that there is an equally likely probability that the spinner will stop at any one of the eight numbers. What is the probability of getting?

   a) 1 or 5
   b) An even number or 7
   c) An even number or 4
   d) An odd number or a number greater than 4

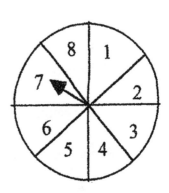

SOLUTION

The possible outcomes are 1, 2, 3, 4, 5, 6, 7, and 8. Find the probability for each event and then use the addition rule.

a) Event A is getting a 1, so $A = \{1\}$ and $P(A) = \dfrac{1}{8}$

Event B is getting a 5, so $B = \{5\}$ and $P\{B\} = \dfrac{1}{8}$

$A \cap B = \{\ \}$, therefore $P(A \cap B) = 0$.

$P(A \cup B) = P(A) + P(B) - P(A \cap B)$

$\quad = \dfrac{1}{8} + \dfrac{1}{8} - 0 \qquad = \dfrac{2}{8} = \dfrac{1}{4} = .25$

b) Event A is getting an even number, so $A = \{2,4,6,8\}$ and $P(A) = \dfrac{4}{8}$

Event B is getting a 7, so $P(B) = \dfrac{1}{8}$

$A \cap B = \{\ \}$ therefore $P(A \cap B) = 0$

$P(A \cup B) = P(A) + P(B) - P(A \cap B)$

$\quad = \dfrac{4}{8} + \dfrac{1}{8} - \dfrac{0}{8} \qquad = \dfrac{5}{8} = .625 = .63$

c) Event A is getting an even number, so $A = \{2,4,6,8\}$ and $P(A) = \dfrac{4}{8}$

Event B is getting a 4, so $B = \{4\}$ and $P(B) = \dfrac{1}{8}$

$A = \{2,4,6,8\}$ and $B = \{4\}$, therefore, $A \cap B = \{4\}$ and $P(A \cap B) = \dfrac{1}{8}$

$P(A \cup B) = P(A) + P(B) - P(A \cap B)$

$\quad = \dfrac{4}{8} + \dfrac{1}{8} - \dfrac{1}{8} \qquad = \dfrac{5}{8} - \dfrac{1}{8} = \dfrac{4}{8} = .5$

Observe that 4 was common to both events, therefore, we subtracted out the probability of getting a 4 to avoid double counting.

d) Event A is getting a odd number, so $A = \{1,3,5,7\}$ and $P(A) = \dfrac{4}{8}$

Event B is getting a number greater than 4, so $B = \{5,6,7,8\}$ and $P(a) = \dfrac{4}{8}$

$A = \{1,3,5,7\}$ and $B = \{5,6,7,8\}$, thus, $A \cap B = \{5,7\}$ and $P(A \cap B) = \dfrac{2}{8}$

$$P(A \cup B) = P(A) + P(B) - P(A \cap B)$$
$$= \frac{4}{8} + \frac{4}{8} - \frac{2}{8} = \frac{6}{8} = \frac{3}{4} = .75$$

Multiplication Rule Is Used For "And"

> The probability of event A occurring and event B occurring is given by the formula:
>
> $$P(A \text{ and } B) = P(A)P(B) \text{ or } P(A \cap B) = P(A)P(B)$$
>
> Where A and B are independent events.

By **independent**, we mean that the occurrence of one event does not affect the outcome of the other event. For example, rolling a six-sided die and tossing a coin are independent events, since the occurrence of one does not affect the outcome of the other.

When trying to determine whether the multiplication rule applies, the word "and" will not always occur. However, words such as both, the same, all, etc. generally suggest "and."

EXAMPLE 3

The two wheels below are spun once. Find the probability that the first spinner stops on a number less than 3 and the second spinner stops on an even number.

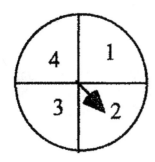 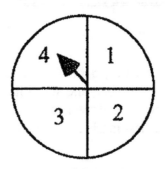

SOLUTION

The sample space for the experiment is

1,1	1,2	1,3	1,4
2,1	2,2	2,3	2,4
3,1	3,2	3,3	3,4
4,1	4,2	4,3	4,4

We need to find the elements in which the first number is less than 3 and the second number is even. The elements satisfying both conditions are

1,2   1,4   2,2   and   2,4

Since there are 4 elements, $P(A \text{ and } B) = \dfrac{4}{16} = \dfrac{1}{4} = .25$.

By the multiplication rule, we find the probability for each event separately, and then multiply.

Let A be the event first number less than three, then:

$A = \{(1,1),(1,2),(1,3),(1,4),(2,1),(2,2),(2,3),(2,4)\}$ and $P(A) = \dfrac{8}{16} = \dfrac{1}{2}$

Let B be the event the second number is even, then:

$B = \{(1,2),(1,4),(2,2),(2,4),(3,2),(3,4),(4,2),(4,4)\}$ and $P(B) = \dfrac{8}{16} = \dfrac{1}{2}$

$$P(A \cap B) = P(A)P(B) = \left(\dfrac{1}{13}\right)\left(\dfrac{1}{13}\right) = \dfrac{1}{169} = .0059 = .006.$$

Conditional Probability

> The **conditional probability** of B and A, denoted $P\left(\dfrac{B}{A}\right)$ refers to the probability of obtaining B when we know the outcome of A.

EXAMPLE 4

A bag contains ten marbles, 5 black, 3 red, and 2 white. An experiment consists of drawing marbles out of the bag one at a time, but not replacing a marble after it has been drawn. If two marbles are drawn, what is the probability of drawing a white marble second, given that the first marble drawn was red?

SOLUTION

Once the red marble is drawn, only 9 marbles are left, therefore, P(selecting a white given that the first was red) = P(white/red) = $\dfrac{2}{9}$.

> For events involving conditional probability, the multiplication rule becomes
>
> $$P(A \text{ and } B) = P(A)P\left(\dfrac{B}{A}\right)$$

EXAMPLE 5

Two cards are drawn from a standard deck of 52 playing cards, without replacement. Find the probability that they are both diamonds.

SOLUTION

There are 13 diamonds in the deck. On the first draw there are 13 diamonds of 52 cards. Therefore $P(\text{selecting a diamond}) = \dfrac{13}{52}$.

Assuming that a diamond was selected the first time, on the second draw, there are 12 diamonds of 51 cards. So, $P(\text{selecting a diamond again}) = \dfrac{12}{51}$.

The word "both" suggests the use of the multiplication rule. The events are dependent, therefore,

$$P(A \text{ and } B) = P(A)P\left(\frac{B}{A}\right) = \left(\frac{13}{52}\right)\left(\frac{12}{51}\right) = \frac{156}{2652} = .059 = .06.$$

Odds

Sometimes probability statements are given in terms of odds. Odds are commonly quoted, rather than probabilities, in horse racing, lotteries, and other gaming situation.

> **Odds** is the ratio that the probability of the event will occur to the probability that the event will not occur.

EXAMPLE 6

Find the odds in favor of rolling a 4 with a single die.

SOLUTION

$P(\text{rolling a 4}) = \dfrac{1}{6}$, $P(\text{not rolling a 4}) = \dfrac{5}{6}$.

Therefore the odds of rolling a $4 = P(\text{rolling a 4}) \div P(\text{not rolling a 4})$

$$\frac{1}{6} \text{ to } \frac{5}{6} \text{ or 1 to 5.}$$

Obtaining Probabilities From Odds

Sometimes we are given the odds in favor of an event, and we want to obtain the probability that it will occur.

> If the odds favoring an event E are m and n, then $P(E) = \dfrac{m}{m+n}$

EXAMPLE 7

On a game show, your odds of selecting the door with the grand prize are 1 to 2. What is the probability of your selecting the right door?

SOLUTION

For the given odds, $m = 1$ and $n = 2$, so, P(selecting right door) $= \dfrac{1}{1+2} = \dfrac{1}{3}$.

EXERCISE 10.5

1) Three coins are tossed, what is the probability of getting all three heads?

2) A pair of dice are tossed, what is the probability of getting
   a) a sum of 7 or 11
   b) a sum greater than 9 or an odd sum
   c) a sum of 9 or an even sum
   d) a sum of 5 or 3

3) A jar contains 4 black balls, 6 white balls, and 5 red balls. If a ball is drawn, what is the probability of getting
   a) a black ball
   b) a red ball
   c) a white ball
   d) a red or white ball
   e) a black or white ball

4) If A and B are events with $P(A) = 7$, $P(B) = 4$, and $P(A \text{ and } B) = .2$, find $P(A \text{ and } B)$.

5) At a certain university, 25% of the freshmen make A's in mathematics, 30% make A's in English, and 14% make A's in both mathematics and English. What is the probability that a freshman makes an A in mathematics or English?

6) If you randomly guess on a five-question true-false test, what is the probability that you will get all of the questions correct?

7) Two cards are drawn from a standard deck of 52 playing cards with replacement. What is the probability of obtaining the following?
   a) a jack or a spade
   b) a face card (jack, queen, king) or club
   c) two aces
   d) two hearts
   e) a 5 and a jack
   f) one jack and one queen
   g) no face cards

8) Two cards are drawn from a standard deck of 52 playing cards. The first card is drawn but not replaced. What is the probability of obtaining the following?
   a) a spade second, given a spade as the first card
   b) two aces
   c) two hearts
   d) a 5 and a jack
   e) one jack and one queen

9) Find the odds in favor of drawing a heart from an ordinary deck of cards.

10) A die is rolled. What are the odds that a 3 will occur?

11) On a multiple question with 4 possible answers, what are the odds that you will answer the question correctly by merely guessing?

12) The odds of answering a question correctly on a test is 1 to 4. What is the probability of answering the question correctly?

13) The odds that it will rain today are 1 to 3. What is the probability that it will rain.

14) The number of symbols on each of the three dials of a standard slot machine is shown in the table.

Symbol	Dial 1	Dial 2	Dial 3
Cherries	2	4	4
Oranges	5	4	5
Plums	6	3	4
Bells	2	4	3
Melons	2	2	2
Bars	2	2	1
7's	1	1	1

What is the probability of getting
a) an orange on the first dial
b) plums on all three dials
c) bars on all three dials
d) 7 on all three dials to win the jackpot

15) Akilah's problem: There are eight cars, 3 Tauruses and 5 Altimas. Three of the cars are green and five of the cars are black. What is the probability of selecting a green Taurus? A black Altima?

CHAPTER 10 REVIEW EXERCISES

List two of each of the following.

1) Measures of central tendency

2) Measures of variability

3) Branches of statistics

4) Characteristics of the normal curve

5) Properties of probability

Define the following terms.

6) Statistics

7) Inferential statistics

8) Descriptive statistics

9) Statistic

10) Percentile

Solve the following problems.

11) When money was collected for a retirement gift, 10 people each contributed $8.00, 12 each contributed $6.00, and 8 people each contributed $4.00. What was the mean contribution of all 30 people?

12) Given the distribution 24, 28, 45, 50, 54, 24, 48 and 40, find
    a) mean b) median c) mode
    d) range e) standard deviation f) Q_1
    g) P_{40}

13) For two events A and B, $P(A) = .76$, $P(B) = .13$, and $P(A$ and $B) = .12$, find $P(A$ or $B)$

14) One card is drawn from a standard deck of 52 playing cards. Find the probability that it is a
    a) spade b) spade or jack c) queen and a heart

15) The mean amount spent by a family of four on food per month is $500 with a standard deviation of $75. Assuming that the food expenditures are normally distributed, what is the proportion of families that spend less than $410?

16) A multiple choice question on a test has 5 possible options. If you randomly guessed, what is the odds that you guessed correctly?

17) Given Data Set I: 2, 9, 1, 8, 6, 4 and Data Set II: 17, 1, 7, 9, 13, 6. Compare Data Sets I and II. Which data set is more variable? What measure of central tendency would you use to report the average of Data Set I? Data Set II? Justify your answer statistically.

18) If you surveyed 10 of your friends and determined the television network (NBC, CBS, ABC, Fox) that they watched the most, what measure of central tendency would you use to report your findings?

19) Determine the heights of 10 of your friends. What average would you report and why?

CHAPTER 10 TEST

1) The range is a measure of
   a) central tendency
   b) variability   c) correlation
   d) none of these

2) In general, the standard deviation is associated with the use of the
   a) median   b) mean
   c) mode   d) none of these

3) Arithmetic mean is to central tendency as standard deviation is to
   a) average   b) variability
   c) SD   d) none of these

4) A company has 5 employees. The president of the company makes $85,000 annually. The two supervisors each made $20,000 annually and the other employees make $10,000 and $5,000 annually. In reporting the "average" salary for the company, which of the following measures of central tendency would be more appropriate?
   a. mean   b) mode
   c) median   d) none of these

5) Generalizing the results of a study based on a sample to make predictions about the whole population is known as
   a) descriptive statistics
   b) inferential statistics
   c) a and b
   d) none of the above

6) A teacher administered a test to 33 students. After correcting the papers she arranged the scores in order from lowest to highest and counted down to the 17th score

which was 70. This score of 70 was
   a) the median   b) the mean
   c) the mode   d) the range

7) A student made a score of 75 on the first hour exam and 65 on the second hour exam. If all of the exams were equally weighted, what score would the student have to have on the final exam to have an average of 73 in the course?
   a) 79   b) 70
   c) 80
   d) cannot be determined

8) Which measure of central tendency is least affected by extreme scores?
   a) mean
   b) standard deviation
   c) median   d) mode

9) Scores that are clustered closely around the mean will have a standard deviation that is
   a) relatively large
   b) relatively small
   c) both a and b
   d) neither a nor b

Use this distribution to answer questions 10–16: 3, 4, 3, 5, 11, 7, 9, and 6

10) The mean is
    a) 3   b) 5
    c) 6   d) 7

11) The mode is
    a) 3   b) 5
    c) 6   d) 7

475

12) The median is
   a) 3   b) 5
   c) 6   d) 5.5

13) The range is
   a) 8   b) 4
   c) 5   d) 10

14) The standard deviation is
   a) 2.24   b) 2.61
   c) 2.7   d) 7.2

15) The 3rd quartile is
   a) 7   b) 8
   c) 9   d) 11

16) P_{67} is
   a) 7   b) 8
   c) 9   d) 11

Use this information to answer questions 17–18.

Three hundred students took a final examination. The mean score was 70 and the standard deviation was 10.

17) What percentage of scores was between 60 and 80?
   a) all   b) 50
   c) 68   d) none

18) How many students made scores between 60 and 80?
   a) 300   b) 204
   c) 150   d) none

19) Too many students sign up for an algebra class at a given hour. It is decided to divide the group into classes, high and low in aptitude level according to students' scores on a mathematical aptitude test.
   a) mean
   b) percentile rank
   c) median   d) mode

20) Lamont made a score of 86 on his final examination. This score had a percentile rank of 90. How would you interpret this score?
   a) 90% of the students taking the exam performed better than Lamont.
   b) 10% of the students taking the exam scored below Lamont.
   c) 90% of the students taking the exam scored below Lamont and 10% scored higher than Lamont.
   d) none of the above

21) The nine basketball players on the home team were lined up from the shortest to the tallest. The median height is the height of the
   a) ninth player
   b) first player
   c) fifth player
   d) the median height cannot be determined

Use for questions 22–24:

The spinner below is spun once.

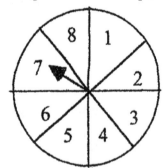

22) What is the probability of getting an odd number greater than 2?
   a) $\dfrac{6}{8}$ b) $\dfrac{3}{8}$
   c) $\dfrac{1}{2}$ d) none of these

23) What is the probability of getting an even number or a number less than 7?

a) $\dfrac{4}{8}$
b) $\dfrac{10}{8}$

c) $\dfrac{7}{8}$
d) none of these

24) What are the odds of the spinner stopping on 4?

a) 1 to 8
b) 4 to 8
c) 1 to 2
d) 1 to 7

Find the indicated probability. Round your answer to 6 decimal places when necessary.

25) Two 6-sided dice are rolled. What is the probability that the sum of the two numbers on the dice will be 4?

a) 3
b) $\dfrac{2}{3}$

c) $\dfrac{1}{12}$
d) $\dfrac{11}{12}$

26) A fair die is rolled. What is the probability of rolling a 3 or a 5?

a) $\dfrac{1}{6}$
b) $\dfrac{1}{36}$

c) 2
d) $\dfrac{1}{3}$

27) You are dealt one card from a 52-card deck. Find the probability that you are not dealt a 7.

a) $\dfrac{12}{13}$
b) $\dfrac{1}{10}$

c) $\dfrac{1}{13}$
d) $\dfrac{9}{10}$

28) When two balanced dice are rolled, there are 36 possible outcomes. What is the probability that the sum of the numbers on the dice is 6 or 10?

a) $\dfrac{4}{9}$
b) $\dfrac{1}{60}$

c) $\dfrac{4}{3}$
d) $\dfrac{2}{9}$

29) A die with 12 sides is rolled. What is the probability of rolling a number less than 11?

a) $\dfrac{5}{6}$
b) $\dfrac{1}{12}$

c) $\dfrac{11}{12}$
d) 10

30) If two fair dice are rolled, what is the probability of not rolling a sum of 12?

a) $\dfrac{5}{6}$
b) $\dfrac{1}{36}$

c) $\dfrac{35}{36}$
d) $\dfrac{1}{6}$

31) One card is selected from a deck of cards. Find the probability of selecting a black card or a king.

a) $\dfrac{1}{26}$
b) $\dfrac{27}{52}$

c) $\dfrac{7}{13}$
d) $\dfrac{15}{26}$

32) A bag contains 5 red marbles, 4 blue marbles, and 1 green marble. What is the probability of choosing a marble that is not blue?

a) 6
b) $\dfrac{2}{5}$

c) $\dfrac{5}{3}$
d) $\dfrac{3}{5}$

33) A class consists of 12 women and 71 men. If a student is randomly selected, what is the probability that the student is a woman?

a) $\dfrac{71}{83}$ b) $\dfrac{1}{83}$

c) $\dfrac{12}{83}$ d) $\dfrac{12}{71}$

34) Two marbles are drawn without replacement from a box with 3 white, 2 green, 2 red, and 1 blue marble. Find the probability that both marbles are white.

a) $\dfrac{9}{56}$ b) $\dfrac{3}{8}$

c) $\dfrac{3}{32}$ d) $\dfrac{3}{28}$

Find the mode(s) for the given sample data.

35) 65, 25, 65, 13, 25, 29, 56, 65
a) 25   b) 65
c) 42.5   d) 42.9

36) 20, 43, 46, 43, 49, 43, 49
a) 41.9   b) 43
c) 49   d) 46

Solve the problem.

37) Find the odds for getting a sum of 4 when two fair dice are rolled.
a) 11 to 1   b) 1 to 12
c) 12 to 1   d) 1 to 11

38) Find the odds for drawing an queen when a card is drawn at random from a normal deck of 52 playing cards.
a) 13 to 1   b) 1 to 13
c) 12 to 1   d) 1 to 12

Find the median for the given sample data.

39) The distances traveled (in miles) to 7 different swim meets are:
10, 24, 34, 46, 61, 73, 86. Find the median distance traveled.
a) 34 miles   b) 61 miles
c) 48 miles   d) 46 miles

40) 8, 14, 17, 21, 30, 30, 49
Find the median for the data.
a) 21   b) 25.5
c) 30   d) 17

Find the range for the given data.

41) The owner of a small manufacturing plant employs six people. The commute distances, in miles, for the six employees are listed below.
2.9 5.7 1.3 4.8 6.2 3.3
a) 1.3 mi   b) 4.9mi
c) 0.4mi   d) 5.7mi

42) Rich Borne is currently taking Chemistry 101. On the five laboratory assignments for the quarter, he got the following scores: 26 40 20 41 60
a) 14   b) 20
c) 60   d) 40

478

11 STATISTICAL GRAPHS AND CHARTS

Graphs are pictures or diagrams that show the relationship between quantities that can be counted or measured. Newspapers, magazines, and textbooks use many different kinds of graphs to present dates or figures quickly and clearly. Common types of graphs are the bar graph, the circle graph, and the line graph. The examples below will illustrate how graphs may be interpreted.

11.1 USES OF GRAPHS

Circle Graph

A **circle graph** is used when a quantity is divided into parts, and we wish to make comparisons of the parts. Thus, quantities that involve proportions or percents would appropriately be presented in a circle graph. The circle represents the whole quantity and the pie-like wedges of various sizes represent the parts. Essential features of a circle graph are (1) a descriptive title, (2) the wedges are proportional to the parts they represent with larger segments representing larger parts, (3) percents are used to label the circle graphs since percents can easily be interpreted to represent a part of the whole, and (4) the entire graph represents 100%.

EXAMPLE 1

Use Figure 11.1 to find the ratio of the annual fuel expense to the total annual cost of operating a car.

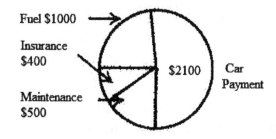

Annual car expenses totaling $4000.
Figure 11.1

479

SOLUTION

Locate the annual fuel expense in the circle graph. Write the ratio of the annual fuel expense to the total annual cost of operating a car in simplest form.

Annual fuel expense: $1000

$$\frac{\$1000}{\$4000} = \frac{1}{4}$$

The ratio is $\frac{1}{4}$.

EXAMPLE 2

The circle graph below describes business expenses totaling $300,000 for a small business. Use the graph to answer the questions below.

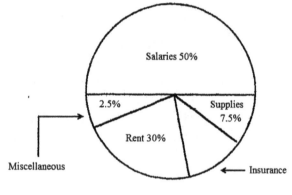

Figure 11.2

a) What percent of the expenses was expended for supplies and salaries?

b) What percent of the expenses was spent for rent?

c) What percent of the expenses was spent on insurance?

d) What were the dollar expenses for salaries, rent, insurance, supplies, and miscellaneous?

SOLUTION

a) To find the percent of the expenses for salaries and supplies, add 50% + 7.5%. The result is 57.5% for the two categories combined.

b) 30%

c) No percentage expenditure for insurance is listed, but the entire graph represents 100%. Summing the percentage expenditures for the other items and subtracting from 100% will give the percentage for insurance.

Insurance Percent = 100% − (Sum of the percentage for Salaries + Rent + Supplies + Miscellaneous)
= 100% − (50% + 30% + 7.5% + 2.5%)
= 100% − 90% = 10% for insurance.

480

d)   To find the dollar expenses for each category, multiply the total expenses of $300,000 by the respective percent for each expense item.

Salaries	50.0%	.50 × 300,000	= $150,000
Rent	30.0%	.30 × 300,000	= 90,000
Supplies	7.5%	.075 × 300,000	= 22,500
Insurance	10.0%	.10 × 300,000	= 30,000
Miscellaneous	2.5%	.0250 × 300,000	= 7,500

EXAMPLE 3

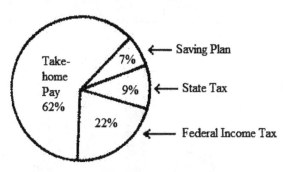

Distribution of gross monthly income of $2000.

Figure 11.3

Use Figure 11.3 to find the take-home pay for the employee.

SOLUTION

To find the take-home pay locate the percent of the distribution which is take-home pay in the circle graph then solve the basic percent equation for amount.

Take-home pay: 62%
Percent × base   = amount
0.62   × $2000 = $1240
The employee's take-home pay is $1240.

Bar Graph

> A **bar graph** uses bars to represent numerical facts. The bars may be horizontal or vertical. Essential features of a bar graph are a descriptive title and a number scale against which the lengths of the bars may be compared. The following example demonstrates how bar graphs are interpreted.

The bar graph in Figure 4 shows the market value of a home for each of five years. For each year, the height of the bar indicates the market value of the home.

The value of the home in 1991 was $90,000.

The value of the home in 1988 was $45,000.

481

Since the 1987 bar is the lowest bar in the graph, the home had the lowest market value in 1987.

A double-bar graph is used to display data for purposes of comparison. The double-bar graph in Figure 11.5 shows the quarterly profits for a company for the years 1999 and 1990.

The profit in the second quarter of 1999 was $14,000.

The profit in the second quarter of 1990 was $16,000.

In 1999 the company had the largest quarterly profit during the third quarter.

EXAMPLE 4

Use Figure 11.5 to find the difference between the company's fourth quarter profits for 1990 and 1999.

SOLUTION

To find the difference read the bar graph to find the fourth quarter profit for each year then subtract to find the difference between the profits.

1990 fourth quarter profit: $15,000
1999 fourth quarter profit: $13,000
$15,000 − $13,000 = $2000
The difference is $2000.

EXAMPLE 5

Title: Company Profit From Five Projects

Use the graph to answer the following

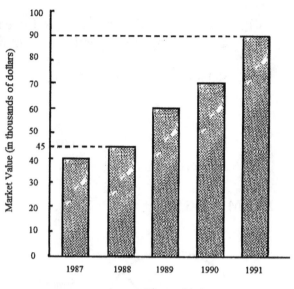

Figure 11.4

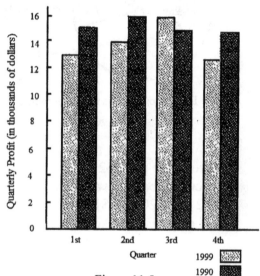

Figure 11.5

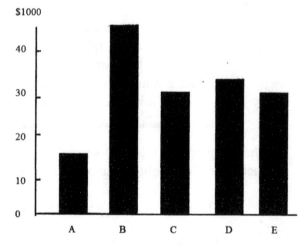

482

questions:

a) What is the maximum profit from any one project? What project?
b) What project yields the smallest profit?
c) Which projects netted approximately the same amount of profit?
d) By what amount did the project with maximum profit exceed the project with the smallest profit?
e) What was the average profit on all five projects?

SOLUTION

a) To find the maximum profit from any one project, locate the tallest bar. Project B has the tallest bar associated with it; therefore, the maximum profit was made on this project. Reading across to the vertical line, this bar corresponds to 45 or $45,000.

b) To find the project that yielded the smallest profit, locate the shortest bar. Project A has the shortest bar associated with it, therefore, the smallest profit was made on this project.

c) To find two projects that netted about the same amount of profit, locate two bars that are approximately the same height. Bars associated with projects C and E are about the same height. The profit on both of these projects was about $30,000.

d) The maximum profit of $45,000 was made on project B while the smallest profit of $15,000 was made on project A. Subtracting, $45,000 − $15,000 = $30,000 (the amount by which the project with the maximum profit exceeded the project with the smallest profit).

e) To find the average profit on all five projects, estimate the profit for each and then find the mean.

Project	Profit
A	$15,000
B	$45,000
C	$30,000
D	$35,000
E	$30,000
Total	$155,000

$$\text{Mean} = \frac{155,000}{5} = \$31,000$$

Line Graph

When the relationship between two quantities is being compared, a **line graph** is usually most informative. Essential features of a line graph are a descriptive title and the peaks of the line give the information for the intervals indicated.

The line graph in Figure 11.6 shows the number of new car sales by an automobile manufacturer for a five-week period. The height of each dot indicates the number of new car sales for each week.

The number of cars sold during Week 1 was 20,000.

The number of cars sold during Week 3 was 15,000.

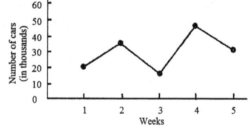

Figure 11.6

Since the highest dot in the graph is above Week 4, the greatest number of cars was sold during Week 4.

Two line graphs are often shown in the same figure for purposes of comparison. Figure 11.7 shows the number of private trucks and commercial trucks parked at an airport for a five-year period.

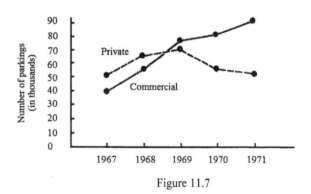

The number of private trucks parking in 1970 was 55,000.

The number of commercial trucks parking in 1970 was 80,000.

Figure 11.7

The greatest number of private trucks parking occurred in 1969.

484

EXAMPLE 5

Use Figure 11.7 to find the difference between the number of private trucks parking and commercial trucks parking in 1968.

SOLUTION

To find the difference read the line graphs to find the number of each type of truck that parked at the airport in 1968. Subtract to find the difference in parking.

Private truck parking: 65,000
Commercial truck parking: 55,000
$65,000 - 55,000 = 10,000$.
The difference is 10,000.

EXERCISE 11.1

The circle graph in Figure 1 shows how a college's annual income of $22,450,000 is budgeted.

1) Find the total amount of money budgeted for administrative costs and teacher salaries.

2) Find the amount of money budgeted for administrative costs.

3) Find the amount of money budgeted for supplies.

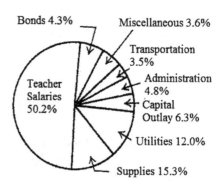

College budget of $22,450,000

Figure 1

The circle graph in Figure 2 shows the annual income groups of 60,000,000 wage earners in the United States.

4) How many people have an annual income between $3000 and $5999?

5) How many people have an annual income between $9000 and $11,999?

6) How many people have an annual income of over $15,000?

7) How many people have an annual income of less than $3000?

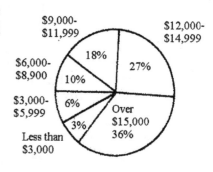

Annual income of 60 million U.S wage earners.

Figure 2

The circle graph in Figure 3 shows how a business spends its annual income of $30,000,000.

8) What is the ratio of operational costs to amount spent on salaries?

9) What is the ratio of operational costs to total income?

10) What is the ratio of renovations costs to total income?

11) What is the ratio of the amount spent on salaries to total income?

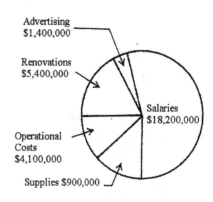

Business expenses totaling $30,000,000

Figure 3

The double-bar graph in Figure 4 shows the number of cars sold by an automobile manufacturer for the last six months of 1980 and 1981.

12) What is the difference between the last three months' sales for 1981 and 1980?

13) What was the total number of cars sold during July, August, and September of 1981?

14) What is the difference between October sales for 1980 and 1981?

15) How many cars were sold in November 1981?

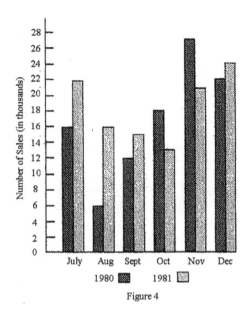

Figure 4

The double-bar graph in Figure 5 shows the number of watches sold at a store for the first six months of 1980 and 1981.

16) What is the difference between the January sales for 1981 and 1980?

17) In 1981, during which month were sales lowest?

18) In 1980, during which month were sales lowest?

19) Find the number of watches sold in February 1981.

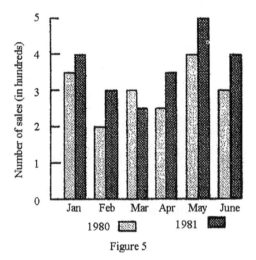

Figure 5

The bar graph in Figure 6 shows the number of shares of a stock traded on the New York Stock Exchange for each of the five days.

20) On which day was the greatest number of the shares traded?

21) How many shares of the stock were traded on Thursday?

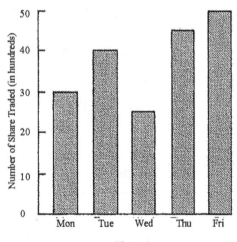

Figure 6

488

22) How many shares of the stock were traded on Wednesday?

23. How many shares of the stock were traded on Monday?

The double-line graph in Figure 7 shows the number of business calls and the number of residential calls made each hour from 9 A.M. 5 P.M. of an 8-hour business day in a small city.

24) What is the difference between the number of business calls and the number of residential calls made between 2 and 5 P.M.?

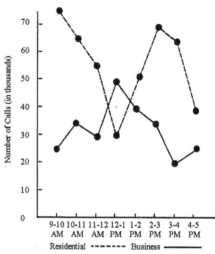

Figure 7

25) How many business calls were made between 9 A.M. and noon?

26) What is the difference between the number of business calls and the number of residential calls made between 11 A.M. and noon?

The double-line graph in Figure 8 shows the quarterly income for John for the years 1980 and 1981.

27) Find the difference between John's third-quarter incomes for 1981 and 1980.

28) In 1981, during which quarter did John have the highest income?

29) Find the income for the second quarter of 1980.

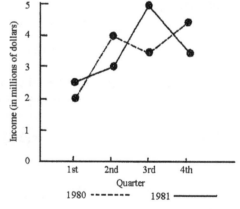

Figure 8

The line graph in Figure 9 shows a company's profits for each of the first six months of a year.

30) During which month was profit the lowest?

31) What was the company's profit for January?

32) What was the company's profit for March?

Use the graphs given to answer each question.

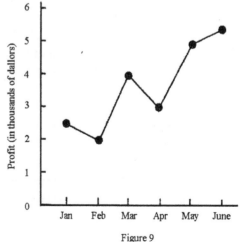

Figure 9

489

The bar graph on the right shows the average class size per year in the Mathematics Department of a medium size institution.

33) What was the average class size in 1992?

34) During which year was the average class the largest?

35) What was the average class size over the four-year period?

This circle graph represents the distribution of a college's enrollment composed of freshmen, sophomores, juniors, and seniors.

36) What percent of the college's enrollment is freshmen?

37) What is the ratio of the number of seniors to the number of freshmen?

38) If there are 3000 students in the college, how many are sophomores? How many are upperclassmen (juniors and seniors)?

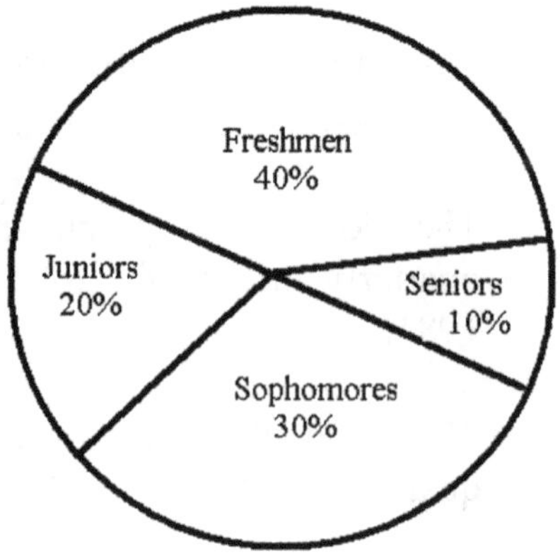

The graph below shows the total production of typewriters at the manufacturing company for 6 months.

39) Rank the months from lowest to highest in production.

40) What was the difference in production between the highest and the lowest months?

41) What was the total production over the six-month period?

42) What was the percent decrease in production from February to March?

43) In which two months was the production about the same?

44) February's production represented what percent of

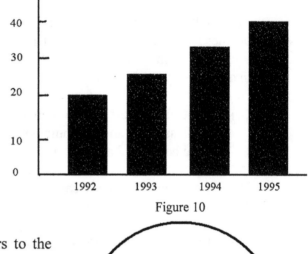

Figure 10

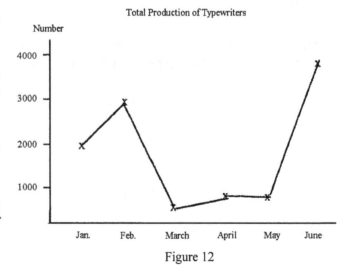

Figure 11

Figure 12

490

the total six months' production?

45) What was the average production per month?

The line graph below shows the sales of minivans over the period 1989 to 1993.

46) What were the total sales of ᵣ

47) Which year had the highest ˢ

48) What was the percent increase i

49) What were the average sales ᵣ

Sales of Minivans

Sales in thousands

Source: Ward's Automotive Yearbook

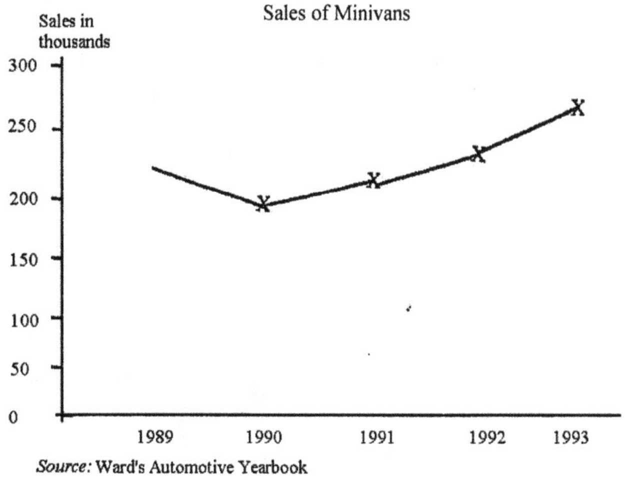

Figure 13

CONSTRUCTING GRAPHS

Sometimes we have data that we have collected from a survey or we have a class report and would like to include a graph to display the data. In this section we will present the basics of constructing a bar graph, a line graph, and a circle graph. The type of graph that you decide to use will depend upon what message you are trying to convey. It will also depend upon whether the data collected are in categories, you are showing changes over time, or you are showing the proportions of some whole. To help you make this decision, let us review the three basic graphs and their description from Section11.1.

> A bar graph is used to show the relationships in a set of quantities. The set of quantities are usually summarized into categories.

> A circle graph is used when a quantity is divided into parts, and we wish to make comparisons of the parts.

> A line graph is used to show the amount of change over time. In line graphs, the horizontal axis (*x*-axis) usually has some unit of time.

Constructing Bar Graphs

The letter grades made by a group of students on a mathematics test are shown in the table below.

Letter Grade	Number
A	4
B	6
C	12
D	5
F	3

To make a bar graph of this information:

1) Draw and name or label a vertical axis on the left side of the paper.
   Draw and label a horizontal axis at a right angle to the vertical axis. The point of intersection is labeled 0. The number or frequency is placed on the vertical axis and the name or categories are listed on the horizontal axis.

2) Select a suitable scale. Be certain the scale is uniform.

3) Draw a bar to represent each amount. Each bar should be the same width and the same distance apart.

4) Select a title for the graph. For example, "Distribution of Letter Grades for a Mathematics Test."

The following is a sample bar graph for the data of this problem.

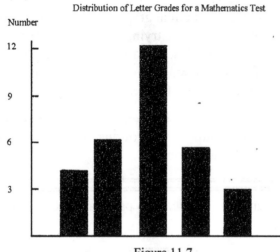

Distribution of Letter Grades for a Mathematics Test

Figure 11.7

Constructing Line Graphs

The data below show the percentage of students meeting all prerequisites for college admissions. Construct a line graph for the data.

1990	1991	1992	1993	1994
67%	71%	74%	79%	84%

To make a line graph of these data set:

1) Draw and label the vertical and horizontal axes on the paper. The point of intersection is labeled 0. The vertical axis should contain the frequency or percentages. The horizontal axis should contain the years or unit of time.

2) Select a suitable scale. Be certain it is uniform.

3) Draw a dot to represent each amount. Connect each dot in order with a continuous line.

4) Choose a title for the graph.

494

Percent   Percentage of students Meeting
                       All College Admissions Requirements

Source: S. C. Commission on Higher Education.

Figure 11.8

Constructing Circle Graphs

The data set below was collected from a survey of 150 randomly selected high school students. Students were asked, "Do you like most of the athletes at your school?" Of the 150 responding, 120 said yes, 24 said no, and 6 gave no response. Prepare a circle graph to display the responses.

To make a circle graph:

1) Represent the data as a fraction or percent.

2) Multiply the fractional parts or the percents by 360°.

Response	Fractional Part	Computation	Degrees
Yes	$\frac{120}{150} = .80$	$.80 \times 360 = 288$	288
No	$\frac{24}{150} = .16$	$.16 \times 360 = 57.6$	58
No Response	$\frac{6}{150} = .04$	$.04 \times 360 = 14.4$	14

3) Draw a circle. Use a protractor to construct the angles in the circle.

495

4) Choose a title.

Do You Like Most of the Athletes at Your School?

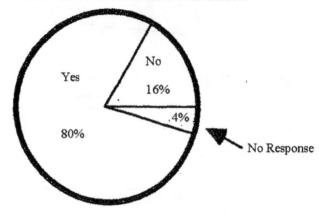

Figure 11.9

EXERCISE 11.2

For each set of data, construct the type of graph indicated.

1) In the mathematics quiz bowl, of the 50 questions asked, four high schools gave the following number of correct answers: High School A, 48; High School B, 46; High School C, 35; and High School D, 31. Make a horizontal bar graph to represent this information.

2) The Ward's Automotive Yearbook reported the sales in thousands for two midsize cars over the period 1989 to 1993.

Year	Car Type A	Car Type B
1889	350	350
1990	425	325
1991	400	300
1992	375	400
1993	325	350

Make a double line graph for the data.

3) A company spent the following amounts for the categories indicated:

Salaries	$4.8 million
Office Supplies	2.0 million
Rent	1.6 million
Insurance	.6 million
Miscellaneous	1.0 million

Construct a circle graph to depict this data set.

4) A *USA Today* survey of 1000 travelers who stayed at a hotel indicated what travelers considered important in selecting food. The results were as follows:

Taste	700
Fat Content	180
Nutritional Content	100
Calories	20

Construct a circle graph of the responses.

5) A Today/CNN/Gallup Poll of a random group of immigrants reported their reasons for coming to America. The percentage response for reasons is as follows:

Better Job	26%
Family	23%
School	19%
Political/Religious Freedom	15%

Construct a bar graph to display the results.

11.3 SCATTER PLOT

Sometimes we have two sets of data and want to know whether they are related. For example, we might have the SAT scores and first semester GPA of a group of students, or the number of members in a family and their mean annual expenditures for food. Are these data related? One way to determine whether a relationship exists is to graph the data. The graph of **bivariate** (two variables analyzed simultaneously) data is called a **scatter plot**.

EXAMPLE 1

There appears to be a relationship between the family size and annual expenditure for food. Suppose the following data set was collected for a sample of six families.

Family Size x	5	6	5	2	3	3
Annual Food Expenditures (in hundreds of dollars) y	19	20	15	6	11	17

Prepare a scatter plot of the data.

SOLUTION

Graph family size on the horizontal axis (x) and food expenditures on the vertical axis (y).

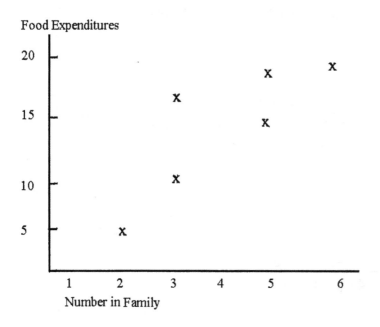

REGRESSION

The scatter plot in Example 1 indicates that a relationship exists between the data sets. That is, the points appear to approximate a straight line. As family size gets larger, food expenditures go up. When a linear relationship exists between data points, we can investigate this relationship using a statistical technique called **regression**.

Regression is based on the concept of a straight line. Recall that the general form of the equation of a straight line is $y = mx + b$ where m is the slope and b is the y-intercept or constant. Real-life examples such as the charge for rental of a car with a base rate of \$18 a day and 6 cents a mile can be expressed with an equation of the form $y = mx + b$.

SLOPE OF A LINE

To write the equation when we have points that all lie on a straight line, we may select any two pair of points and find the rate of change, the slope, using the following formula:

$$m \text{ (slope)} = \frac{\text{difference in } y\text{-coordinates}}{\text{difference in } x\text{-coordinates}} = \frac{y_2 - y_1}{x_2 - x_1}$$

EXAMPLE 2

Find the slope:

    a) $A\,(1, 2)$ and $B\,(5, 8)$ b) $A\,(4, 1)$ and $B\,(-2, 3)$

SOLUTION

a)   Name one pair of points with subscript 1 and the other with subscript 2.

$$A(1,2) = \overset{x_1}{1} \quad \overset{y_1}{2} \qquad\qquad B(5,8) = \overset{x_2}{5} \quad \overset{y_2}{8}$$

Substitute in formula: $m \text{ (slope)} = \dfrac{\text{difference in } y\text{-coordinates}}{\text{difference in } x\text{-coordinates}} = \dfrac{y_2 - y_1}{x_2 - x_1}$

$$= \frac{8-2}{5-1} = \frac{6}{4} = \frac{3}{2}$$

b) $A(4,1) = \overset{x_2}{4} \quad \overset{y_2}{1} \qquad\qquad B(-2,3) = \overset{x_1}{-2} \quad \overset{y_1}{3}$

$$m \text{ (slope)} = \frac{y_2 - y_1}{x_2 - x_1} = \frac{1-3}{4-(-2)} = \frac{-2}{4+2} = -\frac{2}{6} = -\frac{1}{3}$$

From an equation, the slope can be found by using algebra to write the equation in the form $y = mx + b$. In this form, the slope (m) is the coefficient of x and b is the y-intercept or constant term.

EXAMPLE 3

Find the slope and y-intercept of the equations.

    a) $3x + y = 3$ b) $3x - 2y = 4$

SOLUTION

    a) Solve $3x + y = 3$ for y

            $y = -3x + 3 = 3$ Subtract $3x$ from both sides of the equation.

        The slope is -3 or $m = -3$. The y-intercept is 3 or $b = 3$.

    b) Solve $3x - 2y = 4$ for y

           $-2y = -3x + 4$ Subtract $3x$ from both sides of the equation.

$$y = \frac{3}{2}x - 2 \qquad \text{Divide both sides by } -2$$

        The slope is $\frac{3}{2}$ or $m = \frac{3}{2}$ and the y-intercept is -2 or $b = -2$.

WRITING EQUATIONS OF LINES

To write an equation of a line, we need a slope and the y-intercept. The example below shows the steps used to write an equation of a line.

EXAMPLE 4

Write the equation of the line for the following points:

    a) A (2, 1) and B (3, 4)   b) $A(-1, 2)$ and $B(1, -3)$

SOLUTION

a)   To write the equation of the line, follow these steps:

        Step 1:   Find the slope of the points.

$$m \text{ (slope)} = \frac{y_2 - y_1}{x_2 - x_1} = \frac{4 - 1}{3 - 2} = \frac{3}{1} = 3$$

Step 2: Pick either point and substitute these points and the slope in $y = mx + b$ and solve for b. Let us use the point (2, 1).

$$y = mx + b$$
$$1 = 3(2) + b$$
$$1 = 6 + b$$
$$-5 = b$$

Step 3: Substitute the values of m and b in the formula $y = mx + b$.

$$y = 3x - 5$$

b) Follow the same procedure as above.

Step 1: Find the slope.

$$m \text{ (slope)} = \frac{y_2 - y_1}{x_2 - x_1} = \frac{-3 - 2}{1 - (-1)} = -\frac{5}{2}$$

Step 2: Pick the point $(1, -3)$ and use the slope $-\frac{5}{2}$ to find b.

$$y = mx + b$$
$$-3 = -\frac{5}{2}(1) + b$$
$$-3 = -\frac{5}{2} + b$$
$$-\frac{6}{2} + \frac{5}{2} = b$$
$$-\frac{1}{2} = b$$

Step 3: Substitute the values for m and b in the formula $y = mx + b$.

$$y = -\frac{5}{2}x - \frac{1}{2}$$

USING THE CALCULATOR TO FIND THE REGRESSION EQUATION

In some instances, there may be a relationship represented by a number of ordered pairs which, when graphed, approximate a linear equation, that is, all the points do not lie on a straight line. Regression is a technique that uses a least-square criterion to find which one of many possible lines best fits a set of data. The general form of the regression equation is $y = mx + b$ where

m = the slope
b = the y-intercept
x is the explanatory, predictor, or independent variable
y is called the dependent variable or the predicted variable or value.

For a regression equation, the **coefficient of determination**, r^2, tells you how good the equation is (i.e., how well the line it draws fits the data). An r^2 of .70 or 70% or higher generally represents a good-fitting equation. [Note: For simple regression (only one predictor variable x) the r^2 is the same as the correlation coefficient (r) when squared.]

EXAMPLE 5

For the data in Example 1 find:
a) a linear regression equation
b) determine how good the equation is
c) use the equation to predict the food expenditure for a family of 4, that is $x = 4$.

SOLUTION

The data points (x, y) are (5, 19) (6, 20) (5, 15) (2, 6) (3, 11) (3, 17)

a) Using the TI-35X, press [3rd] [STAT2] to place the calculator in the statistical mode.

To clear any existing data, press [2nd] [CSR]

To enter the pair (5, 19), press 5[$x \triangleleft \triangleright y$] 19[Σ+]

To enter (6, 20), press 6[$x \triangleleft \triangleright y$] 20[Σ+]

Do the same for the other points.

When the points have been entered, to find the slope press [2nd] [SLP]. Look on key 5 for [SLP]. The calculator displays 2.83.

To find the y-intercept, press [2nd] [ITC], look on key 4 for ITC. Calculator displays 3.33.

Substituting in $y = mx + b$ the equation is $y = 2.83x + 3.33$.

b) To find the coefficient of determination, r^2, press [3rd] [corr].
The calculator displays .825 = r. $r^2 = (.825)^2 = .68$. Since r^2 is less than .70, but pretty close, we say that the equation is fairly good.

503

c) To predict the food expenditure when the family size is 4, substitute $x = 4$ in the regression equation.

$$y = 2.83x + 3.33 = 2.83(4) + 3.33 = 14.65$$

EXAMPLE 5 – SOLUTION Using the IT-81

Steps

a) Place calculator in statistical mode: [2nd] [STAT]

b) Clear calculator of any previous data: $[\Rightarrow][\Rightarrow]$[DATA][2], for CLR STAT, [ENTER]

c) Enter Data: Quit   [2nd] [CLEAR]

   a. Repeat Step 1   b. Press $[\Rightarrow][\Rightarrow]$[DATA][1]

Calculator displays   DATA
$X1=$
$Y1 = 1$

Now enter the ordered pair, (5, 19). For $X1$ type 5, move cursor down $[\Downarrow]$ to $Y1$, type 19, move cursor down to $X2$, etc. Repeat for the other ordered pairs.

d) After entering all the data, Quit [2nd] [Clear]

e) Compute the slope and y-intercept. Note: On the IT-81, the slope is b and the y intercept is a.

   Quit then return to STAT MODE. Press [CALC] [2] [ENTER]
   Display reads: $a = 3.333...$ $b = 2.8333...$ $r = .82559...$

f) Compute the regression equation.
   Substituting in the general form, $y = mx + b$, the regression equation is

   $$Y = 2.83x + 3.33.$$

   To determine how good the equation is, square r = .82559, $r^2 = .68$.

   To predict the food expenditures for a family of 4, substitute $x = 4$ in the regression equation.

   $$y = 2.83x + 3.33 = 2.83(4) + 3.33 = 14.65$$

504

EXERCISE 11.3

Find the slope for each pair of points.

1) A (4, 2) and B (3, 4)

2) A(2, 1) and B (3, 4)

3) A (1, 3) and B (5, −3)

4) A (5, −2) and B (1, 0)

Write an equation in the form $y = mx + b$ for each point.

5) A (4, 2) and B (3, 4)

6) A (2, 1) and B (3, 4)

7) A (1, 3) and B (5, −3)

8) A (5, −2) and B (1, 0)

Identify the slope and y-intercept in each equation.

9) $3x − 2y = 6$

10) $3x − y = 4$

11) $3x + 2y = 4$

12) $5x − 2y = 8$

13) The amount of natural gas required to heat a home depends on the outdoor temperature. Below is the monthly gas consumption in hundreds of cubic feet for a six-month period during the heating season. The table also contains the average heating degree day for each of the months. One heating degree day is accumulated for each degree a day's average temperature falls below 65°F. For example, an average temperature of 50° corresponds to 15 degree days. The predictor variable (x) is the heating day and the dependent variable (y) is the gas consumption.

	November	December	January	February	March	April
X (Temperature)	6.8	9.5	23.28	22.45	7.8	3.32
Y (Gas Consumption)	30	64	144	159	70	30

a) Generate a scatter plot.
b) Find the regression equation for the data.
c) How good is the equation?
d) Predict the gas consumption when the average temperature is 18.9.

14)   Earnings are related to level of education. Below is information reported in the U.S. Department of Education Educational Statistics for 1993 (p. 303) on the median annual earnings of males ages 25–34 by educational level.

Educational Level	Less than 8th	9–11	High School	Some College	Bachelor's
x (average years of school)	7	10	12	14	16
y (median annual income)	14257	17876	22375	26352	35024

a) Generate a scatter plot.
b) Find the regression equation for the data.
c) Predict the annual earnings for a person with 11 years of education.

CHAPTER 11 REVIEW EXERCISES

1) List the three basic graphs.

2) To display how a government agency spends its money, what type of graph would be used?

3) To display the number of students making grades of A, B, C, D, and F on a test, what type of graph would be used?

4) Adrian has a stock that she has been tracing for 4 months. What type of graph would she use to show the performance of the stock?

5) Use the graph below to answer the following questions.

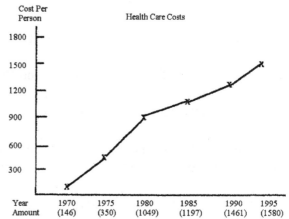

a) What is the percent increase in the cost per person for health care from 1970 to 1995?

b) During which period was the single largest increase experienced?

c) What was the average cost per person for health care over the period given?

6) In 1994 the federal government listed the following categories as the sources of federal income and the percentage from each.

Social Security, Medicare, unemployment, and other retirement taxes	31%
Personal income taxes	38%
Borrowing to cover deficit	18%
Corporate income taxes	8%
Excise, customs, estate, and miscellaneous	5%

Construct a circle graph to display the data.

7) Find the slope of the points A (2, 4) and B (4, −1).

8) Find the slope of the equation $4x - 5y = 20$.

9) Write an equation for the pair of points A $(2, 3)$ and B $(-2, 5)$.

10) The age of five cars in years and the prices paid for the cars in hundreds of dollars are recorded in the following table.

Age, x	2	5	3	6	3
Price, y	27	22	31	18	32

a) Graph the points.

b) Find the regression equation.

c) How good is the equation?

d) Use the equation to predict the price of a 4-year-old car.

11) Find the slope of the points: A(5, 7) and B(3, 4).

Refer to the following information to answer Questions 12 through 16.

Mrs. Pat is an office manager. She has prepared this circle graph to show the workers in the clerk-typist pool how their time will be spent in her office.

12) On which task will the clerk-typists spend of their time?

13) What percent of their time do clerk-typists spend on mailings and data entry?

14) What is the ratio of time spent entering data to time spent answering telephones?

15) Mark works 40 hours per week as a clerk-typist. About how many hours per week should Mark spend doing data entry?

16) Sarah works 36 hours per week. To the nearest whole hour, how many hours per week should she expect to spend on customer service and telephone tasks?

CLERK-TYPIST TASKS

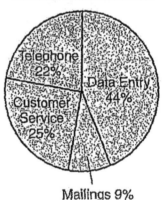

Mailings 9%

Questions 17 through 23.

A high school has received $56,000 in donations and grants to build a new computer lab. The following graph shows how the school plans to spend the money.

508

17) What percent of the money will be spent on construction and electrical upgrade?

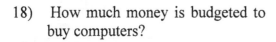

COMPUTER LAB BUDGET

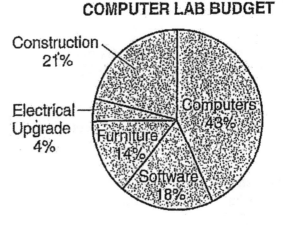

18) How much money is budgeted to buy computers?

19) How much money is budgeted to buy furniture?

20) What percent of the money will not be spent on either computers or software?

21) Which two items combined equal about $\frac{1}{3}$ of the budget?

22) For which item has the school budgeted a little more than $10,000?

23) How much money is budgeted for construction and electrical upgrading?

Refer to the following graph to answer Questions 24 through 29.

24) Netco sells its products online. The graph to the right shows the number of orders placed during the final six months of the year. In which month did the number of orders decline from the previous month?

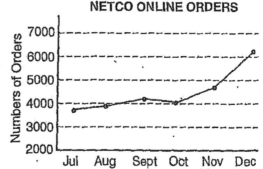

25) About how many more orders were therein December than in September?

26) The rate of increase from December to January is expected to equal the rate of increase from October to November. Estimate the number of orders for January.

27) The average number of orders per month during the six-month period was about 4400. Which two months were closest to the average?

28) Netco reported that the average sale during November was $35. Find November's gross sales by multiplying $35 by the number of sales.

Refer to the following graph to answer Questions 29 through 33.

29) What was the approximate dollar amount in sales for Year 3?

30) Which year showed a decrease in sales?

31) Estimate the difference in sales from Year 4 to Year 5.

32) Approximate the ratio of sales in Year 5 to sales in Year 1.

33) Approximate the average yearly sales for the five-year period.

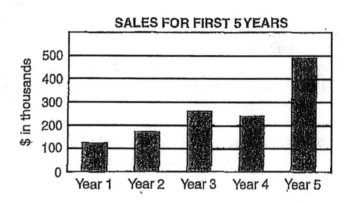

CHAPTER 11 TEST

1) You are requested to prepare a circle graph to show that $\frac{1}{18}$ of the expenses of a business is spent on miscellaneous items. How big an angle must you measure at the corner of the circle to show this amount?

2) Which type of graph is used to display changes over time?
   a) bar
   b) circle
   c) frequency polygon
   d) line

3) Which type of graph would be best for representing the value of a stock on the stock market over time?
   a) line
   b) bar
   c) circle
   d) none of these

4) Which type of graph would be useful in showing how a family spends its monthly budget?
   a) line
   b) bar
   c) circle
   d) none of these

Use the graph above to answer questions 5–8.

5) What percent of the income was spent for supplies and wages combined?
   a) 20%
   b) 30%
   c) 55%
   d) 75%

6) What percent was profit?
   a) 12%
   b) 100%
   c) 33%
   d) 30%

7) If total disbursements were $1,000, how much was spent on operating expenses?
   a) $100
   b) $1,000
   c) $330
   d) $500

8) How many degrees would be in the sector representing wages?
   a) 30
   b) 108
   c) 300
   d) cannot be determined from the information given

511

9) What is the slope of the points?
    A (5, 3) and B (−1, 7)

    a) $-\dfrac{3}{2}$

    b) $-\dfrac{2}{3}$

    c) 1

    d) $\dfrac{3}{4}$

10) An equation for the points
    A (−1, 5) and B (5, 1) is

    a) $y = -\dfrac{10}{3}x + 1$

    b) $y = -\dfrac{2}{3}x + \dfrac{17}{3}$

    c) $y = \dfrac{13}{3}x + 1$

    d) $y = -\dfrac{2}{3}x + \dfrac{13}{3}$

11) The slope of the equation of the
    line $2x + y = 3$ is
    a) 2
    b) −2
    c) 1
    d) 3

Use this information to work problems 12–14.

The following data show the average annual repair cost and the age of the automobile in years.

Age	1	3	5	7	9
Repair	130	320	400	560	700

12) The regression equation is
    a) $y = 77x + 69$
    b) $y = .99x + 69$
    c) $y = 70x + 77$
    d) $y = 77x + .99$

13) The equation is a ? predictor.
    a) very good
    b) fair
    c) poor
    d) cannot be determined

14) The estimated repair cost of a 4-year-old car is
    a) $377
    b) $353
    c) $75.93
    d) $83.93

For Questions 15 to 19, refer to the circle graph below.

15) About what fractional part of the dollar goes for food?

    a) 26   b) $\dfrac{1}{3}$ c) $\dfrac{1}{4}$ d) 2.6   e) $\dfrac{1}{5}$

16) On a family income of $12,000, how much goes for medical expenses?
    a) $10.80   b) $9   c) 108
    d) $900   e) $1,080

17) Together, medical and miscellaneous expenses make up what fractional part of the dollar?

    a) $\dfrac{1}{4}$ b) $\dfrac{1}{5}$ c) $\dfrac{1}{3}$

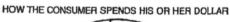

HOW THE CONSUMER SPENDS HIS OR HER DOLLAR

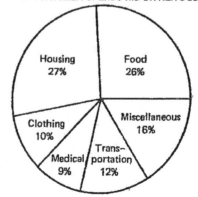

512

d) 25   e) 2.5

18) On an income of $12,000, how much will the average family spend on food and housing?
a) $63.60   b) $636   c) $6,360   d) $53   e) $5,300

19) With a family income of $14,000, how much more is spent on housing than on transportation?
a) $15   b) $21   c) $210   d) $2,100   e) $21,000

Questions 20 to 24 refer to the bar graph below.

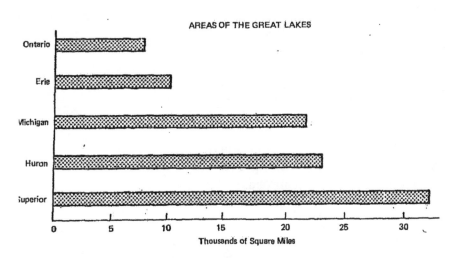

20) About how much larger is Lake Superior than Lake Erie?
a) 32,000 sq mi   b) 22,000 sq mi   c) 10,000 sq mi
d) 15,000 sq mi   e) 14,000 sq mi

21) Lake Michigan is about how many times as large as Lake Ontario?
a) 3   b) 30   c) 300   d) 15,000
e) 15

22) What is the total area of the Great Lakes?
a) 95 sq mi   b) 95,000 sq mi   c) 950,000 sq mi   d) 32,000 sq mi
e) 320,000 sq mi

23) What is the ratio of the area of Lake Erie to that of Lake Superior?
a) $\dfrac{5}{16}$ b) $\dfrac{16}{5}$ c) $\dfrac{1}{4}$ d) $\dfrac{1}{5}$ e) $\dfrac{1}{32}$

24) Which lake is $2\dfrac{1}{3}$ times as large as Lake Erie?

a) Lake Ontario   b) Lake Michigan   c) Lake Huron

513

d) Lake Superior   e) None of these

Questions 25 to 29 refer to the table below.

TRANSPORTATION: ACCIDENT DEATH RATES			
For a Recent Year			
Kind of Transportation	Passenger Miles	Passenger Deaths	Death Rate (per 100,000 miles)
Passenger cars	1,190,000,000	26,800	2.3
Passenger cars on turnpikes	28,000,000	330	1.2
Buses	55,700,000	90	0.16
Railroads	20,290,000	20	0.10
Air transport	35,290,000	123	0.35

25) What form of transportation appears to be least hazardous?
    a) Passenger cars   b) Passenger cars on turnpikes
    c) Buses   d) Railroads
    e) Air transport

26) Which form of transportation resulted in the greatest number of deaths?
    a) Passenger cars   b) Passenger cars on turnpikes
    c) Buses   d) Railroads
    e) Air transport

27) About how many times as great is the death rate for passenger cars as for railroads?
    a) 23   b) 2.3   c) 230   d) 120   e) 12

28) The number of passenger miles by passenger cars is approximately how many times that of railroads?
    a) 6   b) 60   c) 600   d) 6,000   e) 1,340

29) The total number of deaths in forms of transportation other than passenger cars was
    a) 0,61   b) 11,280,000   c) 233   d) 563   e) 27,333

Questions 30 to 34 refer to the line graph below.

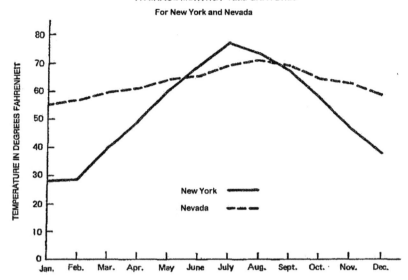

30) In what month was the average temperature highest in New York?
   a) May   b) June   c) July   d) August
   e) September

31) What is the difference between the highest and lowest average monthly temperatures in New York?
   a) 34°   b) 48°   c) 54°   d) 72°   e) 28°

32) What is the range of temperature in Nevada from January to June?
   a) 5°   b) 10°   c) 20°   d) 30°   e) 40°

33) What is the difference between the lowest average monthly temperatures in the two cities?
   a) 20°   b) 37°   c) 17°   d) 7°   e) 27°

34) The lowest temperature in either of the two cities during year was
   a) 28°   b) 55°   c) 34°   d) 57
   e) cannot be determined from the information given.

Appendixes

I. FORMULAS

Simple Interest	Intrest $= $ Principal $\times$ Rate $\times$ Time or $I = PRT$ where Principal (P) is the amount borrowed, Rate (R) is the annual cost given in percent, and Time (T) is the length of time in years.
Compound Interest	$A = P\left(1+\dfrac{r}{t}\right)^{nt}$ where P = the initial principal t = number of compounding periods during the year r = interest rate n = number of years (round calculations to 4 decimal places, answer to 2.)
Effective Annual Rate	$EAR = \left(1+\dfrac{r}{k}\right)^{k} - 1$ where k = number of compounding periods per year and r = annual interest rate.
Amount of Annuity or Future Value of Annuity	$S = R\left[\dfrac{(1+i)^n - 1}{i}\right]$ where S = amount of the annuity, R = periodic payment, i = rate per period divided by the number of compounding periods per year n = number of payments over the period.
Monthly Payments and Mortgages	$PMT = \dfrac{P \times \left(\dfrac{APR}{n}\right)}{\left[1-\left(1+\dfrac{APR}{n}\right)^{(-nY)}\right]}$ where PMT = regular payment amount P = starting loan principal (amount borrowed) APR = annual percentage rate n = number of payment periods per year Y = loan term in years
n^{th} term of an arithmetic sequence **n^{th} term of a geometric sequence**	$a_n = a_1 + (n-1)d$, where a_1 is the first term and d is the common difference. $a_1 r^{n-1}$, where a_1 is the first term and r is the common ratio.
Sum of the first n terms of an arithmetic series	$S_n = \dfrac{n}{2}(a_1 + a_n)$ where a_1 is the first term and a_n is the nth term.
Sum of the first n terms of geometric series	$S_n = \dfrac{a_1(1-r^n)}{(1-r)}$ where a_1 is the first term, $r \neq 1$ and r is the common ratio.

517

Perimeter of Geometric Figures

Rectangle
$P = 2L + 2W$

Square
$P = 4S$

Triangle
$P = a + b + c$

Areas of Geometric Figures

Rectangle
$A = LW$

Square
$A = S^2$

Parallelogram
$A = bh$

Triangle
$A = \dfrac{1}{2}bh$

Trapezoid
$A = \dfrac{1}{2}(B + b)h$

Circle Area
$A = \pi r^2$

Circle Circumference
$C = \pi d$ or $C = 2\pi r$

Volume of Geometric Figures

Rectangular Solid
$V = LWH$

Cube
$V = e^3$

Cylinder
$V = \pi r^2 h$

Sphere
$V = \dfrac{4}{3}\pi r^3$

Pythagorean Theorem
$c^2 = a^2 + b^2$

II. TABLE OF MEASURES
English Measures

Length

1 foot (ft or ') = 12 inches (in or ")
1 yard (yd) = 36 inches
1 yard = 3 feet
1 rod (rd) = 16½ feet
1 mile (mi) = 5280 feet
1 mile = 1760 yards
1 mile = 320 rods

Liquid Measure

1 cup (c) = 8 fluid ounces (fl oz)
1 pint = 16 fluid ounces (oz)
1 quart (qt) = 2 pints
1 gallon (gal) = 4 quarts

Weight

1 pound (lb) = 16 ounces (oz)
1 ton (T) = 2000 pounds

Dry Measure

1 quart = 2 pints (pt)
1 bushel (bu) = 4 pecks

Area

1 square foot (ft^2) = 144 square inches (in^2)
1 square yard (yd^2) = 2 square feet

Volume

1 cubic foot (ft^3 or cu ft) = 1728 cubic inches
1 cubic yard (yd^3 or cu yd) = 27 cubic feet
1 gallon = 231 cubic inches

Table of English-Metric Conversions

1 inch = 2.54 centimeters
1 yard = .9 meter
1 mile = 1.6 kilometers
1 ounce = 28 grams
1 pound = 454 grams
1 fluid ounce = 30 milliliters
1 fluid quart = .95 liters

Table of Metric-English Conversions

1 centimeter = .39 inches
1 meter = 1.1 yards
1 kilometer = .6 miles
1 kilogram = 2.2 pounds
1 liter = 1.06 liquid quarts

III. Table of Areas under the Normal Curve

z-score	Area below	z-score	Area below	z-score	Area below
−3.0	.0013	−1.0	.1587	1.1	.8643
−2.9	.0019	−0.9	.1841	1.2	.8849
−2.8	.0026	−0.8	.2119	1.3	.9032
−2.7	.0035	−0.7	.2420	1.4	.9192
−2.6	.0047	−0.6	.2743	1.5	.9332
−2.5	.0062	−0.5	.3085	1.6	.9452
−2.4	.0082	−0.4	.3446	1.7	.9554
−2.3	.0107	−0.3	.3821	1.8	.9641
−2.2	.0139	−0.2	.4207	1.9	.9713
−2.1	.0179	−0.1	.4602	2.0	.9773
−2.0	.0227	0.0	.5000	2.1	.9821
−1.9	.0287	0.1	.5398	2.2	.9861
−1.8	.0359	0.2	.5793	2.3	.9893
−1.7	.0446	0.3	.6179	2.4	.9918
−1.6	.0548	0.4	.6554	2.5	.9938
−1.5	.0668	0.5	.6915	2.6	.9953
−1.4	.0808	0.6	.7258	2.7	.9965
−1.3	.0968	0.7	.7580	2.8	.9974
−1.2	.1151	0.8	.7781	2.9	.9981
−1.1	.1357	0.9	.8159	3.0	.9987
		1.0	.8413		

Answers to Selected Exercises

CHAPTER 1

EXERCISE 1.1
1) {8, 9, 10, 11, 12}
3) {1, 3, 5, 7, 9, 11, 13, 15, 17, 19}
5) {10, 12, 14, 16, 18, 20, 22, 24}
7) {x|x is an even number}
9) {x|x is a number between 1 and 60, inclusive}
11) {0, 1, 2, 3, ..., 10}
13) {}
15) {0, 1, 2, 3, ..., 10}
17) False
19) False
21) {2, 3, 4, 5, 6, 7, 8}
23) {4, 5, 6, 7, 8}
25) Equivalent
27) Equivalent
29) a) 3, 7   b) 0, 3, 7, −9
    c) $-\sqrt{3}$, $\sqrt{5}$
    d) all except $-\sqrt{3}$, $\sqrt{5}$
    e) all   f) 0, 3, 7
31) False
33) True
35) True
37) $\notin$

EXERCISE 1.2
1) Commutative
3) Associate
5) Identity
7) Inverse
9) Closure
11) Commutative
13) Inverse
15) 7(5) + 7(2)
17) 9(6) + 4(6)
19) Inverse
21) Distributive
23) Closure
25) Commutative

EXERCISE 1.3
1) 2
3) 3, 5, 15
5) 3
7) None
9) 3
11) 3, 5, 15
13) 2, 3, 5, 9, 15
15) 2, 3
17) None
19) 2

EXERCISE 1.4
1) Prime
3) Prime
5) Composite
7) Composite
9) Prime
11) Prime
13) Composite
15) Composite
17) Composite
19) $4 \times 7, 2 \times 14$
21) $1 \times 16, 4 \times 4$
23) $1 \times 9, 3 \times 3$
25) $1 \times 35, 5 \times 7$
27) $1 \times 98, 2 \times 49$
29) $1 \times 135, 5 \times 27$
31) $1 \times 49, 7 \times 7$
33) {19, 38, 57, ...}
35) {35, 70, ...}
37) {40, 80, ...}
39) {20, 40, ...}
41) 2×7
43) 5^4
45) $2^3 \times 3 \times 5$
47) $3 \times 2 \times 7$
49) 3×41
51) $5^2 \times 7$
53) $2 \times 3 \times 17$
55) 2×43
57) $2^4 \times 5$
59) $2 \times 3 \times 5$

EXERCISE 1.5

1) 4
3) 17
5) 10
7) 3
9) 6
11) 2
13) 8
15) 4
17) 12
19) 25
21) 1
23) 1
25) 35
27) RP
29) RP
31) RP
33) RP
35) RP
37) 105
39) 216
41) 200
43) 196
45) 210
47) 726
49) 2250
51) 120
53) 120
55) 270
57) 4410
59) 3375
61) GCD: 1;   LCM: 45
63) GCD: 2;   LCM: 252
65) GCD: 1;   LCM: 1, 225
67) GCD: 8;   LCM: 336
69) GCD: 15;   LCM: 675

CHAPTER 1 REVIEW EXERCISES

1) True
3) False
5) False
7) {August, April}
9) {2, 3, 6, 8, 9, 10}
11) {}
13) Real numbers
15) {}

17) a) False   b) True
    c) True
19) Commutative
21) Identity
23) Distributive
25) False
27) True
29) False
31) False
33) True
35) {3, 5, 9, 15}
37) {1, 7, 29}
39) a) 4, $\sqrt{16}$ b) 0, 4, $\sqrt{16}$
    c) 0, 4, $\sqrt{16}$, -4
    d) $\sqrt{2}$

CHAPTER 1 TEST

1) c
3) d
5) c
7) a
9) c
11) a
13) a
15) b
17) b
19) a
21) c
23) c
25) a
27) c
29) c
31) a
33) c
35) e
37) c
39) d
41) c

CHAPTER 2

EXERCISE 2.1

1) 1130 (1130)
3) 610 (615)
5) 10060 (10057)

7) 900 (903)
9) 2850 (2854)
11) 86000 (85034)
13) 6000 (5989)
15) 13000 (13401)
17) 178000 (177798)
19) 16000 (16770)
21) 252000 (272620)
23) 320000 (326560)
25) 360000 (390241)
27) 120 (112.6)
29) 10 (6.786)
31) 460 (333.59)
33) 450 (455.42)
35) $300($272.32)
37) 22000
39) 11000
41) 4700
43) 2100
45) $15
47) $20000
49) 1500
51) 350000
53) 80000
55) 4500
57) 800
59) 85
61) 2200
63) 15 miles per gallon

EXERCISE 2.2

1) 5^4
3) 5^7
5) 5^4
7) 7^9
9) $\dfrac{1}{2^3 \times 3^2 \times 4^2}$
11) 4
13) 36
15) 98
17) 125000
19) 16
21) 400
23) 3,010,936,384
25) .0013717421
27) 761

29) 156169

EXERCISE 2.3

1) <
3) >
5) <
7) <
9) >
11) <
13) ≤
15) ≤
17) ≤
19) ≥
21) <, ≤
23) <, ≤
25) <, ≤
27) >, ≥
29) >, ≥
31) ≤, ≥
33) >, ≥
35) <, ≤
37) $0 \geq 0$
39) $12 \neq 5$
41) $3 < \dfrac{50}{5}$
43) $7 = 5 + 2$
45) True
47) False
49) False
51) False
53) False
55) False
57) True
59) True
61) True
63) True

EXERCISE 2.4

1) 60
3) 0
5) 2
7) 16
9) 66
11) $\dfrac{10}{3}$
13) 15

525

15) 65
17) $3 \geq 3$, true
19) $2 \geq 3$, false
21) $66 > 72$, false
23) $45 \geq 46$, False
25) $0 \geq 0$, true
27) $61 \leq 60$, false
29) $16 \leq 16$, true
31) $(8 - 2^2) \cdot 2 = 8$
33) $(6 + 5) \cdot 3 = 99$
35) No parentheses needed
37) $(3 \cdot 5 + 2) \cdot 4 = 68$
39) $(2^3 + 4) \cdot 2 = 24$
41) No parentheses needed
43) $3 \cdot (5 - 4) = 3$
45) No parentheses needed
47) $10 - (7 - 3) = 6$

EXERCISE 2.5A
1) 6.7×10^4
3) 2.18×10^3
5) $6..28 \times 10^{18}$
7) 7.49×10^{-6}
9) 8.92×10^{-6}
11) 9.3×10^7
13) 2.28×10^{-5}
15) 1.0×10^{-9}
17) 6.6×10^7
19) 1.64×10^5
21) 4.7×10^1
23) 1.0×10^6
25) 6.85×10^5
27) 1.1×10^4

EXERCISE 2.5B
1) b
3) a
5) b
7) d
9) a
11) d
13) d
15) a

17) a
19) b
21) c

EXERCISE 2.6A
1) 69500
3) .0000564
5) 2.15
7) .375
9) 181000
11) .0145
13) 3.3
15) 898000
17) 769000
19) .0000103
21) .0386

EXERCISE 2.6B
1) b
3) c
5) b
7) c
9) d
11) c
13) a
15) b
17) d
19) a

EXERCISE 2.7-2.8A
1) 0.0362
3) 726
5) 3613
7) 0.017432
9) 8.2×10^5
11) 2.9648×10^{-1}
13) 7.917×10^6
15) 2.16×10^1
17) 1.023×10^2
19) 2×10^1
21) 4.36×10^{-2}
23) 3.4×10^{-5}
25) 1.22×10^{-2}

1) a
3) b
5) a
7) d
9) d

CHAPTER 2 REVIEW EXERCISES
1) 14
3) 4
5) 125
7) 216
9) 6.4×10^{-5}
11) >
13) <
15) 70 < 68, false
17) $\frac{8}{9} < 1$, true
19) $\frac{1}{8}$
21) 3
23) 7
25) 2
27) d
29) d
31) b
33) a
35) a

CHAPTER 2 TEST
1) c
3) b
5) b
7) a
9) c
11) c
13) b
15) d
17) b
19) c

CHAPTER 3

EXERCISE 3.1A
1) $\frac{23}{4}$
3) $\frac{34}{3}$
5) $\frac{23}{5}$
7) 7
9) $\frac{51}{4}$
11) $\frac{16}{3}$
13) $2\frac{1}{2}$
15) 7
17) $2\frac{9}{11}$
19) $5\frac{6}{7}$
21) $3\frac{1}{5}$
23) $2\frac{4}{5}$
25) $\frac{5}{6}$
27) $\frac{3}{7}$
29) $\frac{3}{5}$
31) $\frac{4}{5}$
33) $\frac{3}{8}$
35) $\frac{3}{8}$
37) 24
39) 28
41) 102
43) 28
45) 64
47) 28

EXERCISE 3.1B

1) $\dfrac{3}{4}$

3) $\dfrac{9}{16}$

5) $\dfrac{8}{13}$

7) $\dfrac{7}{9}$

9) $\dfrac{20}{21}$

11) $\dfrac{11}{3}$

13) 25

15) 8

17) 81

19) 33

21) 153

23) 17

25) $\dfrac{61}{4}$

27) $\dfrac{22}{9}$

29) 11

31) $\dfrac{161}{7}$

33) $\dfrac{90}{7}$

35) $\dfrac{56}{3}$

37) $\dfrac{93}{5}$

39) $\dfrac{96}{7}$

EXERCISE 3.2

1) $\dfrac{3}{8}, \dfrac{1}{2}, \dfrac{2}{3}, \dfrac{5}{6}$

3) $\dfrac{3}{4}, \dfrac{5}{6}, \dfrac{11}{12}$

5) $\dfrac{7}{16}, \dfrac{1}{2}, \dfrac{5}{8}$

7) $\dfrac{3}{4}, \dfrac{4}{5}, \dfrac{4}{3}, \dfrac{3}{2}$

9) $\dfrac{73}{88}, \dfrac{74}{88}, \dfrac{75}{88}, \dfrac{76}{88}$

11) $\dfrac{1}{3}$

13) $\dfrac{1}{3} + \dfrac{1}{4}$

15) $\dfrac{3}{4}$

17) $\dfrac{3}{4}$

EXERCISE 3.3A

1) $\dfrac{3}{7}$

3) 4

5) $\dfrac{23}{60}$

7) $1\dfrac{5}{72}$

9) $2\dfrac{5}{72}$

11) $10\dfrac{1}{12}$

13) $18\dfrac{8}{9}$

15) $15\dfrac{29}{60}$

17) $\dfrac{9}{16}$

19) $\dfrac{23}{80}$

21) $8\dfrac{6}{9}$

23) $6\dfrac{32}{45}$

25) $8\dfrac{37}{44}$

27) $68\dfrac{1}{60}$

EXERCISE 3.3B

1) 2

3) $\dfrac{7}{60}$

5) 3

7) $2\dfrac{2}{15}$

9) $\dfrac{9}{10}$

11) $\dfrac{1}{3}$

13) $\dfrac{3}{40}$

15) $8\dfrac{7}{12}$

17) $\dfrac{3}{40}$

19) $2\dfrac{3}{8}$

21) $1\dfrac{1}{4}$

23) $\dfrac{39}{70}$

25) $\dfrac{11}{12}$

27) $\dfrac{1}{6}$

29) $8\dfrac{11}{12}$

31) $\dfrac{7}{12}$

33) $8\dfrac{1}{15}$

EXERCISE 3.4

1) 1

3) $2\dfrac{1}{3}$

5) $\dfrac{5}{12}$

7) $\dfrac{1}{4}$

9) $1\dfrac{1}{20}$

11) $\dfrac{2}{3}$

13) $\dfrac{7}{12}$

15) $1\dfrac{2}{3}$

17) $\dfrac{4}{25}$

19) $\dfrac{7}{32}$

21) $\dfrac{11}{13}$

23) $1\dfrac{1}{18}$

25) $17\dfrac{2}{3}$

27) $2\dfrac{6}{13}$

29) 3

31) $\dfrac{1}{10}$

33) $\dfrac{1}{10}$

35) 3
37) 12

39) $\dfrac{7}{12}$

EXERCISE 3.5A

1) $4\dfrac{1}{5}$

3) $1\dfrac{1}{10}$

5) $\dfrac{1}{14}$

7) $\dfrac{31}{50}$

9) $\dfrac{5}{18}$

EXERCISE 3.5B

1) $\dfrac{63}{19}$

3) $\dfrac{429}{365}$

5) $\dfrac{76}{135}$

7) $\dfrac{33}{4}$

9) $\dfrac{25}{21}$

11) $\dfrac{111}{31}$

13) $\dfrac{1}{5}$

15) $\dfrac{5}{7}$

17) $\dfrac{11}{70}$

19) $\dfrac{15}{7}$

EXERCISE 3.6A

1) $\dfrac{1}{6}$

3) 4

5) 16

7) $\dfrac{1}{16}$

9) 8000

11) 8000

EXERCISE 3.6B

1) $9\dfrac{1}{60}$ hours

3) $2{,}053\dfrac{29}{80}$ miles

5) $49\dfrac{97}{240}$ miles

7) $7\dfrac{33}{80}$ inches

9) $\dfrac{3}{5}$ hours

11) $22\dfrac{11}{88}$ feet above

13) 177 miles

15) $\dfrac{27}{46}$

17) $\dfrac{89}{216}$

19) $\dfrac{353}{630}$

EXERCISE 3.7A

1) 150.1065
3) 5.627
5) 15.7865
7) .355
9) 73.636
11) 2.678
13) 1.275
15) 56.37

EXERCISE 3.7B

1) 44.6516
3) 276.096
5) 72.31
7) 7.55
9) 2.3
11) 7.148
13) 118.333
15) 3.9974
17) 4.7974
19) 64.947
21) 15.89
23) 1.44
25) 155.073
27) 1,143.665

EXERCISE 3.8

1) 3
3) 1
5) .094
7) 10.79
9) 18.6
11) .025
13) .42
15) .28
17) .84419

19) .943
21) .002
23) 20
25) 2,056
27) .08
29) 9.9
31) 2.34
33) 57.69 mph
35) 309.6
37) 55.93
39) 28.4

EXERCISE 3.9
1) .17
3) .125
5) .833
7) 0.56
9) .083
11) .38
13) .12
15) .8
17) .06
19) .42
21) $3\frac{3}{25}$
23) $6\frac{1}{2}$
25) $\frac{13}{1000}$
27) $\frac{1}{40}$
29) $1\frac{41}{50}$
31) $2\frac{1}{8}$
33) $\frac{14}{25}$
35) $\frac{7}{10}$
37) .011, .1001, .101, 1.01, 1.1
39) 1.708
41) 500
43) 3

CHAPTER 3 REVIEW EXERCISE A
1) $\frac{2}{7}$
3) $\frac{109}{15}$
5) $\frac{1}{5}, \frac{7}{15}, \frac{2}{3}$
7) $7\frac{13}{15}$
9) $8\frac{43}{48}$
11) $\frac{9}{10}$
13) 1.7904
15) 36.2513
17) 6.875
19) $\frac{3}{5}$
21) Possible answers: 1.57, 1.52, etc
23) 24 inches
25) 12500

CHAPTER 3 REVIEW EXERCISE B
1) Improper
3) Improper
5) Proper
7) $4\frac{2}{5}$
9) 0
11) 60
13) $\frac{5}{4}$
15) $\frac{0}{4}$
17) $5\frac{23}{36}$
19) $9\frac{1}{8}$
21) $\frac{19}{20}$
23) $\frac{353}{480}$

25) $\dfrac{33}{4}$

27) nine hundred and five tenths

29) 200.17

31) 3,003.003

33) 73.00

35) 3.9623

37) 17.79

39) 55.26066

41) 0.733

43) $340

45) $99

47) $203.10

CHAPTER 3 TEST

1) b

3) d

5) b

7) b

9) b

11) a

13) b

15) a

17) c

19) d

21) d

CHAPTER 4

EXERCISE 4.1A

1) $\dfrac{2}{7}$

3) $\dfrac{2}{3}$

5) $\dfrac{4}{5}$

7) Yes

9) Yes

11) No

13) $2\dfrac{2}{3}$

15) 96

17) 4

19) 22

21) 29.7 yards

23) 4.48 Kg

25) 13.5 miles

27) 5.45 hours

29) 45

31) $3\dfrac{1}{3}$

33) 63 g

35) 8 fictions and 32 non-fictions

37) 25

EXERCISE 4.1B

1) $\dfrac{6}{5}$

3) $\dfrac{1}{48}$

5) $\dfrac{4}{7}$

7) $\dfrac{1}{4}$

9) not true

11) not true

13) true

15) true

17) true

19) not true

21) 3

23) 6

25) 9

27) 4

29) 8

31) 2.22

33) 6.67

35) 10.67

37) 21.33

39) 7 units

41) $1\dfrac{1}{4}$ units

43) $8\dfrac{3}{4}$ cups

45) $4.85

47) 1,500 gallons

49) $41.67

51) $11\frac{1}{4}$ hours

53) 259,000 revolutions

55) $420

57) 29 stores

59) 10 cookies, 15 crackers, 20 pieces of candy

61) 23,400 miles

63) $437\frac{1}{2}$ grams

EXERCISE 4.2A

1) 95.6%

3) 7543%

5) 590%

7) 214%

9) $33\frac{1}{3}$%

11) 155.55%

13) 71.42%

15) 128.57%

17) 280.00%

19) 57.14%

21) 1557.14%

23) 0.375

25) 0.0268

27) 0.153

29) 1

31) 0.2725

33) 0.008

35) 4.82

37) 0.15625

39) 0.0125

41) 0.0005

43) 2.13

45) 2.1

47) $\frac{4}{125}$

49) $3\frac{3}{4}$

51) $\frac{17}{25,000}$

53) $\frac{163}{1000}$

55) $\frac{3}{400}$

57) $\frac{1}{125}$

59) 4

EXERCISE 4.2B

1) $\frac{23}{50}$

3) $\frac{47}{1000}$

5) $\frac{153}{2500}$

7) $\frac{5}{32}$

9) .23

11) .6725

13) 52

15) .00009

17) .448

19) .049

21) 80%

23) 260%

25) 367.8%

27) 120%

29) 435%

31) 1476%

33) 62.5%

35) 17.5%

37) 166.7%

39) 120%

EXERCISE 4.3A

1) 8

3) 71.43

5) 26.67

7) 311.11

9) 3

11) 32

13) 28.57

15) 180

EXERCISE 4.3B

1) 7

3) 150

5) 20%

7) $33\frac{1}{3}\%$

9) $23\frac{17}{21}$

11) 12.5

13) 2.109

15) $25\frac{1}{3}$

17) 900%

19) 6

21) $32

23) 2.8%

25) 400 cavities

27) Rent – 20%, food – 40%, Taxes – 15%

29) 20 problems

31) $15\frac{15}{49}\%$

33) 750%

EXERCISE 4.4A

1) 23.08%

3) 5.6%

5) Original price: $10.00; sales price $7.00

7) 11%

9) 141

11) 1213.63

13) 90%

15) $700

17) $9185

EXERCISE 4.4B

1) 20%

3) a) $2860   b) $28,860

5) $33\frac{1}{3}\%$

7) $16\frac{2}{3}\%$

9) a) $19.04   b) $243.04

11) 25%

13) $552.50

15) a) 36¢   b) $1.44

17) $4.20

CHAPTER 4 REVIEW EXERCISE A

1) $\frac{5}{9}$

3) $\frac{15}{16}$

5) 57

7) 25

9) True

11) True

13) $9,120

15) $1.95

17) 40,000

19) 575%

21) 120%

23) 60%

25) 85%

27) $\frac{1}{6}$

29) 25

31) $\frac{33}{50}$

33) $30

35) $25,400

37) $4,044

39) 1

41) 50

43) 25%

45) 14.91

47) a) .0075   b) 3.5   c) .15

49) a) 300%   b) 200%   c) 19.5
    d) 180   e) 100   f) 80%
    g) 400%   h) 180   i) 7
    j) 16000   k) 40.85   l) 90

51) $3.60

53) $318.50

CHAPTER 4 REVIEW EXERCISE B

1) a) 36   b) 42

3) a) 9   b) $\frac{56}{9}$
   c) 90   d) 5

5) 166

7) 78.57

9) $8.15

CHAPTER 4 TEST A

1) c
3) b
5) d
7) c
9) c
11) c
13) a
15) b
17) b
19) c
21) a
23) d
25) c
27) b

CHAPTER 4 TEST B

1) a
3) a
5) d
7) b
9) d
11) a
13) d
15) d
17) a
19) d
21) c
23) b
25) e
27) c
29) b
31) d
33) b

CHAPTER 5

EXERCISE 5.1

1) $476.11
3) $59.20
5) $216.67
7) $103.67
9) $1606.50
11) $513.48
13) $7429.24

15) a) $90   b) $360   c) $50
    d) $4   e) $14
17) $6
19) a) 20   b) 45   c) 44
    d) 12   e) 45   f) 90
21) $57000
23) a) $198.40   b) $41.60
25) $1024
27) $13500
29) a) $72000   b) $6187.50
31) $953.69
33) $2468.41

EXERCISE 5.2

1)

	Total amount to be paid	Total interest	Effective interest rate
a)	$416	$41.60	29%
b)	$481.20	$81.70	40%
c)	$800	$50	41%
d)	$540	$80	14%
e)	$1800	$300	13%
f)	$536	$41	13%
g)	$12000	$2800	24%
h)	$13400	$1400	6%
i)	$720	$22	2%
j)	$48	$3	23%

3) $4295.57
5) $7429.74

EXERCISE 5.3

1) a) $241.26   b) $1261.28
   c) $1449.17
3) a) $166.07   b) 5978.52
   c) $978.52
5) a) $477.83   b) $86099
   c) $36099
7) $283.53
9) $446.71
11) a) 6370   b) $57330
    c) $1719.90   d) $518.96
13) a) 2380   b) $45220
    c) $2261   d) $595.09

EXERCISE 5.4A
1) 8, 12, 16, 20, 24, 28, 32, 36, 40, 44
3) 27, 32, 37, 42, 47, 52, 57, 62, 67, 72
5) 19, 23, 27, 31, 35, 39, 43, 47, 51, 55
7) 80, 73, 66, 59, 52, 45, 38, 31, 24, 17
9) $12, 11\frac{1}{3}, 10\frac{2}{3}, 10, 9\frac{1}{3}, 8\frac{2}{3}, 8, 7\frac{1}{3},$

$\quad 6\frac{2}{3}, 6$

11) 13, 17, 21, 25, 29, 33, 37, 41, 45, 49
13) 17, 20, 23, 26, 29, 32, 35, 38, 41, 44
15) 7, 10, 13, 16, 19, 22, 25, 28, 31, 34
17) 38, 45, 52, 59, 66, 73, 80, 87, 94, 101
19) $d = 4$
21) Not arithmetic
23) $d = 11$
25) Not arithmetic

EXERCISE 5.4B
1) 51
3) 144
5) 291
7) 11
9) 51

EXERCISE 5.4C
1) 21
3) 20
5) $a_1 = 6$, $a_n = 246$, $d = 8$, $n = 31$
7) $a_1 = 5$, $a_n = 62$, $d = 3$, $n = 20$

EXERCISE 5.4D
1) 678
3) 650
5) 516
7) 959
9) 45100
11) 63
13) 174
15) 750
17) 1275

EXERCISE 5.5A
1) 1, 3, 9, 27, 81, 243
3) 8, 24, 72, 216, 648, 1444
5) 11, 33, 99, 297, 2673
7) 1024, 256, 64, 16, 4, 1
9) −9, 18, −36, 72, −144, 288
11) 6, 18, 54, 162, 486, 1458
13) −4, −8, −16, −32, −64, −128
15) −11, 22, −44, 88, −176, 352
17) $r = 4$
19) Not a geometric
21) $r = \dfrac{1}{2}$
23) 768
25) 2048
27) $r = \dfrac{1}{768}$

EXERCISE 5.5B
1) 252
3) 7651
5) 7875
7) 305
9) 2457
11) $a_1 = 2$, $r = 2$, $n = 5$, $s_5 = 62$
13) $a_1 = 7$, $r = 2$, $n = 5$, $s_5 = 217$
15) $a_1 = 1$, $r = 2$, $n = 8$, $s_8 = 255$
17) $a_1 = 2$, $r = 3$, $n = 7$, $s_7 = 2186$

CHAPTER 5 REVIEW EXERCISE A
1) $175.5
3) $66.67
5) a) $421.34
   b) $467.43
7) .3%
9) Discount = $45
   New proceeds = $1,455
11) 45 days
13) Effective rate of interest = 58%
15) 120 days

CHAPTER 5 REVIEW EXERCISE B
1) 9, 15, 21, 27, 33
3) 3, 11, 19, 27, 35
5) 6, 12, 24, 48, 96
7) −10, 20, −40, 80, −160
9) 76

11) 16
13) 460
15) 325
17) 500500
18) 484

CHAPTER 5 TEST

1) a
3) c
5) a
7) b
9) d
11) c
13) a
15) c

CHAPTER 6

EXERCISE 6.1

1) $-1\frac{1}{2}$

3) $2\frac{1}{2}$

5) -2

7) 1

9) 12

11) -6

13) -1.7

15) $6\frac{7}{8}$

17) 22

19) -4

21) $-3\frac{4}{5}$

23) 13.08

EXERCISE 6.2

1) $\frac{3}{16}$

3) $-7\frac{5}{6}$

5) $-1\frac{11}{18}$

7) $\frac{-5}{24}$

9) 35
11) -52
13) -21
15) 17
17) -6
19) -4

EXERCISE 6.3

1) 9
3) 22
5) 3
7) 2
9) -24

11) $3\frac{11}{24}$

13) $\frac{3}{4}$

15) $\frac{9}{28}$

17) $-1\frac{7}{24}$

19) $\frac{-19}{60}$

21) $-1\frac{3}{8}$

23) -2.96
25) -17.97
27) 6.78

EXERCISE 6.4

1) $-6\frac{3}{4}$

3) $\frac{-5}{6}$

5) $\frac{3}{20}$

7) -112.97
9) -7.84
11) 28.14
13) -168
15) 0
17) -126
19) 36
21) -72

23) 0
25) 13.57
27) 23.94
29) -11.5
31) -26.25
33) undefined
35) -2.4
37) 8
39) -9

EXERCISE 6.5
1) -17
3) -2
5) -15
7) 6
9) 30
11) 20
13) -24
15) 3
17) 13
19) -4
21) -6
23) -10
25) 10
27) -12

EXERCISE 6.6
1) 1
3) -1
5) 35
7) -9
9) like terms
11) like terms
13) $-14p^3 + 5p^2$
15) $9x^2$
17) $23x$
19) $-17 + x$
21) $14 + 3m$
23) $2x + 6$
25) $-4y$
27) $7k + 15$
29) $13b$
31) $-6a$
33) $17y$
35) $7x + 14$

37) $5w + 6$
39) $y - 6$
41) $6z + 14$
43) $10x + 4$
45) $-2y - 8$
47) $y + 6$
49) 0
51) 4
53) 6
55) 52
57) 30
59) -31
61) 11
63) $\dfrac{-2}{3}$
65) 27
67) -12
69) 1

EXERCISE 6.7A
1) 1
3) -16
5) $1\dfrac{1}{8}$
7) $1\dfrac{1}{3}$
9) $\dfrac{-2}{5}$
11) 10
13) -22
15) $3\dfrac{1}{4}$
17) $-3\dfrac{1}{2}$
19) $6\dfrac{1}{2}$
21) -3
23) $\dfrac{-3}{4}$
25) 1
27) 3
29) -3
31) 2
33) -21

35) $3\frac{2}{3}$

37) 1

39) 3

41) -3

43) 2

45) -21

47) $1\frac{11}{12}$

49. -1.525

EXERCISE 6.7B

1) $\frac{1}{2}$

3) 12

5) $2\frac{1}{8}$

7) $\frac{2}{5}$

9) 24

11) 2

13) -7

15) 1

17) 2

19) -8

21) 12

23) 2

25) -2

27) 5

29) 13

31) 12

33) 5

EXERCISE 6.8

1) $(x-3)x$

3) $(x+3)+\frac{x}{2}$

5) $7(z+8)$

7) $5m-m$

9) $\frac{3}{4}x$

11) $\frac{x}{20}$

13) $x+2x+11$

15) $2x+7+8$

17) $10-\frac{x}{2}$

19) $3+y-6$

21) $x+7$

23) $x-12$

EXERCISE 6.9

1) 4

3) 15

5) -15

7) 10

9) 24

11) 3

13) 12

EXERCISE 6.10

1) $96

3) 500 bicycles

5) $1,750

7) $200

9) $3750

11) $2.47

13) $249,577.78

15) 38,400 bushels

17) 5°

19) $25

EXERCISE 6.11

1) $x>3$

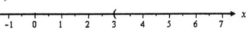

3) $m\geq 10$

5) $x>15$

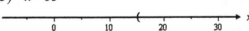

7) $z\geq -6$

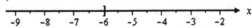

9) $-7<x<7$

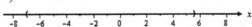

539

11) $k > 3$

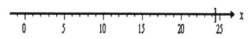

13) $m \le -11$

15) $-6 < m < -\dfrac{13}{3}$

17) $x < -2$

19) $-\dfrac{14}{3} \le x \le 2$

21) $-\dfrac{1}{2} \le x \le \dfrac{35}{2}$

23) $k \le 24.5$

25) No solution

27) $-14 \le k \le 10$

29) $-5 \le x \le 6$

CHAPTER 6 REVIEW EXERCISE A

1) coefficient
3) variable
5) term
7) $3n - 9p + 18$
9) $4x - 16$
11) $x + 6$

13) -38
15) $\$137.50$
17) -2
19) 56
21) 26
23) 0
25) $-3\dfrac{1}{2}$
27) $5(x + 3)$
29) 9

CHAPTER 6 REVIEW EXERCISE B

1) $11m$
3) $16p^2 + 2p$
5) $-2m + 29$
7) $m = 6$
9) $k = 7$
11) $r = 11$
13) $k = 5$
15) $p = 5$
17) $k = \dfrac{64}{5}$
19) $m = -\dfrac{9}{2}$
21) 6
23) 12 miles
25) 3 meters
27) 10 fives
29) 24.1 liters
31) $m \ge -2$

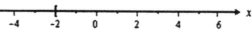

33) $-5 \le p < 6$

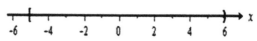

35) $y \ge -7$
37) $k \ge 53$
39) $-8 \le x \le -2$
41) $k \ge -3$
43) $p < 5$
45) $-\dfrac{3}{2} \le m \le 3$
47) $y = 7$

49) $z < 4$
51) $k = -6$

CHAPTER 6 TEST A
1) b
3) d
5) b
7) a
9) c
11) d
13) e
15) c
17) b
19) a
21) b
23) a
25) c
27) b
29) c
31) b

CHAPTER 6 TEST B
1) b
3) a
5) d
7) a
9) a
11) d
13) a
15) a
17) a
19) d
21) b

CHAPTER 7

EXERCISE 7.1
1) Yes
3) Yes
5) No
7) Yes
9) Yes
11) No
13) 11
15) 29

17) -4
19) 3
21) $(3, 7), (0, 1), (-1, -1)$
23) $(2, 2), (0, 8), (-3, 17)$
25) $(0, 9), (3, 3), (12, -15)$
27) $(-4, 6), (-4, 2), (-4, -3)$
29) $(-9, 8), (-9, 3), (-9, 0)$
31) $(-2, 6), (0, 0), (3, -6)$

EXERCISE 7.2
1) $(0, 5), (5, 0), (2, 3)$

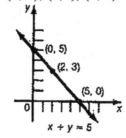

3) $(0, 4), (-4, 0), (-2, 2)$

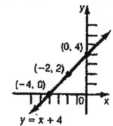

5) $(0, -6), (2, 0), (3, 3)$

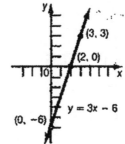

7) $(0, 4), (10, 0), (5, 2)$

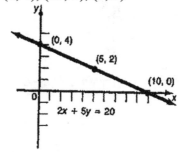

541

9) $(-5,2)$, $(-5, 0)$, $(-5, -3)$

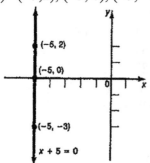

11) x-intercept: 3; y-intercept: 2

13) x-intercept: 3; y-intercept $-\dfrac{9}{5}$

15) x-intercept: 0; y-intercept: 0

17) x-intercept: -4; y-intercept: none

19)

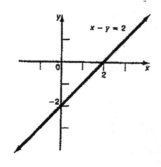

21)

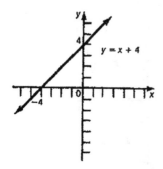

23)

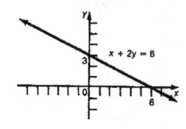

25)

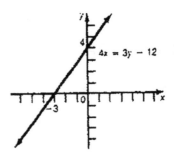

27)

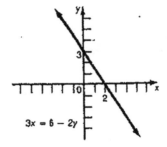

29)

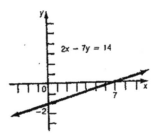

31)

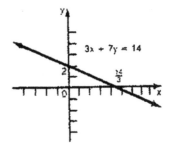

33)

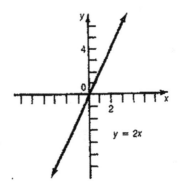

542

35)

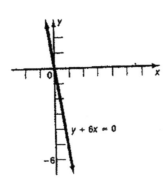

37)

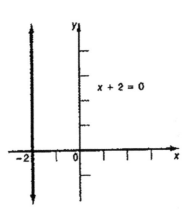

39)

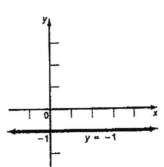

41)

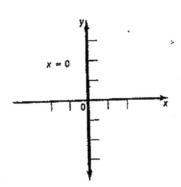

EXERCISE 7.3

1) $\dfrac{7}{6}$

3) $\dfrac{9}{2}$

5) $-\dfrac{5}{8}$

7) 0

9) undefined

11) -1.020

13) -1

15) -5

17) -2

19) $\dfrac{3}{2}$

21) $-\dfrac{2}{5}$

23) 0

25) Parallel

27) Parallel

29) Perpendicular

31) Perpendicular

33) Neither

35) Neither

37) $\dfrac{8}{15}$

39) -5

EXERCISE 7.4

1) $y = 3x + 5$

3) $y = -x - 6$

5) $y = \dfrac{2}{5}x - \dfrac{1}{4}$

7) $y = -5$

9) $y = 4.61x - 2.38$

11) $2x + y = -4$

13) $2x - 3y = -9$

15) $3x + 4y = -17$

17) $8x + 11y = 48$

19) $3x - 5y = -11$

21) $3x + 5y = -17$

23) $2x + 3y = 6$

543

25) $2x + y = -1$

CHAPTER 7 REVIEW EXERCISE
1) Yes
3) No
5) Yes

7) $(-1, -5), (0, -2), (\frac{7}{3}, 5)$

9) $(-4, -3), (-4, 0), (-4, 5)$
11)

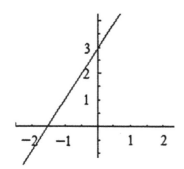

13)

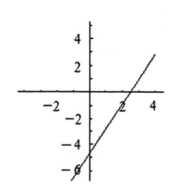

15)

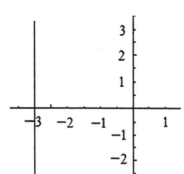

17) $x = \frac{5}{2}, y = -5$

19) $x = 0, y = 0$

21) 2

23) 3

25) Undefined

27) Parallel
29) Perpendicular
31) $3x - y = 2$
33) $2x - 3y = -15$
35) $2x - 3y = -14$
37) $x + 4y = 6$

CHAPTER 7 TEST
1) c
3) a
5) c
7) a
9) b
11) d
13) a
15) c
17) d
19) c
21) b
23) a

CHAPTER 8

EXERCISE 8.1
1) parallel lines
3) parallelogram
5) rectangle or square
7) sphere
9) ray
11) 180°
13) two
15) hexagon
17) right
19) straight
21) obtuse
23) right
25) isosceles
27) equilateral

EXERCISE 8.2
1) 360°
3) 18°
5) 52°
7) 16°

9) Complement $\Rightarrow 67\dfrac{1}{2}$

Supplement $\Rightarrow 157\dfrac{1}{2}$

11) 53°
13) 33°
15) 77°
17) $y = 104^o$, $x = 56^o$
19) $3x = 60$, $4x + 60 = 140$, $3x + 20 = 80$

EXERCISE 8.3
1) 62cm
3) 23.3 inches
5) 29 feet
7) 36 yards
9) 192 inches
11) 16
13) 24

EXERCISE 8.4
1) 165 square inches
3) 25 square feet
5) 25 square inches
7) 108
9) $577.50
11) 17.4
13) $1,760
15) 45 carpet titles
17) 17,400 square inches

EXERCISE 8.5
1) Area = 113.10 square feet)
   Circumference = 37.7 feet
3) Area = 254.46 square inches
   Circumference = 56.55 inches
5) Area = 888 square feet
7) Area = 105.5 square feet
9) Area = 88 square meters
11) Area = 1.72 units

EXERCISE 8.6
1) 1.728 cubic inches
3) 22 cubic meters
5) 1620 cubic inches
7) 113.10 cubic inches
9) 6 inches

11) 3060 cubic inches
13) 2411.52 cubic inches
15) 305.61 cubic centimeters
17) 1,215 cubic feet

EXERCISE 8.7
1) 3.6
3) 6.40
5) Yes, $(15)^2 + (36)^2 = (39)^2$
7) 10.91 ft.
9) 7.416 meters

EXERCISE 8.8
1) $\dfrac{1}{3}$

3) $\dfrac{1}{2}$

5) Yes
7) Yes
9) 4.9 feet
11) 7.2 cm
13) 49 m^2
15) 12 inches
17) 12 meters

CHAPTER 8 REVIEW EXERCISES
1) scalene
3) trapezoid
5) perpendicular
7) 180°
9) d
11) 12.8 meters
13) $96
15) 94.2 ft
17) 20°
19) 64 sq. inches
21) right triangle
23) 4.59 feet

CHAPTER 8 TEST
1) a
3) d
5) e
7) b
9) c
11) e

13) b
15) c
17) b
19) b
21) a
23) b
25) c
27) d
29) d
31) b
33) d
35) d
37) b
39) c
41) c

CHAPTER 9

EXERCISE 9.1

1) .003 kl
3) 0.005261 km
5) 2500 liter
7) 0.02591 kg
9) 978000000 or 9.78×10^8 mm
11) .68 dm
13) 2.6×10^4
15) <
17) =
19) <
21) 1235 m
23) 4.43644 m
25) 10300000 or 1.03×10^7 m

EXERCISE 9.2A

1) 45 ft
3) 9.67 yards
5) 11 qts
7) 120 ft
9) $1\frac{1}{9}$ yards
11) 96 inches
13) 4 gals
15) 21 ft
17) 36 qts
19) 8.5 galsm

21) >
23) <
25) <
27) <
29) >
31) b
33) b
35) b
37) c
39) meter
41) meter
43) kilograms
45) liters
47) kilometers
49) meter
51) mg
53) meter

EXERCISE 9.2B

1) $1\frac{5}{6}$ yd
3) 180 in.
5) $1\frac{1}{2}$ yd
7) $13\frac{1}{2}$ ft
9) $5\frac{1}{3}$ ft
11) 108 in.
13) 28 qt
15) $2\frac{1}{2}$ gal
17) 6 qt
19) $2\frac{1}{2}$ pt
21) 20 fl oz
23) 6 c
25) 42 oz
27) $5\frac{5}{8}$ lb
29) 2,500 lb
31) $3\frac{1}{2}$ tons
33) 112 oz.

35) $2\frac{1}{4}$ lb

EXERCISE 9.2C
1) 0.854
3) 183
5) 16.39
7) 2.46
11) 4.73
13) 11
15) 2.72
19) 6.44
21) 4.03
23) 46.6
25) 4.73
27) 20.32
29) 39 cents per-liter
31) $3.72/kg
33) 48.3 km/h
35) 8.89 meters
37) 53.961
39) 3.771
41) 61.36 kg

EXERCISE 9.3
1) 9 lbs. 3 oz
3) 9 gal 2 qt
5) 21 lbs 13 oz
7) 5 ft 7 in
9) 1 hr 40 min
11) 4 lbs 10 oz
13) 5 lbs
15) 16 ft 3 in.
17) 23 gal 1 pt
19) 2 ft 3 in
21) 5 hrs 8 min
23) 9 yrs. 11 mo
25) 134 ft 8 in

EXERCISE 9.4
1) a) 72°C   b) 68°F   c) 50°F
    d) -22 ^{o}F e) 21.1 ^{o}C
3) a) -223.15^oC   b) -33.15^oC
    c) 283.15^{o}K
5) -108.4^oF
7) a) 5°C b) 100°C

9) 37°C

EXERCISE 9.5
1) 2
3) $3\frac{3}{8}$
5) 1
7) 3
9) 12' by 12'
11) 20 cm
13) 165 cm
15) 6 cm

CHAPTER 9 REVIEW EXERCISE
1) gram, meter, liter
3) .001, .1, 100
5) $2.18
7) 680 miles
9) a) 45 ft   b) 3 yards
    c) 1.2 gal d) 2.9375 lbs
11) a) 10 ft 4 in   b) 13 ft 9 in
    c) 2 lbs 7 oz d) 2 lbs 4 oz
13) 117
15) b
17) a
19) b
21) 35.56
23) a) .75 b) 0.7925
25) 24.8 mph
27) 25°C
29) .250
31) $\frac{1}{10}$
33) 100
35) .92
37) 82
39) 900

CHAPTER 9 TEST
1) b
3) a
5) a
7) b
9) d
11) c
13) d

15) b
17) b
19) a
21) a
23) d
25) b
27) b
29) c
31) d
33) b
35) b
37) a
39) c

CHAPTER 10

EXERCISE 10.1

	Mean	Median	Mode
1)	3	3	None
3)	105	105	None
5)	91	95	95
7)	6	2.5	1
9)	11	9	none
11)	19	19	none

13) Mean: 13; Median: 11; Mode: 10
15) mean = 6.05; median = 6.5; mode = 8; the mode is the least representative.
17) a) mean = 9; median = 9
    b) mean = 24.2; median = 9
    c) mean = 11; median = 9
19) Travil is correct. He took account of the number of students making each score, and Derrick did not.
21) 60
23) $84 per-week
25) 40
27) 10.1 gallons
29) 272 hot dogs
31) $131,257.50
33) 28 requests
35) $36.20

EXERCISE 10.2

	Range	Standard Deviation
1)	4	1.58
3)	18	7.21
5)	4	1.29
7)	4	1.58
9)	25	9.11

11) $Q_1 = 135$ $Q_2 = 17$ $Q_3 = 20$
    $P_{45} = 17$ $P_{89} = 25$

EXERCISE 10.3

1) a) 584.85 b) 725 c) 2%
3) 41.92%
5) 49.01%
7) 11.79%
9) 19.15%
11) 49.25%
13) 17.72%
15) 34
17) 91.92%
19) 31
21) 1
23) 25

EXERCISE 10.4

1) a) $\dfrac{4}{8} = \dfrac{1}{2}$
   b) $\dfrac{6}{8} = \dfrac{3}{4}$
   c) 1
3) a) $\dfrac{1}{8}$
   b) $\dfrac{3}{8}$
   c) $\dfrac{1}{8}$
   d) $\dfrac{1}{8}$
5) a) $\dfrac{1}{5}$
   b) $\dfrac{4}{5}$
7) a) .1333
   b) .40

548

c) .90

9) a) $\dfrac{4}{54} = \dfrac{1}{13}$

b) $\dfrac{13}{54} = \dfrac{1}{4}$

c) $\dfrac{13}{54} = \dfrac{1}{4}$

d) $\dfrac{26}{54} = \dfrac{1}{2}$

e) 1

EXERCISE 10.5

1) $\dfrac{1}{8}$

3) a) $\dfrac{4}{14} = .27$

b) $\dfrac{5}{15} = .333$

c) $\dfrac{6}{15} = .40$

d) $\dfrac{11}{15} = .733$

e) $\dfrac{10}{15} = .67$

5) .41

7) a) $\dfrac{12}{51} = .24$

b) $\dfrac{22}{52} = .42$

c) .0059

d) .0625

e) .006

f) .006

g) .77

9) 1 to 3

11) 1 to 3

13) $\dfrac{1}{4} = .25$

15) a) .140

b) .391

CHAPTER 10 REVIEW EXERCISE

1-10 See definitions in textbook

11) $6.13

13) .77

15) 11.51%

17) Data set II is more variable. Its S.D. is 3.42 compared to 2.94 for set I. Use the mean for set I and median for set II.

19) mean

CHAPTER 10 TEST

1) b

3) b

5) b

7) a

9) b

11) a

13) a

15) b

17) c

19) c

21) c

23) c

25) c

27) d

29) a

31) c

33) c

35) b

37) d

39) d

41) b

CHAPTER 11

EXERCISE 11.1

1) $12,347,500

3) $3,434,850

5) 10,800,000 people

7) 1,800,000 people

9) $\dfrac{41}{300}$

11) $\dfrac{91}{150}$

13) 53,000 cars

15) 21,000 cars
17) March
19) 300 watches
21) 4,500 shares
23) 3000 shares
25) 90,000 calls
27) $1,500,000
29) $4,000,000
31) $2,500
33) 30
35) 27.5
37) 1 to 4
39) March, April, May, Jan., Feb., June
41) 11,500
43) April, May
45) 1916.67
47) 1993
49) 235

Exercise 11.2
1)

High
School

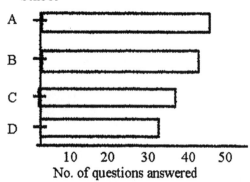

No. of questions answered

3)

5)

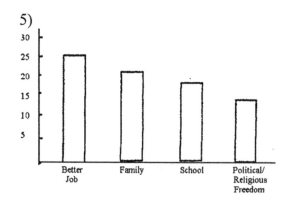

EXERCISE 11.3
1) -2

3) $-\dfrac{3}{2}$

5) $y = -2x + 10$

7) $y = -\dfrac{3}{2}x + \dfrac{9}{2}$

9) $m = \dfrac{3}{2}$ and $b = -3$

11) $m = -\dfrac{3}{2}$ and $b = 2$

13) a)
Gas Consumption

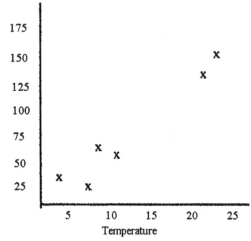

b) $y = 6.4x + 4.8$
c) $r = .97, r^2 = .94$, excellent
   equation
d) 125.76

CHAPTER 11 REVIEW EXERCISE

1) Line, bar, circle
3) bar
5) a) 982.19%
   b) $963.83

7) $-\dfrac{5}{2}$

9) $y = -\dfrac{1}{2}x + 4$

11) $\dfrac{3}{2}$

13) 53%
15) between 17 and 18 hours
17) 25%
19) $7840
21) furniture and software
23) $14000
25) about 2000
27) September and November
29) $260000
31) between $250000 and $270000
33) between $230000 and $260000

CHAPTER 11 TEST

1) c
3) a
5) c
7) c
9) b
11) b
13) a

15) $\dfrac{1}{4}$

17) $\dfrac{1}{4}$

19) d
21) 3

23) $\dfrac{5}{16}$

25) d
27) a
29) d
31) b
33) e

Index

Absolute value, 230
Acute angle, 334, 336
Addend, 45
Additive inverse, 4, 8
Algebraic expressions, 229, 253, 254
Algebra, 229
Amount, 157
Angle (s)
 complementary, 339
 supplementary, 339
Annuities, 195, 196
Area
 of circles, 355
 of polygons, 349
 of quadrilaterals, 349
 of triangles, 349
Associative property
 of Addition, 9
 of Multiplication, 9
Average, 432

Bar graph, 481
Base, 157, 254
Buying a house, 203

Calibrated scales, 415
Central tendency, 431, 435
Centiliter, 388
Circle, 356
Circle graph, 479, 493
Circumference, 355
Closing costs, 204
Closure, 9
Commission, 172
Common factors, 29
Common multiple, 30
Commutative property, 9
Compound interest, 185, 188
Composite
 figures, 335, 356
 numbers, 23
Congruent triangles, 374
Constructing graphs, 493
Cube, 363
Cylinder, 363

Decimal
 ordering, 126, 127
Decrease, percent, 165
Degree, 254
Denominate numbers, 393, 405
Denominator, 87, 89
Diameter, 355
Decimal point, 115

Decimals
 conversion/to fractions, 125
 division of, 121
 multiplication of, 121
 and percent, 152
 place values, 115
 rounding, 127
 subtraction of, 95

Discount, 159
Distributive property, 9
Divisibility, 17, 18
Division
 of decimals, 159
 of fractions, 101
 of mixed numbers, 101
 of signed numbers, 243
 of whole numbers, 17, 45
Divisor, 29
Down payment, 203, 207

Elements, 1, 6
Empty set, 2
English system, 395
Equations, 261
Estimating
 sums, 41
 differences, 41
 product, 41, 42
 quotient, 43
Exponent, 49, 50, 254

Factor
 tree, 24
Factorization, 24, 25

Finite, 2
Fractions
    addition, 95
    and decimals, 115, 125
    division of, 101
    improper, 85
    mixed numbers, 86, 96
    multiplication of, 101
    ordering, 91
    and percent, 158
    subtraction of, 95
Fundamental counting principle, 86

Geometry
    angles, 334
    area, 349
    composite geometric figures, 335, 356
    lines, 333
    perimeter, 345
    Pythagorean, 369
    similar triangles, 373
    volume, 363
Gram, 388
Graphing, 285, 303

Identity property, 10
Integers
    addition, 235
    division, 243
    multiplication, 243
    subtraction, 239
Interest
    compound, 18
    simple, 185
Intersection, 5
Inverse property, 10

Kilogram, 389

Least common denominator, 30
Line graph
    segment, 333
Liter, 388

Mean, 432, 453
Measurements, 340, 387, 393

Meter, 388
Metric conversion, 394
Metric system
    prefixes, 388
    base units, 388
Mixed numbers, 86
Mode, 432
Monthly payment, 199, 204
Normal curve
    percentages, 450
    properties, 450
Numbers
    integers, 4, 5
    irrational, 4
    mixed, 86
    prime, 23
    rational, 4, 5
    whole, 4
Number line, 230
Numerator, 85

Obtuse angle, 334
Ordered pairs, 304
Order of operations, 61, 247, 248

Parallelogram, 349
Part, 157
Percent, 51
Pi, 355
Polygon
    regular, 334
Prime factor, 24, 29
Prime factorization, 24
Prime number, 23
Properties of equality
    addition and subtraction, 261
    multiplication and division, 261
Proportions, 139
Pythagorean Formula, 369

Quadrilateral, 335

Radius, 355
Range, 441
Rational numbers, 4, 85
Right angle, 334
Right triangle, 369

Rounding, 43, 44

Scientific notation, 41, 65
Set
        braces, 1
        disjoint, 3
        empty, 2
        subset, 3
Similar triangles, 373
Simple interest, 185
Solving proportions, 139
Sphere, 364
Square
        area of, 349
        perimeter, 345
Standard deviation, 441, 442, 449
Statistics, 431
Straight angle, 334

Subset, 3
Symbols of equality, 55, 57

Trapezoid, 349
Triangle
        area, 349

Union, 3
Unlike terms, 253

Variability, 441
Variable, 253
Volume
        of a box, 363
        of a cylinder, 363
        of a sphere, 364

Word problems, 142, 269, 277